MECHANICAL ENGINEERING EXAM PREP

MECHANICAL ENGINEERING EXAM PREP

Problems and Solutions

Layla S. Mayboudi, PhD

MLI Exam Prep Series
Sarhan Musa, PhD
Prairie View A&M
(Series Editor)

MERCURY LEARNING AND INFORMATION

Dulles, Virginia
Boston, Massachusetts
New Delhi

Publisher: David Pallai
MERCURY LEARNING AND INFORMATION
22841 Quicksilver Drive
Dulles, VA 20166
info@merclearning.com
www.merclearning.com
800-232-0223

L. S. Mayboudi. *Mechanical Engineering Exam Prep: Problems and Solutions.*
ISBN: 978-1-683921-34-9

The publisher recognizes and respects all marks used by companies, manufacturers, and developers as a means to distinguish their products. All brand names and product names mentioned in this book are trademarks or service marks of their respective companies. Any omission or misuse (of any kind) of service marks or trademarks, etc. is not an attempt to infringe on the property of others.

Library of Congress Control Number: 2020952431

212223321   Printed on acid-free paper in the United States of America.

Our titles are available for adoption, license, or bulk purchase by institutions, corporations, etc.
For additional information, please contact the Customer Service Dept. at 800-232-0223(toll free).

All of our titles are available in digital format at *academiccourseware.com* and other digital vendors.

To scientists, mathematicians, engineers, and physicians, the warriors without borders, who honored their professional oaths, defended the rights to human empowerment, knowledge, and freedom, who were shunned for being the outliers, worked ethically and diligently, failed but persevered, who lost their lives but lived happily forever and ever after, the trailblazers . . .

CONTENTS

Preface   *ix*

Acknowledgments   *xv*

Chapter 1   Mathematics and Physics   **1**

Chapter 2   Probability and Statistical Analysis   **25**

Chapter 3   Engineering Mechanics   **73**

Chapter 4   Structure of Materials   **101**

Chapter 5   Dynamics and Control   **137**

Chapter 6   Thermodynamics   **159**

Chapter 7   Heat Transfer   **199**

Chapter 8   Fluid Mechanics   **215**

Chapter 9   Mechanical Engineering Drawing   **249**

    Orthogonal Views   249

    Isometric Views   252

    Auxiliary Views   253

    Section Views   254

    Dimensional and Working Drawings   260

    CAD with Solid Edge®   265

Good Practices and Lean Six Sigma Implementation   **271**

Assessment Tools   **275**

Appendix: Thermodynamic Tables and Diagrams   **279**

Bibliography   *285*

PREFACE

Introduction to Mechanical Engineering

Mechanical engineering is a broad field that, like other engineering disciplines, is founded on the fundamentals of physics, mathematics, and materials science. This basic knowledge is applied to design, analysis, manufacture, and maintenance of mechanical systems. These systems can be as simple as a ramp or as complex as a jumbo jet, built from a few million individual parts. Use of simple machines has been reported in medieval documents and ancient societies. These were levers, wheels and axels, pulleys, inclined planes, wedges, and screws. A force applied to a wedge, for example, is multiplied in magnitude and changed in direction. Some of these devices have been attributed to the Greek philosopher Archimedes, who lived around the third century BCE; however, similar machines were also widely used in the construction of monuments and cities such as Persepolis in ancient Persia or the Great Pyramids in ancient Egypt. As technology developed over centuries, these simple machines were combined to become building blocks of much more complex mechanical systems.

Use of the wedge and ramp has been reported since prehistoric times. Wheel and axle mechanisms were invented in Mesopotamia during the fifth millennium BCE. Lever mechanisms appeared in the same region in 3,000 BCE and were used in balance scales as well as in shadoof water-lifting devices, the first crane. Use of pulleys has been reported in Mesopotamia in the early second millennium BCE. The *Saqiyah*, or the Persian wheel, was a mechanical device used to scoop water by means of jars or buckets. Its use has been traced to the ancient Nubians in the Kingdom of Kush, where animals were employed to drive this wheel. The works of Archimedes in the third century BCE were among the influencing factors affecting future Western inventions. The first steam-powered device was invented in Roman Egypt by Heron of Alexandria in the first century CE. Use of water-powered machines, waterwheel and watermill, were first reported in Persia (Iran) in the early fourth century.

The Islamic Golden Age, between the seventh and fifteenth centuries, witnessed the greatest contributions made by Muslim inventors in the field of mechanical technology. Al-Jazari, a Mesopotamian polymath and mechanical engineer from Jazira, wrote *The Book of Knowledge of Ingenious Mechanical Devices* in 1206, describing 100 mechanical devices as well as instructions for how to make them. He is considered the originator of the flush toilet as well as devices such as crankshafts and camshafts.

Dedicated Persian polymaths and engineers such as Muhammad Ibn Musa al-Khwarizmi, the inventor of algebra in the ninth century, focused on the technical side of inventions as much as their artisanry and diligently developed and employed mathematical relations. An example is developing and employing thermal management principles, both for heating and cooling, centuries ago. In fact, the industrial revolution is not responsible for the invention of heating, ventilation, and cooling in buildings, nor for the invention of refrigerators and freezers. Persian engineers used the *Yakhchal*, meaning *ice container* in Farsi, as early as 400 BCE, to store during the summer months the ice created in winter. The Iranian aqueduct, also known as a *Qanat*, was responsible for the transfer of water to the *Yakhchal's* tall conical-shaped structure, where it was cooled to the freezing point by its surroundings. To make the heat transfer more efficient, the flow was directed through the northern wall to use its shadow and keep the water cool, and an additional eastern-western wall was constructed to protect the wall from the Sun's radiation. The same word is still used for refrigerator and freezer in modern-day Iran.

The *Abanbar*, a Persian cistern, is another thermal engineering example where a water reservoir with an insulating structure built below ground level was able to manage water temperature and the ventilating effect due to the installation of windcatchers while avoiding the creation of mold and mildew due to stagnation. Windcatchers were ventilating systems used in the old Iran which are still in use to this day. If you wonder what material was used to make such an insulating structure, think of baking a low-fat sand cubed cake with two-meter sides: egg white, sand, clay, lime, goat hair, and ash.

The mechanical engineering field experienced accelerated growth during the industrial revolution in Europe that started in the eighteenth century. Many breakthroughs that still form the foundations of today's mechanical engineering practice were made in England, with the development of calculus and laws of motion by Sir Isaac Newton. Another major contribution to calculus was made in Germany by Gottfried Wilhelm Leibniz in the seventeenth century. The early nineteenth century witnessed significant advancements in mechanical machines and tools in England, Scotland, and Germany, primarily in support of the development and refinement of the steam engine technology that powered the industrial revolution. This is the point where mechanical engineering was officially recognized as an independent field within engineering, as it focused on the development of manufacturing machines and engines.

If you think of a steam engine, you will see that it connects to all the founding disciplines of mechanical engineering. There are heat transfer and fluid mechanics of the heated water and steam, thermodynamics, and combustion in conversion of the fuel energy to useful work, statics and dynamics of the rotating shafts and linkages, solid mechanics of the pressure vessels, and machine design of the bearings and gears. All of these lead to improvements of the engine's design and are made into a working machine through manufacturing technologies.

The first British Professional Society of Mechanical Engineers was founded in 1847, thirty years after civil engineers formed their own, the first one of its kind. In the United States, the American Society of Mechanical Engineers (ASME) was founded in 1880. The United States Military Academy was the first school in the United States to offer an engineering education in 1817; they were followed by the school that later became known as Norwich University (1819), and by Rensselaer Polytechnic Institute (1825).

In the nineteenth century, developments in physics led to the growth of mechanical engineering science. The field is also continually evolving to incorporate new advancements; today mechanical engineers are pursuing developments in the recently established areas of composites, mechatronics, and nanotechnology. Many mechanical engineers also collaborate with medical researchers in what can be broadly called biomedical engineering. This encompasses biomechanics, transport phenomena, bio-mechatronics, bio nanotechnology, and modeling of biological systems. Additionally, climate change concerns increased importance in the studies of the renewable energy field, such as wind and solar energy and fuel cell technologies.

Education in mechanical engineering is based on a strong foundation of mathematics and science. Most countries set standards in order to ensure the students graduating from their mechanical engineering programs meet the requirements for safe professional practice. The ultimate purpose is to ensure they have the needed skills, both technically and professionally. The fundamental subjects are mathematics and statistics, physical sciences, strength of materials and solid mechanics, statics and dynamics, materials sciences, control and vibration, machine design, manufacturing, thermofluids (e.g., thermodynamics, heat transfer, fluid mechanics, and combustion), and engineering design.

Introduction to This Book

This book can be considered a companion to the author's previous publications [1,2,3,4]. The intention is to provide a review of the mechanical engineering curriculum, covering topics in thermal fluid sciences, solid mechanics, manufacturing, and other specialized topics such as statistical analysis. Students of engineering, both on undergraduate or graduate levels, as well professors and scientists, can benefit from the content. The book can help to review this material when taking recency tests, preparing for upcoming job interviews, or for on-the-job training.

Readers can follow the material in the sequence presented or focus on the selected chapters. If sitting for a PhD comprehensive test, one can review the multiple-choice questions, contemplating the whys and the associated scenarios to promote in-depth understanding of the material, and also work diligently on the more comprehensive problems provided at the end of each section.

The purpose is to provide a broad perspective of the subject matter, and then continue to learn in greater depth in the fields of engineering graphics, mathematics, thermal sciences, mechanics, dynamics, material sciences, and statistics. All the artwork, diagrams, and graphs in this publication are the work of the author, created in CAD software or using Microsoft Office® applications. Much effort has been expended to create original problems. This is to promote the importance of original thinking but also to express the problems as such and come up with methods to develop the questions further and find innovative solutions. It is not simply another test book in mechanical engineering. Several topics in this publication are investigated in detail in the author's other publications referenced previously [1–4]. Asking questions, learning, and trying to understand things are the most basic elements of humanity. If there is one single lesson that the author would like the reader to take from this publication, it is to never stop questioning and learning; these are the privileges given to all, rich or poor, old or young. There exists nothing such as *are we too curious!* Scientific curiosity is good for all—not only for those destined to be scientists or the selected few.

As the closing remarks of this work are written, humans are being launched to the ISS from North American soil for the very first time. The science and knowledge required to make this launch a reality can be traced back centuries, when bright Middle Eastern scientists diligently laid the foundations for modern mathematics, physics, and science. Their work was carried on by Western scientists such as Galileo Galilei, who defended their scientific observations against the ones who had launched the inquisition against humanity during the dark ages. This effort and opening to the new era are coincident with the COVID-19 pandemic. Innovation will be needed to tame this pandemic, encompassing fields such as genetics, medicine, social sciences, humanity, and even engineering, where specialized gear is being designed to keep people safe from harm brought by this widespread but invisible enemy.

The Road Map

The book's primary focus is on mechanical engineering, but some chapters will also be more generally relevant. This section provides an outline of this work, a road map for the reader on this journey. The author encourages the reader to review the *Solution Guide* and to work through the presented problems.

Chapter 1 is related to mathematics and physics, both the foundations of all applied sciences and thus also of mechanical engineering. Problems are given related to mathematical concepts such as numbers, limits, functions, derivatives, integrals, and transformations. Physics and mathematics are closely intertwined; therefore, any of the mathematical operations reviewed herein are directly applicable to physical phenomena. Think of predicting the trajectory of a heavenly body under gravitational forces or designing a propulsion system that generates thrust for a rocket.

Chapter 2 presents probability and statistical analysis. Probability and statistics are widely used in engineering fields when conducting tests, identifying a system's parameters, and administering performance evaluations. These tools are also much employed in other diverse fields, from aerospace to medicine to sports. Selecting the best manufacturing method, process parameters, and material properties can be a challenging task. It can be facilitated by implementing mathematical techniques such as regression analysis and valid methods of comparison in order to make a sound decision. The application of statistical methods can help you to determine whether it is the *curse* of appearing on the cover page of a famous sport magazine that is the reason for the subsequent poor performances of an athlete. Problems involving different methods of regression analysis in order to predict data trends, either by moving the average or over time, are presented. Problems are given related to the statistical observations made during the current pandemic, threatening the lives of many people, of any age group and socioeconomic background. It is recommended that the presented questions and methods are carefully studied not only for the specialized tests but also for gaining deeper understanding of the analysis method in lab settings and manufacturing applications. Time regression methods are among the techniques that are used when correlating experimental and theoretical data. Empirical methods often require an in-depth understanding of data analysis, data trending, and design of experiments, which rely on selecting good sets of data in order to make conclusions based on scientific principles. Probability distribution diagrams are used in the presented problems. These diagrams demonstrate the practical use of data analysis in process improvement fields such as Lean Six Sigma, where the status of processes and products is investigated in order to identify its deviation from the targeted criteria. This is closely related to all stages of the product life cycle, including software-human interactions.

Chapter 3 covers problems in engineering mechanics, such as those relying on the balance of forces and free body diagrams. Examples of mechanical structures from bridges to cars and airplanes are used. Problems involving stationary or transient analysis, with differing loads and initial conditions, are listed.

Chapter 4 is about the structure of materials and includes questions related to crystallography, with emphasis on its mathematics aspect. Although some may consider the study of materials to be a field distinct from that of mechanical engineering, a good understanding of these concepts will be of help to anyone working within the latter field. The problems take into account thermofluids as well as mechanical concepts and should be suited for readers of varying backgrounds, including ones with the desire to specialize in thermofluids or solid mechanics. In today's world, thermoplastics are widely used in the manufacturing of parts due to being easily formable into complex shapes, while keeping production costs low, and from the environmental perspective, due to being easily recyclable. It would be beneficial to be familiar with their structure and the variation of their properties under different loads or temperatures. For most engineering materials that have a crystalline structure, change of the structure, which has a known shape and undertakes a variety of symmetry shapes, can be predicted. Knowledge of vector algebra is a must for these problems. Identifying types of symmetry is an important topic that relates mathematics and material properties. The concept of material strain covered here is also shared with the solid mechanics topic.

Chapter 5 focuses on dynamics and control, including problems related to the concepts of proportional, integral, and derivative control systems. The application of control parameters both in terms of implementing control systems and also in the form of varying parameters within the physics problems are discussed. For example, a question may ask to determine the appropriate flow rate in order to control the flow level in a tank with inlet and outlet flows. Case studies are presented examining the sensitivity of critical variables to the variation of physical parameters. Additionally, more comprehensive problems are included with the purpose to awake the sense of understanding and application of the control theory to a variety of areas not necessarily confined to the field of electrical engineering.

The next three chapters deal with thermofluids sciences. It is often challenging to treat these sciences separately from other ones such as applied mathematics or even mechanics, since some problems require the understanding of several disciplines. Obviously, when assessment of the retained knowledge in the engineering fields is desired, basic evaluation with particular emphasis on related analytical methods is important. Methods such as the separation of variables and the Fourier transform may be occasionally seen in higher level or graduate studies. Before electronic computers, human "calculators" were employed by NASA in the 1950s to perform the necessary computations such as launch trajectories. Although modeling some of these physical phenomena involves heavy calculations, they may also be assessed with some degree of approximation using more basic techniques for relating their descriptive parameters; for example, the variable X is inversely or linearly related to the parameter b. Concepts such as conservation of mass, momentum, energy, and continuity are further clarified by means of practical examples used in daily life or more complex aerospace applications.

Chapter 6 introduces thermodynamics. This field of study is concerned with the interactions among temperature, pressure, and volume, and generation of heat due to molecular motion. Thermodynamics laws are reviewed in the form of problems and occasional hints as suggested approaches to solutions. Open and closed systems and related concepts are presented. Problems associated with thermodynamic systems, where matter or energy enters and leaves the system, are offered. If the reader struggled with understanding of some of the thermodynamic concepts such as temperature and heat, the presented problems will help to address that, assuming that the reader diligently follows through with them. Problems related to the Carnot, Rankine, Stirling, Brayton, Diesel, and Otto cycles are presented along with the associated diagrams, making this chapter a comprehensive study and problem-solving guide. The author recommends that the reader review the problems carefully before attempting them or focusing on the provided solution hints, and to solve them on a step-by-step basis. Problems are given involving real and ideal thermodynamic processes. Examples for cases involving isobaric, isothermal, isochoric, or a combination of conditions are presented as well.

Chapter 7 focuses on heat transfer, another major thermal-science subject. Problems cover fundamentals of heat transfer and modeling thermal problems by incorporating modes of heat transfer (conduction, convection, and radiation), energy balance equations, and a material's thermal properties. Problems based on case studies showing the relation between heat and work generation and transfer are presented. Successful completion of problems presented in this chapter will help the reader to gain a good understanding of thermal models of multilayer systems, such as those for walls or slab layers with different geometrical configurations and thermophysical properties. Phase change is a concept that is an integral part of the thermofluids sciences, especially thermodynamics and heat transfer, and thus it is covered in the relevant two chapters. Renewable energy and its harvesting in the form of wind and solar are good examples for heat transfer applications. Thermal management methods such as use of extended surfaces rely on conduction and convection heat transfer modes, and the related problems are presented in this chapter.

Chapter 8 is about fluid mechanics. A variety of topics in this field, from the basic concepts to the more advanced ones, are presented herein. Fluid mechanics is an important subject in mechanical engineering as well as other fields such as civil and chemical engineering. The complexity of the associated problems in the mechanical engineering field is more obvious and therefore requires detailed attention, given that it covers the elementary to more advanced topics during the undergraduate and graduate studies as well as being the prerequisite for a number of undergraduate and graduate courses. Topics such as flow inside a pipe are among the problems presented. Although heated flow as moved inside the channels can also be investigated in heat transfer problems, fluid moving at different velocities inside the channels with varying cross sections are best treated as fluid mechanics problems.

Chapter 9 focuses on engineering drawing and engineering graphics concepts. This chapter is presented last as it relates to manufacturing and design, which in turn are a culmination for application of all the previously covered subjects. There are six different sub-topics: orthographic, isometric, auxiliary, section views, dimensioning and working drawings, and CAD with SolidEdge®. Engineering graphics is vital due to the extensive use of its concepts in daily life as well as in technical fields. Even if a person is not working as a designer, where mastering a CAD tool is essential, being able to interpret engineering drawings accurately is necessary to suggest design improvements. Nowadays, engineering graphics has found ways into other fields such as art and culture in the form of new designs and creations that are not only aesthetically pleasing but also practical. Its application in 3D printing is particularly important; the latest developments in this field have made it possible to design innovative medical equipment, such as eyeglasses and shoes, and parts that are either used as part of an assembly or are built as a single unit for weight-sensitive applications in the aerospace industry.

ACKNOWLEDGMENTS

My family, spouse, teachers, and publisher—I am infinitely grateful to them all for their generous support and the positive influence on my life that they have had.

Layla S. Mayboudi
January 2021

CHAPTER 1

MATHEMATICS AND PHYSICS

1. Mathematics is only used in scientific and engineering applications. (T/F)
2. Round off 312.14567 to figures with 5, 4, and 3 significant figures, respectively:
    (a) 312.15 / 312.1 / 312
    (b) 312.14 / 312.1 / 312.0
    (c) 312.14 / 312.2 / 312.1
    (d) 312.14 / 312.2 / 312.2
3. What is the slope of line $f(x) = ax + ba + cb$, where a, b, and c are constants?
    (a) b
    (b) c
    (c) a
    (d) $ba + cb$
4. The derivative of function $f(x) = x^n$ is:
    (a) $(n-1) \times x^{n-1}$
    (b) $n \times x^n$
    (c) $n \times x^{n-1}$
    (d) $(n-1) \times x^n$
5. The integral of function $f(x) = x^n$ is:
    (a) $x^{n-1}/(n-1)$
    (b) $(n+1) \times x^{n+1}$
    (c) $(n-1) \times x^{n-1}$
    (d) $x^{n+1}/(n+1)$
6. The average velocity (V) is calculated by (X is distance, t is time, and i and f are indices indicating the initial and final states):
    (a) $V = \dfrac{\Delta X}{\Delta t} = \dfrac{X_f - X_i}{t_f - t_i}$
    (b) $V = \dfrac{\Delta V}{\Delta t} = \dfrac{V_f - V_i}{t_f - t_i}$
    (c) $V = \dfrac{\Delta X}{\Delta V} = \dfrac{X_f - X_i}{V_f - V_i}$
    (d) $V = \dfrac{\Delta V}{\Delta x} = \dfrac{V_f - V_i}{X_f - X_i}$
7. The average acceleration (a) is calculated by (X is distance, t is time, V is velocity, and i and f are indices indicating the initial and final states):
    (a) $a = \dfrac{\Delta X}{\Delta t} = \dfrac{X_f - X_i}{t_f - t_i}$
    (b) $a = \dfrac{\Delta V}{\Delta t} = \dfrac{V_f - V_i}{t_f - t_i}$
    (c) $a = \dfrac{\Delta X}{\Delta V} = \dfrac{X_f - X_i}{V_f - V_i}$
    (d) $a = \dfrac{\Delta V}{\Delta x} = \dfrac{V_f - V_i}{X_f - X_i}$
8. Derive from first principles the derivative of the function $f(x) = ax^3 + bx^2 + cx + d$, where a, b, c, and d, are constants. Present the function y and its derivative in one diagram.
    (a) $3ax^2 + 2bx + c$
    (b) $ax^2 + bx + c$
    (c) $a + b + c$
    (d) $2ax^2 + bx + c$

Solution Guide

In order to derive the relation for the derivative, the limit mathematical definition is employed, which represents the change in value in a function when its independent variable varies by a small amount. This perturbance in the independent variable is very small so that it can be assumed to be approaching zero after the mathematical simplifications are made. See Figure 1.

$$f(x) = y(x) = ax^3 + bx^2 + cx + d$$

$$f(x + \Delta x) = y(x + \Delta x) = a(x + \Delta x)^3 + b(x + \Delta x)^2 + c(x + \Delta x) + d$$

$$y' = f'(x) = \frac{df(x)}{dx} = \lim_{\Delta x \to 0} \left(\frac{f(x + \Delta x) - f(x)}{\Delta x} \right)$$

$$\therefore y' = f'(x) = \lim_{\Delta x \to 0} \frac{1}{\Delta x} \left(\begin{array}{l} a(x^3 + 3x^2\Delta x + 3x\Delta x^2 + \Delta x^3) + b(x^2 + 2x\Delta x + \Delta x^2) + \\ c(x + \Delta x) + d - (ax^3 + bx^2 + cx + d) \end{array} \right)$$

$$= \lim_{\Delta x \to 0} \frac{1}{\Delta x} (3ax^2\Delta x + 3ax\Delta x^2 + a\Delta x^3 + 2bx\Delta x + b\Delta x^2 + c\Delta x) = 3ax^2 + 2bx + c$$

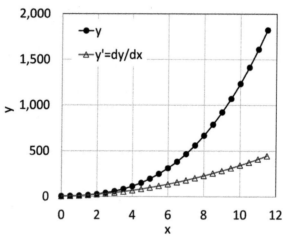

FIGURE 1. Function y and its derivative.

9. Change of distance with time is described by:
   (a) speed   (b) acceleration
   (c) deceleration   (d) length

10. Acceleration (m/s^2) is given by $a = 3t^2 + 2t + 1$, where t is time. Starting from the initial time and velocity of zero, the speed reached after 2 s is:
    (a) 16 m/s   (b) 14 m/s   (c) 12 m/s   (d) 10 m/s

11. Acceleration (m/s^2) is given by $a = 3t^2 + 2t + 1$, where t is time. Starting from the initial time and velocity of zero, the distance traveled after 3 s is:
    (a) 60.3 m   (b) 33.8 m   (c) 50.5 m   (d) 52.3 m

12. An object initially located at $X = 0$ starts moving along X from rest at time zero, with distance traveled varying according to a quadratic relationship. If the distance traveled after 4 s is 50 m, the speed reached after 2 s is 2 m/s, and the acceleration is constant and is equal to 2 m/s^2, the distance can be described by the function:
    (a) $X = -t^2 + 2t - 42$ (b) $X = -t^2 - 2t + 42$
    (c) $X = t^2 + 2t - 42$ (d) $X = t^2 - 2t + 42$

13. The second and third derivatives of function $f(x) = 5t^2 + 6t + 1$ are:

    (a) 5, 4   (b) 10, 4   (c) 10, 0   (d) 5, 0

14. The integral of function $f(x) = 5t^2 + 4t + 3$ $(2 \leq t \leq 3)$ is:

    (a) 44.7   (b) 44.3   (c) 44.5   (d) 44.9

15. A plot of velocity versus time is shown in Figure 2. Calculate the distance traveled from the initial time $t = 0$ to 3 s after the start of the motion (x and y are abscissa and ordinate, respectively).

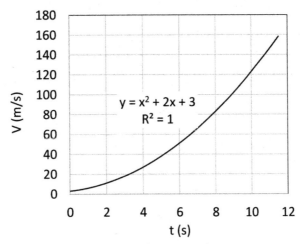

FIGURE 2. Velocity versus time.

    (a) 12   (b) 20   (c) 27   (d) 30

Solution Guide

$$x = \int_0^3 V \, dt = \int_0^3 (t^2 + 2t + 3) dt = \left(\frac{t^3}{3} + t^2 + 3t \right)\Big|_0^3$$

16. A plot of velocity versus time is shown in Figure 3. Calculate the acceleration reached after 100 s from the start of the motion (x and y are abscissa and ordinate, respectively).

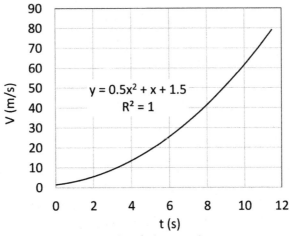

FIGURE 3. Velocity versus time.

    (a) 101   (b) 102   (c) 100   (d) 103

Solution Guide

$$a = \frac{dV}{dt} = (t+1)_{t=100\,s}$$

17. A plot of acceleration versus time is shown in Figure 4. Calculate the change in velocity from the initial time $t = 3$ to 5 s after the start of the motion (x and y are abscissa and ordinate, respectively).

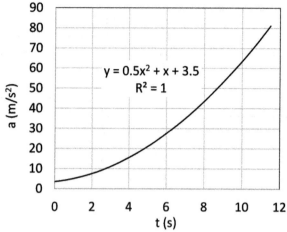

FIGURE 4. Acceleration versus time.

(a) 35.2   (b) 31.5

(c) 35.3   (d) 31.3

Solution Guide

$$V = \int_{0}^{3} a\,dt$$

$$= \int_{3}^{5} (0.5t^2 + t + 3.5)\,dt$$

$$= \left(\frac{0.5t^3}{3} + \frac{t^2}{2} + 3.5t \right)\Bigg|_{3}^{5}$$

18. For the distance versus time plot in Figure 5, select the appropriate velocity and acceleration diagrams from Figure 6:

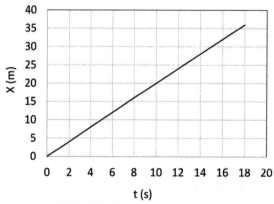

FIGURE 5. Distance versus time plot.

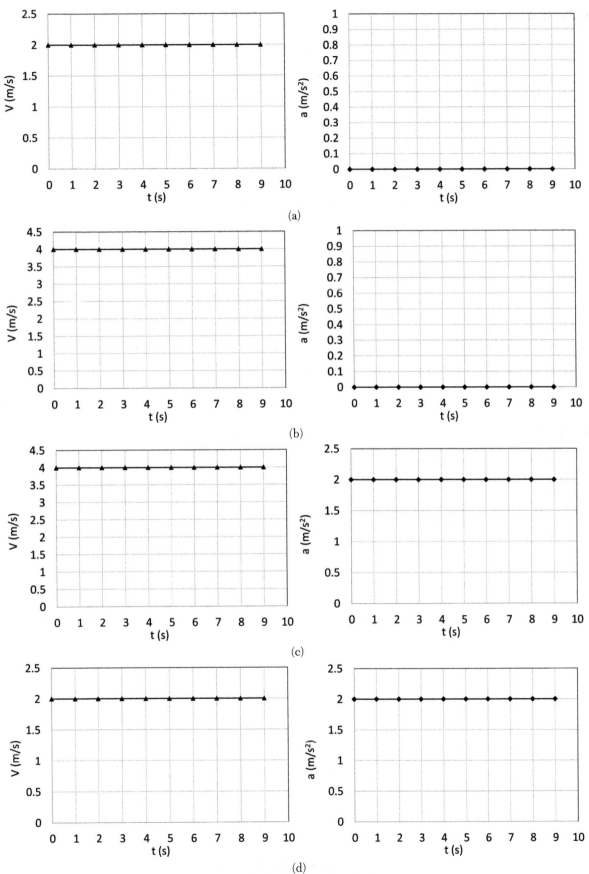

FIGURE 6. Velocity and acceleration plots.

19. For the distance versus time plot in Figure 7, the matching velocity and acceleration diagrams in Figure 8 are:

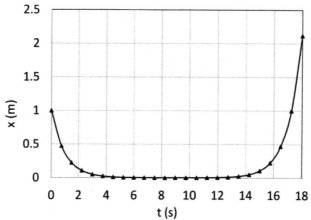

FIGURE 7. Distance versus time plot.

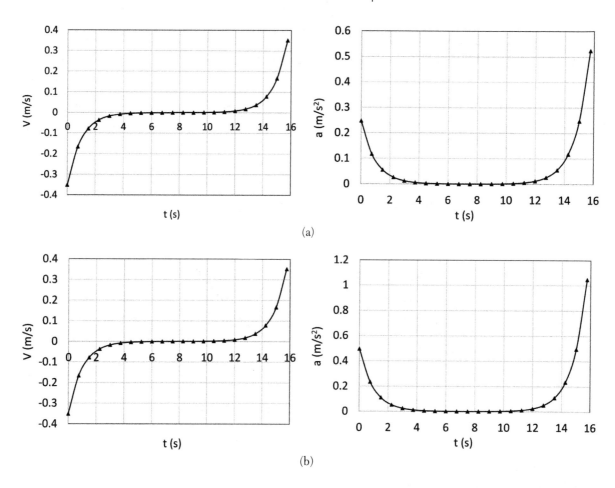

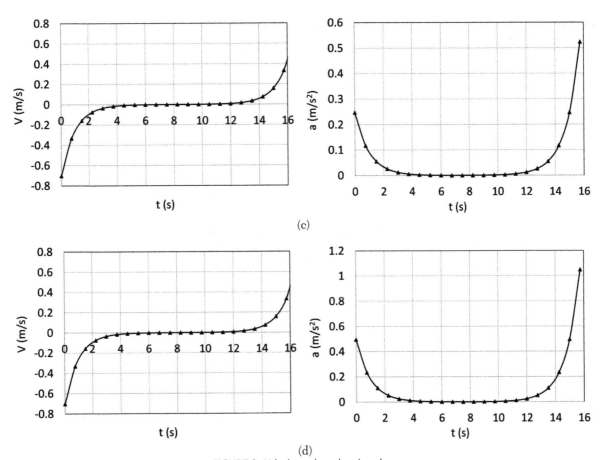

(c)

(d)

FIGURE 8. Velocity and acceleration plots.

20. Displacement versus time is given in Figure 9. Velocity:

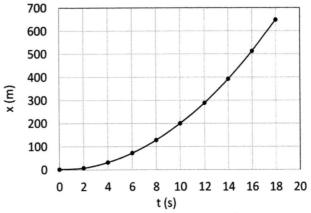

FIGURE 9. Displacement versus time plot.

(a) increases linearly with time
(b) remains constant with time
(c) increases quadratically with time
(d) increases exponentially with time

21. Velocity versus time is given in Figure 10. Acceleration:

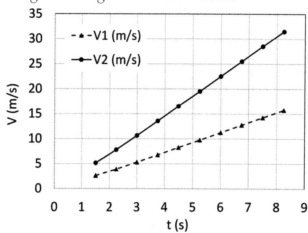

FIGURE 10. Plots of two velocities.

(a) increases linearly for V_1 but decreases linearly for V_2 with time
(b) decreases linearly for V_1 but increases linearly for V_2 with time
(c) remains constant for V_1 and V_2
(d) is zero for V_1 and V_2

22. Velocity versus time is given in Figure 11. Acceleration:

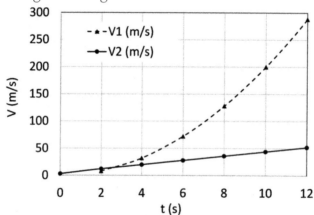

FIGURE 11. Plots of two velocities.

(a) increases linearly for V_1 but remains constant for V_2 with time
(b) decreases linearly for V_1 but increases linearly for V_2 with time
(c) remains constant for V_1 and V_2
(d) is zero for V_1 and V_2

23. Velocity (V) as a function of time (t) for constant acceleration (a) and initial velocity V_0 is given by:

(a) $V = -at + V_0$ (b) $V = -at - V_0$
(c) $V = at - V_0$ (d) $V = at + V_0$

24. The acceleration (a) and distance (X) traveled over a time segment ($6 \leq t \leq 10s$) (as shown in Figure 12) are:

(a) $a = 0.5$ m^2/s, $X = 16$ m   (b) $a = 1$ m^2/s, $X = 12$ m
(c) $a = 0.5$ m^2/s, $X = 4$ m   (d) $a = 1$ m^2/s, $X = 20$ m

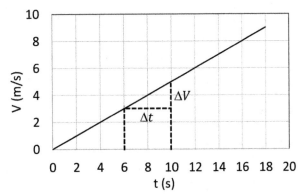

FIGURE 12. Plot of velocity versus time.

Solution Guide

$$X = \frac{1}{2} \times \Delta V \times \Delta t + V_6 \times \Delta t = \frac{1}{2} \times (V_{10} - V_6) \times (t_{10} - t_6) + V_6 \times (t_{10} - t_6)$$

NOTE: $a = \dfrac{\Delta V}{\Delta t} = \dfrac{V_{10} - V_6}{t_{10} - t_6}$

$\therefore (V_{10} - V_6) = a \times (t_{10} - t_6)$

$\therefore X = \dfrac{1}{2} \times a \times (t_{10} - t_6)^2 + V_6 \times (t_{10} - t_6) = \dfrac{1}{2} \times a \times \Delta t^2 + V_0 \times \Delta t$

25. The general expression for the distance traveled (X) over a time segment (as shown in Figure 12) as a function of the initial (V_i) and final (V_f) velocities, and acceleration (a) during this time segment is (assume the initial time is zero):

(a) $X = \dfrac{1}{2a} \times \left(-V_f^2 + V_i^2\right)$

(b) $X = \dfrac{1}{2a} \times \left(V_f^2 + V_i^2\right)$

(c) $X = \dfrac{1}{2a} \times \left(-V_f^2 - V_i^2\right)$

(d) $X = \dfrac{1}{2a} \times \left(V_f^2 - V_i^2\right)$

Solution Guide

$$X = \frac{1}{2} \times \Delta V \times \Delta t + V_i \times \Delta t = \frac{1}{2} \times (V_f - V_i) \times (t_f - t_i) + V_i \times (t_f - t_i)$$

NOTE: $a = \dfrac{\Delta V}{\Delta t} = \dfrac{V_{10} - V_6}{t_{10} - t_6}$

$\therefore \Delta t = (t_{10} - t_6) = (V_{10} - V_6)/a = \Delta V/a$

$\Delta X = X_{10} - X_6 = \dfrac{1}{2} \times a \times \left(t_{10}^2 - t_6^2\right) + V_6 \times (t_{10} - t_6)$

$\therefore$

$\qquad = \dfrac{1}{2} \times a \times \Delta t \times (t_{10} + t_6) + V_6 \times \Delta t$

$\therefore \Delta X = X_{10} - X_6 = \dfrac{1}{2} \times a \times \left(\dfrac{\Delta V}{a}\right) \times (t_{10} + t_6) + V_6 \times \dfrac{\Delta V}{a}$

$\therefore \Delta X = X_{10} - X_6 = \dfrac{1}{2} \times (V_{10} - V_6) \times \left(2t_6 + \dfrac{V_{10} - V_6}{a}\right) + \dfrac{1}{a} \times V_6 \times (V_{10} - V_6)$

$\therefore \Delta X = X_{10} - X_6 = \dfrac{1}{2} \times (V_{10} - V_6) \times \left(2t_6 + \dfrac{V_{10} - V_6}{a}\right) + \dfrac{1}{a} \times V_6 \times (V_{10} - V_6)$

$$\therefore \Delta X = X_{10} - X_6 = t_6 \times (V_{10} - V_6) + \frac{1}{2a} \times (V_{10} - V_6)^2 + \frac{1}{a} \times V_6 \times (V_{10} - V_6)$$

$$\therefore \Delta X = X_{10} - X_6 = t_6 \times (V_{10} - V_6) + \frac{1}{2a} \times (V_{10}^2 - V_6^2)$$

$$\therefore X_{10} = X_6 \times \frac{V_{10} - V_6}{V_6} + \frac{1}{2a} \times (V_{10}^2 - V_6^2)$$

$$\therefore \Delta X|_{t_6=0} = X_{10}|_{t_6=0} = \frac{1}{2a} \times (V_{10}^2 - V_6^2)$$

$$\therefore X = \frac{1}{2a} \times (V_f^2 - V_i^2)$$

26. The general expression for final velocity (V_f) as a function of the initial (V_i) velocity, distance traveled (X), and acceleration (a) during a time segment (as shown in Figure 12) is (assume the initial time is zero):

(a) $V_f = \sqrt{V_i^2 - 2a \times X}$
(b) $V_f = \sqrt{V_i^2 + 2a \times X}$
(c) $V_f = \sqrt{-V_i^2 + 2a \times X}$
(d) $V_f = \sqrt{-V_i^2 - 2a \times X}$

Solution Guide

$$X = \frac{1}{2} \times \Delta V \times \Delta t + V_i \times \Delta t = \frac{1}{2} \times (V_f - V_i) \times (t_f - t_i) + V_i \times (t_f - t_i)$$

NOTE: $a = \dfrac{\Delta V}{\Delta t} = \dfrac{V_{10} - V_6}{t_{10} - t_6}$

$$\therefore \Delta t = (t_{10} - t_6) = (V_{10} - V_6)/a = \Delta V/a$$

$$\therefore \quad \Delta X = X_{10} - X_6 = \frac{1}{2} \times a \times (t_{10}^2 - t_6^2) + V_6 \times (t_{10} - t_6)$$

$$= \frac{1}{2} \times a \times \Delta t \times (t_{10} + t_6) + V_6 \times \Delta t$$

$$\therefore \Delta X = X_{10} - X_6 = \frac{1}{2} \times a \times \left(\frac{\Delta V}{a}\right) \times (t_{10} + t_6) + V_6 \times \frac{\Delta V}{a}$$

$$\therefore \Delta X = X_{10} - X_6 = \frac{1}{2} \times (V_{10} - V_6) \times \left(2t_6 + \frac{V_{10} - V_6}{a}\right) + \frac{1}{a} \times V_6 \times (V_{10} - V_6)$$

$$\therefore \Delta X = X_{10} - X_6 = \frac{1}{2} \times (V_{10} - V_6) \times \left(2t_6 + \frac{V_{10} - V_6}{a}\right) + \frac{1}{a} \times V_6 \times (V_{10} - V_6)$$

$$\therefore \Delta X = X_{10} - X_6 = t_6 \times (V_{10} - V_6) + \frac{1}{2a} \times (V_{10} - V_6)^2 + \frac{1}{a} \times V_6 \times (V_{10} - V_6)$$

$$\therefore \Delta X = X_{10} - X_6 = t_6 \times (V_{10} - V_6) + \frac{1}{2a} \times (V_{10}^2 - V_6^2)$$

$$\therefore X_{10} = X_6 \times \frac{V_{10} - V_6}{V_6} + \frac{1}{2a} \times (V_{10}^2 - V_6^2)$$

$$\therefore \Delta X\big|_{t_6=0} = X_{10}\big|_{t_6=0} = \frac{1}{2a}\times\left(V_{10}^2 - V_6^2\right)$$

$$\therefore X = \frac{1}{2a}\times\left(V_f^2 - V_i^2\right)$$

$$\therefore V_{10} = \sqrt{2a\times X + V_6^2}$$

$$\therefore V_f = \sqrt{V_i^2 + 2a\times X}$$

27. A train slows down from 90 km/h to rest in 1 min. The average acceleration (a) and the total distance (X) traveled are:
    (a) $a = 1.5$ m^2/s, $X = 1.5$ km   (b) $a = 0.42$ m^2/s, $X = 2.25$ km
    (c) $a = -0.42$ m^2/s, $X = 0.75$ km   (d) $a = -1.5$ m^2/s, $X = 4.2$ km

28. A bullet train accelerates at the rate of 0.5 m/s^2 from 100 km/h for 1 min. The final velocity (V) reached in km/h and the distance traveled over this time (X) are:
    (a) $V = 208$ km/h, $X = 2.6$ km   (b) $V = 308$ km/h, $X = 6.9$ km
    (c) $V = 154$ km/h, $X = 1.1$ km   (d) $V = 154$ km/h, $X = 2.1$ km

29. A train accelerates from 10 km/h to 130 km/h for 0.5 min. It then travels at a constant velocity of 130 km/h for t s. It then decelerates in 2 min to a full stop. The total distance traveled is 4 km. The time (t), in seconds, that the train traveled at constant velocity is:
    (a) 10.4 s   (b) 20.8 s   (c) 17.3 s   (d) 34.6 s

30. A train accelerates from 10 km/h to 130 km/h for 0.5 min. It then travels at a constant velocity of 130 km/h for t s. It then decelerates in 2 min to a full stop. The total distance traveled is 4 km. Select the correct distance-time, velocity-time, and acceleration-time diagrams in Figure 13.

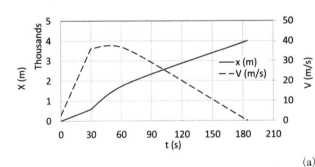 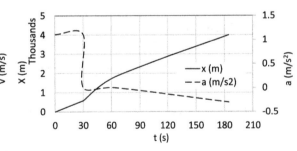

(a)

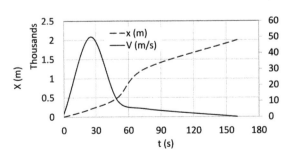 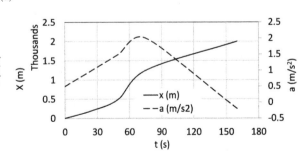

(b)

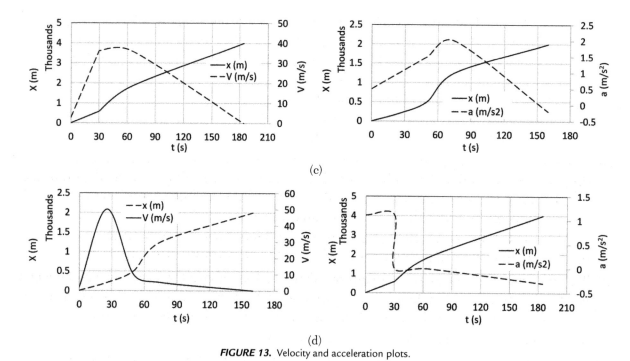

FIGURE 13. Velocity and acceleration plots.

31. An object is launched vertically from the ground into the air with an initial velocity V_i. After t s, it falls back on the ground due to the force of gravity. Ignoring air resistance, the speed of the object at the highest point (V_f) reached is:

(a) $V_f = V_i$ (b) $V_f = 0$ (c) $V_f = V_i/2$ (d) $V_f = 2V_i$

32. An object is launched vertically from the ground into the air with an initial velocity V_i. After t s, it falls back on the ground due to the force of gravity. Ignoring air resistance, the time it takes for the object to reach its highest point (t_{top}) is:

(a) $t_{top} = g/V_i$ (b) $t_{top} = V_i/g$ (c) $t_{top} = V_i/(2g)$ (d) $t_{top} = g/(2V_i)$

33. An object is launched vertically from the ground into the air with an initial velocity V_i. After t s, it falls back on the ground due to the force of gravity. Ignoring air resistance, the maximum height (H_{max}) reached is:

(a) $H_{max} = g/(2V_i^2)$ (b) $H_{max} = V_i^2/(2g)$ (c) $H_{max} = 2V_i^2/g$ (d) $H_{max} = g/V_i^2$

34. An object is launched vertically from the ground into the air with an initial velocity V_i. After t s, it falls back on the ground due to the force of gravity. Ignoring air resistance, the initial speed of the object (V_i), as a function of the total travel time, t, is:

(a) $V_i = gt$ (b) $V_i = 2gt$ (c) $V_i = -gt$ (d) $V_i = gt/2$

35. An object is launched with speed V_i vertically from the Moon's surface. 5 s after launch, it returns back to the surface due to the force of gravity. Assume the Moon's gravity force is 1/6 that on the Earth's surface. The initial speed of the object must have been equal to:

(a) 4.1 m/s   (b) 1.9 m/s   (c) 8.2 m/s   (d) 16.4 m/s

36. An object is launched with speed V_i vertically from the Moon's surface. 10 s after launch, it returns back to the surface due to the force of gravity. Assume the Moon's gravity force is 1/6 that on the Earth's surface. At 2 s after the launch, the object is traveling at V_f speed and is at h height. Ignoring air resistance, speed V_f, height h, and speed V_i are:

(a) $V_f = 4.9$ m/s, $h = 8.2$ m, $V_i = 13.1$ m/s   (b) $V_f = 13.1$ m/s, $h = 8.2$ m, $V_i = 4.9$ m/s

(c) $V_f = 4.9$ m/s, $h = 13.1$ m, $V_i = 8.2$ m/s   (d) $V_f = 13.1$ m/s, $h = 4.9$ m, $V_i = 8.2$ m/s

37. An object is launched with speed V_i vertically from the Moon's surface. 10 s after launch, it returns back to the surface due to the force of gravity. Assume the Moon's gravity force is 1/6 that on the Earth's surface. At 2 s after the launch the object is traveling at V speed and is at h height. Ignoring air resistance, the time it takes the object to cover distances of 10 m (t_{10}) and 20 m (t_{20}) along its path of motion are:

(a) $t_{10} = 1.43$ s, $t_{20} = 4.91$ s   (b) $t_{10} = 1.43$ s, $t_{20} = 7.91$ s
(c) $t_{10} = 2.91$ s, $t_{20} = 4.91$ s   (d) $t_{10} = 2.91$ s, $t_{20} = 7.91$ s

38. Two objects are dropped from an aircraft flying at 1 km above the ground level. Object 1 is dropped first at $t = 0$, followed by object 2, after Δt time delay. Object 2 hits the ground at $t = 20$s. Ignoring air resistance, the maximum speed reached by the objects (V), time it takes for object 1 to reach the ground (t), and Δt, are:

(a) $V = 140$ m/s, $t = 14.3$ s, $\Delta t = 34.3$ s   (b) $V = 14.3$ m/s, $t = 140$ s, $\Delta t = 5.7$ s
(c) $V = 140$ m/s, $t = 14.3$ s, $\Delta t = 5.7$ s   (d) $V = 14.3$ m/s, $t = 140$ s, $\Delta t = 34.3$ s

39. Two objects are dropped from an aircraft flying at 1 km above the ground level. Object 1 is dropped first at $t = 0$, followed by object 2, after a 2-s delay. Ignoring air resistance, the vertical distance traveled by object 1 during the time interval between 10 and 13 s is:

(a) 414.5 m   (b) 338.5 m   (c) 490.5 m   (d) 828.9 m

40. Two objects are dropped from an aircraft flying at 1 km above the ground level. Object 1 is dropped first, followed by object 2, after Δt time delay. Ignoring air resistance, the time it takes for object 2 to travel the last 100 m before hitting the ground is:

(a) 1.43 s   (b) 1.35 s   (c) 0.73 s   (d) 0.67 s

41. Wind is blowing at 25 knots (kt) from 220-degrees south-west. You are sailing to the 310-degrees north-west. The headwind and crosswind components are:

(a) 25 kt from the right, 0 kt   (b) 12.5 kt, 12.5 kt, from the left
(c) 25 kt, 25 kt from the right   (d) 0 kt, 25 kt from the left

42. A missile is moving toward a ship over a long-distance trajectory; speed is 3.4 km/s. The ship is located at latitude 13 N and longitude 144 W. The missile's initial coordinates are (2 N, 158 W). The velocity components (V_x, V_y), the distance between the two coordinates (missile and ship, X and Y), the distance between the line connecting the missile and the ship (d), and the direction of parallels (α), west to east (positive x is along the parallels, left to right, toward the east, and positive y is along the meridians, toward the north) are:

(a) $V_x = -2.8$ m/s, $V_y = 1.9$ m/s, $X = 1{,}070$ km, $Y = -1{,}5567$ km, $d = 1{,}889$ km, $\alpha = 145.5°$
(b) $V_x = 1.9$ m/s, $V_y = -2.8$ m/s, $X = 1{,}070$ km, $Y = -1{,}5567$ km, $d = 1{,}889$ km, $\alpha = -55.5°$
(c) $V_x = 2.8$ m/s, $V_y = -1.9$ m/s, $X = -1{,}070$ km, $Y = 1{,}5567$ km, $d = 1{,}889$ km, $\alpha = -145.5°$
(d) $V_x = -1.9$ m/s, $V_y = 2.8$ m/s, $X = -1{,}070$ km, $Y = 1{,}5567$ km, $d = 1{,}889$ km, $\alpha = 55.5°$

43. Wind is blowing at 30 kt from a compass bearing of 20 degrees. You are sitting in an aircraft on the runway ready for takeoff, with a compass showing 300 degrees. The crosswind (V_x) and headwind (V_y) components are:

(a) $V_x = 3.5$ m/s, $V_y = 39.9$ m/s   (b) $V_x = 15.2$ m/s, $V_y = 2.7$ m/s
(c) $V_x = 49.2$ m/s, $V_y = 8.7$ m/s   (d) $V_x = 8.7$ m/s, $V_y = 49.2$ m/s

44. You are flying on the compass bearing of 350 degrees with an airspeed of 100 kt. You know that the wind is 40 kt and coming from 290 degrees. The crosswind (V_x) and headwind (V_y) components as well as the magnitude of the velocity with respect to the ground (V) are:

(a) $V_x = 49.2$ m/s, $V_y = 8.7$ m/s, $V = 50.1$ m/s
(b) $V_x = 8.7$ m/s, $V_y = 49.2$ m/s, $V = 103.8$ m/s
(c) $V_x = 17.8$ m/s, $V_y = 10.3$ m/s, $V = 91.5$ m/s
(d) $V_x = 8.7$ m/s, $V_y = 49.2$ m/s, $V = 50.1$ m/s

45. The magnitude of vector $\vec{A} = X\vec{i} + Y\vec{j} + Z\vec{k}$ is 10. The X component is 5 and the Y component is 7. The Z component is:

(a) 5.1   (b) 4.5   (c) 6   (d) 2

46. For the vectors $\vec{A} = a\times b\vec{i} + b\times c\vec{j} + c\times d\vec{k}$, and $\vec{B} = -b\times c\vec{i} + a\times b\vec{j} - (a-b)\vec{k}$, identify the relationship between a, b, and c, if the vectors are perpendicular to one another. Note that a, b, c, and d, are all nonzero.

(a) $2\times a\times b^2\times c + (a-b)\times c\times d = 0$, $a = b$
(b) $-2\times a\times b^2\times c + (a-b)\times c\times d = 0$, $c\times d = 0$
(c) $(a-b)\times c\times d = 0$, $a = b$
(d) $(-a-b)\times c\times d = 0$, $c\times d = 0$

47. For the vectors $\vec{A} = a\times b\vec{i} + b\times c\vec{j} + c\times d\vec{k}$ and $\vec{C} = -b\times c\vec{i} + a\times b\vec{j} - (a-b)\vec{k}$, identify vector $\vec{B}$, if $\vec{y} = \ - \ $:

(a) $\vec{B} = b\times(a+c)\vec{i} + b\times(c-a)\vec{j} + (a-b+c\times d)\vec{k}$
(b) $\vec{B} = -b\times(a+c)\vec{i} + b\times(c-a)\vec{j} + (a-b-c\times d)\vec{k}$
(c) $\vec{B} = -b\times(a+c)\vec{i} - b\times(c-a)\vec{j} + (a-b+c\times d)\vec{k}$
(d) $\vec{B} = -b\times(a+c)\vec{i} - b\times(c-a)\vec{j} - (a-b-c\times d)\vec{k}$

48. A roller-coaster car A is located at $A(X, Y, Z)$. The roller-coaster car B is located at $B(U, V, W)$. The straight-line distance between the two cars is:

(a) $\sqrt{(X-V)^2 + (Y-U)^2 + (Z-W)^2}$ (b) $\sqrt{(X-U)^2 + (Y-W)^2 + (Z-V)^2}$

(c) $\sqrt{(X-W)^2 + (Y-V)^2 + (Z-U)^2}$ (d) $\sqrt{(X-U)^2 + (Y-V)^2 + (Z-W)^2}$

49. A missile located at $A(X, Y, Z)$ is traveling in a straight line toward location $B(U, V, W)$. The velocity vector angle (β) relative to the x-axis is:

(a) $\arccos\left(\dfrac{X\times U + Y\times V + Z\times W}{\sqrt{X^2 + Y^2 + Z^2} \times \sqrt{U^2 + V^2 + W^2}}\right)$ (b) $\arcsin\left(\dfrac{X\times U + Y\times V + Z\times W}{\sqrt{(X-U)^2 + (Y-V)^2 + (Z-W)^2}}\right)$

(c) $\arccos\left(\dfrac{X\times U + Y\times V + Z\times W}{\sqrt{(X-U)^2 + (Y-V)^2 + (Z-W)^2}}\right)$ (d) $\arcsin\left(\dfrac{X\times U + Y\times V + Z\times W}{\sqrt{X^2 + Y^2 + Z^2} \times \sqrt{U^2 + V^2 + W^2}}\right)$

50. A roller-coaster car A is located at $A(0, b, c)$. The roller-coaster car B is located at $B(a, 0, 0)$. The straight-line distance between the two cars is:

(a) $\sqrt{a^2 + b^2 + c^2}$ (b) $\sqrt{-a^2 + b^2 + c^2}$

(c) $\sqrt{a^2 - b^2 - c^2}$ (d) $\sqrt{-a^2 - b^2 + c^2}$

51. A missile located at $A(a, 0, c)$ moves toward location $B(0, b, c)$. The missile's angle (β) relative to the x-axis is:

(a) $\arcsin\left(\dfrac{c^2}{\sqrt{a^2+c^2}\times\sqrt{b^2+c^2}}\right)$

(b) $\arccos\left(\dfrac{c^2}{\sqrt{a^2+c^2}\,\sqrt{b^2+c^2}}\right)$

(c) $\arcsin\left(\dfrac{ab+bc+ac}{\sqrt{a^2+c^2}\times\sqrt{b^2+c^2}}\right)$

(d) $\arccos\left(\dfrac{ab+bc+ac}{\sqrt{a^2+c^2}\times\sqrt{b^2+c^2}}\right)$

52. A missile located at $A(a, b, c)$ moves toward location $B(a, b, 0)$. The missile's angle (β) relative to the x-axis is:

(a) $\arcsin\left(\dfrac{a^2+b^2}{\sqrt{a^2+b^2+c^2}\times\sqrt{a^2+b^2}}\right)$

(b) $\arccos\left(\dfrac{a^2+b^2}{\sqrt{a^2+b^2+c^2}\times\sqrt{a^2+b^2}}\right)$

(c) $\arcsin\left(\dfrac{ab+bc+ac}{\sqrt{a^2+c^2}\times\sqrt{b^2+c^2}}\right)$

(d) $\arccos\left(\dfrac{ab+bc+ac}{\sqrt{a^2+c^2}\times\sqrt{b^2+c^2}}\right)$

53. A missile located at $A(b, c, 0)$ moves toward location $B(0, 0, d)$. It takes t s to reach the destination. Assuming that the acceleration is given by $\vec{a} = (4a, 2a)$, and the initial velocity vector is $\vec{V} = (V_x, V_y, V_z)$, the relationship between the distance traveled, initial velocity, and acceleration relative to the x-axis is:

(a) $b\vec{i} + c\vec{j} + d\vec{k} = -\dfrac{a}{2}\times\left(4\vec{i}+2\vec{j}+\vec{k}\right)\times t^2 + \left(V_x\vec{i}+V_y\vec{j}+V_z\vec{k}\right)$

(b) $b\vec{i} + c\vec{j} - d\vec{k} = -a\times\left(2\vec{i}+\vec{j}+\vec{k}\right)\times t^2 + \left(V_x\vec{i}+V_y\vec{j}+V_z\vec{k}\right)$

(c) $b\vec{i} + c\vec{j} + d\vec{k} = \dfrac{a}{2}\times\left(4\vec{i}+2\vec{j}\right)\times t^2 + \left(V_x\vec{i}+V_y\vec{j}+V_z\vec{k}\right)\times t$

(d) $b\vec{i} + c\vec{j} - d\vec{k} = a\times\left(2\vec{i}+\vec{j}\right)\times t^2 + \left(V_x\vec{i}+V_y\vec{j}+V_z\vec{k}\right)\times t$

54. A rock initially at (X, Y, Z) is launched with velocity magnitude V_0 at a positive angle α degrees relative to the horizontal $x - z$ plane. It experiences an acceleration vector $\vec{a} = (0, -g, 0)$ and takes t s to reach its highest point under gravity. If the initial velocity vector is $\vec{V} = (V_x, V_y, V_z)$, the maximum height (H_{max}) reached is:

(a) $H_{max} = \dfrac{V_x^2}{2g} = \dfrac{V_0^2\times\cos^2\alpha}{2g}$

(b) $H_{max} = \dfrac{V_y^2}{2g} = \dfrac{V_0^2\times\sin^2\alpha}{2g}$

(c) $H_{max} = \dfrac{V_x^2}{g} = \dfrac{V_0^2\times\cos^2\alpha}{g}$

(d) $H_{max} = \dfrac{V_y^2}{g} = \dfrac{V_0^2\times\sin^2\alpha}{g}$

55. A rock initially at (X, Y, Z) is launched with velocity magnitude V_0 at a positive angle α degrees relative to the horizontal $x - z$ plane. It experiences an acceleration vector $\vec{a} = (0, -g, 0)$ and takes t s to reach its highest point under gravity. The initial velocity vector is $\vec{V} = (V_x, V_y, 0)$ and the maximum height reached is H_{max}. The range (maximum distance traveled) (R) is:

(a) $R = X + V\times t = X + \dfrac{\times\cos(2\alpha)}{}$

(b) $R = Y + V_x\times t = H_{max} + \dfrac{V_0^2\times\sin\alpha}{2g}$

(c) $R = Y + V_y\times t = H_{max} + \dfrac{V_0^3\times\cos\alpha}{2g}$

(d) $R = X + V_x\times t = X + \dfrac{V_0^2\times\sin(2\alpha)}{g}$

56. A small projectile initially located on the ground surface is launched into the air at an angle of 60 degrees relative to the horizontal plane with the initial velocity of 100 km/h. Assume air resistance can be ignored. Maximum height (H_{max}), horizontal range (R), and total time (t) in the air are:

(a) $H_{max} = 29.5$ m, $R = 68.1$ m, $t = 4.9$ s
(b) $H_{max} = 58.9$ m, $R = 34.1$m, $t = 2.4$ s
(c) $H_{max} = 29.5$ m, $R = 68.1$m, $t = 4.9$ s
(d) $H_{max} = 58.9$ m, $R = 34.1$m, $t = 2.4$ s

57. A small projectile initially located at coordinates (100, 200, 300) m is launched into the air at an angle of 45 degrees relative to the horizontal $x - z$ plane with the initial velocity of 45 m/s. It takes t s to reach its highest point under gravity. Maximum height (H_{max}), horizontal range (R), time (t), and final position vector ($\vec{A}$) are:

(a) $H_{max} = 206.4$ m, $R = 51.6$ m, $t = 4.2$ s, $A = (252, 200, 352)$
(b) $H_{max} = 51.6$ m, $R = 206.4$ m, $t = 4.2$ s, $A = (252, 306, 200)$
(c) $H_{max} = 51.6$ m, $R = 206.4$ m, $t = 3.2$ s, $A = (306, 252, 300)$
(d) $H_{max} = 206.4$ m, $R = 51.6$ m, $t = 3.2$ s, $A = (306, 200, 352)$

58. A small projectile initially at $A(0, 0, 0)$ is launched into the air at an angle α relative to the horizontal plane. x-axis is horizontal and y-axis is vertical. Neglect air resistance. It takes t s to reach its highest point under gravity. Assuming that the initial velocity magnitude is V, find a relation for range (R) versus height reached (H_{max}), and time the maximum height is reached (t):

(a) $R = \left(\dfrac{V \times H_{max}}{t \times g}\right)\sqrt{1 - \left(\dfrac{2H_{max}}{t \times V}\right)^2}$ (b) $R = 2t \times \sqrt{1 - \left(\dfrac{H_{max}}{t \times V}\right)^2}$

(c) $R = \left(\dfrac{V \times H_{max}}{t \times g}\right)\sqrt{1 - \left(\dfrac{H_{max}}{t \times V}\right)^2}$ (d) $R = 4H_{max}t \times \sqrt{\left(\dfrac{V^2}{2g \times H_{max}}\right) - 1}$

59. A ball at location $A(X_1, Y_1)$ is launched, intended to hit location $B(X_2, Y_2)$. Assuming that the initial velocity magnitude is V, find the angle (α) required for the ball to hit point B:

(a) $\arctan\left(\dfrac{Y_2 - Y_1}{X_2 - X_1} + \dfrac{g}{2} \times \dfrac{X_2 - X_1}{V^2}\right)$ (b) $\arcsin\left(\dfrac{Y_2 - Y_1}{2 - 1} + \dfrac{g}{2} \times \dfrac{X_2 - X_1}{2}\right)$

(c) $\arcsin\left(\dfrac{1}{X_2 - X_1} \times \left(Y_2 - Y_1 + \dfrac{g}{2} \times t^2\right)\right)$ (d) $\arctan\left(\dfrac{1}{X_2 - X_1} \times \left(Y_2 - Y_1 + \dfrac{g}{2} \times t^2\right)\right)$

60. A ball at location $A(10, 20)$ is thrown, intended to hit location $B(30, 17)$. Assuming that the initial velocity is 100 m/s, find the angle (α) required for the ball to hit point B (x-axis is horizontal and y-axis is vertical, both pointed upside):

(a) 7 degrees   (b) 9 degrees   (c) 10 degrees   (d) 8 degrees

61. The third Newton's law states that:

(a) For every action there is an equal and opposite reaction.
(b) The force applied to an object is equal to the change of momentum per change in time.
(c) Every object either remains stationary or moves under the influence of a force.

(d) Acceleration of two equal masses under the influence of the same force are the same in all directions.

62. Thrust to drag is like lift to:
    (a) displacement
    (b) acceleration
    (c) reaction force
    (d) weight

63. A jet is taking off. Velocity (V) of the jet with mass m that accelerates at a (at an instant) with air density ρ and drag coefficient C_D (assuming that the projected cross-sectional area of the body perpendicular to flow direction is A and thrust is F) at the given instant, is equal to:

    (a) $\sqrt{\dfrac{2(F - m \times a)}{\rho \times C_D}}$

    (b) $\sqrt{\dfrac{2 \times (F - m \times a)}{\rho \times C_D \times A}}$

    (c) $\sqrt{\dfrac{2a}{m \times C_D \times A \times F}}$

    (d) $\sqrt{\dfrac{2a \times F}{m \times C_D}}$

64. If you throw a fast-spinning ball into the air, the ball's trajectory will be affected by its spin due to the:
    (a) Magnus effect
    (b) Coandă effect
    (c) Bernoulli effect
    (d) Kepler effect

65. The Coandă effect is the tendency of a fluid jet to:
    (a) not stay attached to a convex surface
    (b) stay attached to a concave surface
    (c) not stay attached to a concave surface
    (d) stay attached to a convex surface

66. The tendency of the air flow to remain attached to the ball surface when it is moving through the air is an indication of the:
    (a) force applied to the air by the ball   (b) ball acceleration within the air
    (c) force applied to ball by the air   (d) air acceleration hitting the ball

67. If a cylinder moves through the air while spinning about its axis, if flow velocity is V, the generated Kutta-Joukowski lift (F_L) as a function of the generated vortex with strength G is estimated by (assume cylinder length is L, air fluid density is ρ, rotational speed is V_r, and radius of cylinder is r):
    (a) $F_L = \rho \times V \times G$, $G = 2\pi \times r \times V_r$
    (b) $F_L = \rho \times V^2 \times G \times L$, $G = \pi \times r \times V_r$
    (c) $F_L = \rho \times V \times G \times L$, $G = 2\pi \times r \times V_r$
    (d) $F_L = \rho \times V^2 \times G$, $G = \pi \times r \times V_r$

68. The first Newton's law states that:
    (a) the force applied to an object is equal to the change of momentum per change in time.
    (b) every object either remains stationary or moves under the influence of a net force at constant velocity.
    (c) acceleration of two equal masses under the influence of the same force is the same in all directions.
    (d) for every action there is an equal and opposite reaction.

69. The second Newton's law states that:
    (a) the net force applied to an object is equal to the change of momentum per change in time.

(b) every object either remains stationary or moves under the influence of a force.

(c) acceleration of two equal masses under the influence of the same force is the same in all directions.

(d) for every action there is an equal and opposite reaction.

70. According to the second Newton's law, acceleration is caused by the . . . force applied on an object.

(a) derivative of the   (b) absolute

(c) resultant   (d) horizontal

71. A net force F is applied to a ball of mass m that is initially at rest. After time t, the ball speed (V) will be given by:

(a) $V = \dfrac{F \times m}{t}$ (b) $V = \dfrac{m \times t}{F}$ (c) $V = \dfrac{m \times F}{t}$ (d) $V = \dfrac{F \times t}{m}$

72. An object with mass m_1 moves at constant speed (V_2), and then accelerates to V_1 in t s and remains at a constant speed. Another object with mass m_2, moving in the opposite direction at constant speed (V_2), hits the first object. The objects then join and move together reaching a speed V that is equal to:

(a) $V = \dfrac{m_1 \times V_1 - m_2 \times V_2}{m_1 + m_2}$ (b) $V = \dfrac{V_1 + V_2}{m_1 + m_2}$

(c) $V = \dfrac{m_1 + m_2}{m_1 \times V_1 - m_2 \times V_2}$ (d) $V = \dfrac{m_1 + m_2}{V_1 + V_2}$

73. An object with mass m is stationary on a flat horizontal surface. Contact force that is applied to the object is:

(a) 0   (b) m × g/2   (c) m × g   (d) 2 m × g

74. Person A is standing on the ground having a conversation with his colleague, person B. Assume the third Newton's law did not apply to person A but did apply to person B. Your expectations are:

(a) Person A carries on with his tasks while person B passes through the ground.

(b) Person A and person B both pass through the ground.

(c) Person A and person B do not pass through the ground.

(d) Person A passes through the ground because the ground does not react, and person B remains on the ground due to the gravity force.

75. Two objects with masses m_1 and m_2 connected by a rope are being accelerated upward along a sloping surface with an angle θ and a friction coefficient of μ (Figure 14). Write the relationship between the thrust force (Thrust) required to accelerate the two objects with a positive acceleration (a) and m_1, m_2, θ, μ, and cable tension (T).

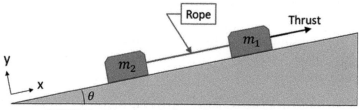

FIGURE 14. Objects on a sloping surface connected by a cable.

1. $\left(\sum F_y\right)_{m_1} = -m_1 \times g \times \cos\theta + N_1 = 0 \rightarrow N_1 = m_1 \times g \times \cos\theta$

2. $\left(\sum F_y\right)_{m_2} = -m_2 \times g \times \cos\theta + N_2 = 0 \rightarrow N_2 = m_2 \times g \times \cos\theta$

3. $\left(\sum F_x\right)_{m_1} = \text{Thrust} - m_1 \times g \times (\sin\theta + \mu \times \cos\theta) - T = m_1 \times a$

4. $\left(\sum F_x\right)_{m_2} = T - m_2 \times g \times (\sin\theta + \mu \times \cos\theta) = m_2 \times a$

5. $\left(\sum F_x\right)_{m_1} = \text{Thrust} + m_1 \times g(\sin\theta - \mu \times \cos\theta) + T = m_1 \times a$

6. $\left(\sum F_x\right)_{m_2} = m_2 \times g \times (\sin\theta - \mu \times \cos\theta) - T = m_2 \times a$

7. $\left(\sum F_x\right)_{m_1} = \text{Thrust} - \mu \times m_1 \times g - T = m_1 \times a$

8. $\left(\sum F_x\right)_{m_2} = T - \mu \times m_2 \times g = m_2 \times a$

(a) (1), (3), (5), and (7)
(b) (2), (4), (6), and (8)
(c) (1), (2), (3), and (4)
(d) (1), (2), (3), and (5)

76. Two objects with masses m_1 and m_2 connected by a rope are being accelerated downward along a sloping surface with an angle θ and a friction coefficient of μ (Figure 15). Write the relationship between the thrust force (Thrust) required to accelerate the two objects with a positive acceleration (a) and m_1, m_2, θ, μ, and cable tension (T).

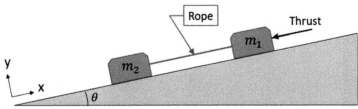

FIGURE 15. Objects on a sloping surface connected by a cable.

1. $\sum F_y = -m_1 \times g \times \cos\theta + N_1 = 0 \rightarrow N_1 = m_1 \times g \times \cos\theta$

2. $\sum F_y = -m_2 \times g \times \cos\theta + N_2 = 0 \rightarrow N_2 = m_2 \times g \times \cos\theta$

3. $\left(\sum F_x\right)_{m_1} = \text{Thrust} - m_1 \times g \times (\sin\theta + \mu \times \cos\theta) - T = m_1 \times a$

4. $\left(\sum F_x\right)_{m_2} = T - m_2 \times g \times (\sin\theta + \mu \times \cos\theta) = m_2 \times a$

5. $\left(\sum F_x\right)_{m_1} = \text{Thrust} + m_1 \times g(\sin\theta - \mu \times \cos\theta) + T = m_1 \times a$

6. $\left(\sum F_x\right)_{m_2} = m_2 \times g \times (\sin\theta - \mu \times \cos\theta) - T = m_2 \times a$

7. $\left(\sum F_x\right)_{m_1} = \text{Thrust} - \mu \times m_1 \times g - T = m_1 \times a$

8. $\left(\sum F_x\right)_{m_2} = T - \mu \times m_2 \times g = m_2 \times a$

(a) (1), (3), (5), and (7)   (b) (2), (4), (6), and (8)
(c) (1), (2), (5), and (6)   (d) (1), (2), (3), and (5)

77. Two masses (m_1 and m_2) are connected by two ropes in a fixed pulley system hanging from the ceiling. This system is situated on the Moon's surface, where gravity force is 1/6 that on the Earth's surface (Figure 16). Write the expression for the tension (T) in the cable. The acceleration expression when $m_1 \gg m_2$ is (assume pulley and rope are massless):

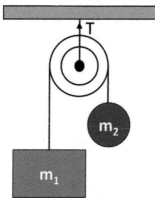

FIGURE 16. Objects connected by two ropes in a fixed pulley system.

(a) $T = \dfrac{m_1 \times m_2}{(m_1 + m_2)} \times \dfrac{2 \times g}{3}$, $a(m_1 \gg m_2) = \dfrac{g}{6}$ (b) $T = \dfrac{(m_1^2 + m_2^2)}{(m_1 + m_2)} \times \dfrac{g}{3}$, $a(m_1 \gg m_2) = \dfrac{g}{3}$

(c) $T = \dfrac{(m_1^2 + m_2^2)}{(m_1 + m_2)} \times \dfrac{2 \times g}{3}$, $a(m_1 \gg m_2) = \dfrac{g}{6}$ (d) $T = \dfrac{m_1 \times m_2}{(m_1 + m_2)} \times \dfrac{g}{3}$, $a(m_1 \gg m_2) = \dfrac{g}{3}$

78. The Neptune's mass is m_N and the Earth's mass is m_E. The acceleration due to gravity on the Neptune's surface is 1.36 times that on the Earth; the Neptune's radius is 3.6 times that of the Earth. The ratio of the Neptune's mass to that of the Earth is:

   (a) $\dfrac{m_N}{m_E} = 23.3z$

   (b) $\dfrac{m_N}{m_E} = 11.6$

   (c) $\dfrac{m_N}{m_E} = 17.2$

   (d) $\dfrac{m_N}{m_E} = 8.61$

79. A three-tier cake is being moved by a motorized cart on a level surface at a steady speed of 1 m/s. The base tier has mass m and each subsequent tier has a mass equal to 1/3 of the previous tier. The cart mass is 10 m. If the cart is generating a thrust of 10 N and the coefficient of friction between the cart and the ground surface is 0.25, the mass of the top cake tier is:

   (a) 40 g   (b) 356 g   (c) 119 g   (d) 4 kg

80. A four-tier cake is being pulled by a caterer on a cart to the event venue. The base tier has mass m and each subsequent tier has a mass equal to 1/2 of the previous tier. The cart mass is 12 m. The cart gets stuck at one point, and the caterer applies extra force to get it moving. The cart accelerates, reaching speed of 3 m/s in 1 s. If the coefficient of friction between the cart and the ground surface is 0.34, the force applied by the caterer (F) and the mass (m_{top}) of the top cake tier are (the threshold force to overcome the friction is 30 N):

   (a) $F = 36$ N, $m_{top} = 324$ g   (b) $F = 36$ N, $m_{top} = 81$ g
   (c) $F = 57$ N, $m_{top} = 162$ g   (d) $F = 57$ N, $m_{top} = 778$ g

81. On an aircraft, the center of gravity (COG) is an important characteristic which determines whether the plane can operate safely. The COG location must be within certain limits. The location is measured along the longitudinal axis of the aircraft and referenced to a firewall that separates the forward engine compartment from the cockpit. For this particular

aircraft, the pilot and the first officer are sitting side-by-side at a distance of 2 m from the firewall, and they each weigh 90 kg. The box of cargo is located 9 m from the firewall and weighs 150 kg. The fueled aircraft without cargo and crew *COG* is 8 m from the firewall and its weight is 2,500 kg. If the safely loaded aircraft *COG* must be between 7.1 and 7.5 m from the firewall, is the aircraft airworthy for the current load, and what is the location of its *COG* relative to the firewall?

(a) No, 7.7 m   (b) Yes, 7.3 m   (c) Yes, 7.4 m   (d) No, 7.6 m

82. A bag of flour is thrown from an aircraft while flying over a hangar. The bag hits a spot on the ground. This spot is located at a horizontal distance d from the point of the bag's release and at a height h from the ground. The angle of the bag's initial velocity (V) vector relative to the horizon is:

(a) $\arctan\left(\dfrac{h}{d}+\dfrac{1}{2}\times g\times\dfrac{d}{V^2}\right)$

(b) $\arcsin\left(\dfrac{h}{d}+\dfrac{1}{2}\times g\times\dfrac{d}{V^2}\right)$

(c) $\arctan\left(\dfrac{1}{d}\times\left(h+\dfrac{g}{2}\times t^2\right)\right)$

(d) $\arcsin\left(\dfrac{1}{d}\times\left(h+\dfrac{g}{2}\times t^2\right)\right)$

83. The impulse can be described as an integral of the resultant force on an object. (T/F)

84. The main difference between impulse and impact is:

(a) Impact is related to momentum force; impulse depends on force.
(b) Impact is equal to force by time change; impulse is equal to mass by velocity change.
(c) Impact is change in momentum; impulse is application of a force over a short time.
(d) Impact is equal to mass by velocity change; impulse is equal to force by time change.

85. An object of 500 g mass resting on a frictionless horizontal surface experiences an impact resulting in a horizontal force of 100 N over 0.1 s. The object velocity after the impact is:

(a) 20 m/s   (b) 10 m/s   (c) 5 m/s   (d) 15 m/s

86. A pitcher throws a ball of mass m to the batter. The ball impacts the bat with speed V_1 and is directed straight back toward the pitcher. Next, a second pitch is thrown with a higher speed V_2 and a ball that is slightly heavier (by Δm) than the first ball. The ball is similarly hit by the batter. The change in hit impulse from the first to the second pitch is:

(a) $m\times(V_1-V_2)+\Delta m\times V_2$

(b) $m\times(V_1+V_2)-\Delta m\times V_2$

(c) $m\times(V_1+V_2)+\Delta m\times V_2$

(d) $m\times(V_1-V_2)-\Delta m\times V_2$

87. In one case, object A with mass of m_A and velocity of V_A is launched directly toward a stationary object C with mass m_C. In another case, object B with mass of m_B and velocity of V_B is also launched directly toward the stationary object C. The inelastic collisions in both cases create the same impacts. Object A is lighter than object B. The relationship between the two velocities is:

(a) $V_A > V_B$ (b) $V_A < V_B$ (c) $V_A = V_B$ (d) Not related

88. A ball is dropped from a height d and it falls due to gravity until hitting the ground. If the ball's mass is m, the work done is W, applied force is F, and the acceleration is g, the relationship between the ball's distance traveled, mass, acceleration, and work is:

(a) $d = W/(m \times g)$ (b) $m \times \Delta V = W$ (c) $g = W/m$ (d) $F = m \times \Delta V/gW$

89. An object is dropped from a height of 50 cm and falls to the ground due to the gravity force. If the object mass is 10 kg, the work done is:

(a) 490 J (b) 4.9 J (c) 49 J (d) 4,900 J

90. A 10-kg object is dropped from a height of 10 m and falls to the ground due to the gravity force. If the object momentum at height h is 100 kgm/s, the impact (J) on the ground and height (h) are:

(a) $J = 15$ kgm/s, $h = 1.4$ m (b) $J = 5$ kgm/s, $h = 4.5$ m
(c) $J = 20$ kgm/s, $h = 9.5$ m (d) $J = 140$ kgm/s, $h = 5.1$ m

91. A 10-kg object is dropped from a height of 5 m and falls to the ground due to the gravity force. If the object momentum at height h is 100 kgm/s, speed (V) of the object when hitting the ground and impact time (t) are:

(a) $V = 10$ m/s, $t = 1$ s (b) $V = 10$ m/s, $t = 0.01$ s
(c) $V = 0.1$ m/s, $t = 1$ s (d) $V = 0.1$ m/s, $t = 0.01$ s

92. A 10-kg object is dropped from a height of 50 m and falls to the ground due to the gravity force. If the object momentum at height h is 200 kgm/s, h is:

(a) 20.4 m (b) 10.2 m (c) 13.6 m (d) 40.7 m

93. If two objects with masses m_1 and m_2 and initial speeds V_1 and V_2, moving in the same direction, hit one another elastically, the speed of the objects after separation (V_1' and V_2'), are:

(a) $V_1' = \left(\dfrac{m_1 + m_2}{m_1 - m_2}\right)V_1 + \left(\dfrac{2m_2}{m_1 - m_2}\right)V_2, V_2' = \left(\dfrac{2m_1}{m_1 - m_2}\right)V_1 + \left(\dfrac{m_1 + m_2}{m_1 - m_2}\right)V_2$

(b) $V_1' = \left(\dfrac{2m_1}{m_1 - m_2}\right)V_1 + \left(\dfrac{m_1 + m_2}{m_1 - m_2}\right)V_2, V_2' = \left(\dfrac{m_1 + m_2}{m_1 - m_2}\right)V_1 + \left(\dfrac{2m_2}{m_1 - m_2}\right)V_2$

(c) $V_1' = \left(\dfrac{m_1 - m_2}{m_1 + m_2}\right)V_1 + \left(\dfrac{2m_2}{m_1 + m_2}\right)V_2, V_2' = \left(\dfrac{2m_1}{m_1 + m_2}\right)V_1 + \left(\dfrac{m_1 - m_2}{m_1 + m_2}\right)V_2$

(d) $V_1' = \left(\dfrac{2m_1}{m_1 + m_2}\right)V_1 + \left(\dfrac{m_1 - m_2}{m_1 + m_2}\right)V_2, V_2' = \left(\dfrac{m_1 - m_2}{m_1 + m_2}\right)V_1 + \left(\dfrac{2m_2}{m_1 + m_2}\right)V_2$

94. Three balls, the first two having the same mass (m) and velocity (V), are traveling in the positive direction and along one line. Ball 3 is moving toward balls 1 and 2 in the negative direction at a velocity that is double the velocity of the first two. Ball 2 is ahead of ball 1 by a time interval of t s. Ball 2 first hits ball 3 and comes to a full stop. Ball 3 continues moving in the positive direction. In an instant ball 1 collides with balls 2 and 3 and the three balls join, moving in the positive direction at a velocity of 0.5 V. Assume a one-dimensional Newtonian collision. The mass of ball 3 is:

(a) 2m/5 (b) 5m/2 (c) 2m/3 (d) 5m/3

95. Three balls, the first two having the same mass (m) and velocity (V), are traveling in the positive direction and along one line. Ball 3 is moving toward balls 1 and 2 in the negative direction at a velocity that is double the velocity of the first two. Ball 2 is ahead of ball 1 by a time interval of t s. Ball 2 first hits ball 3 and comes to a full stop. Ball 3 continues moving in a positive direction. In an instant ball 1 collides with balls 2 and 3 and the three balls join, moving in the positive direction at a velocity of 0.5 V. Assume a one-dimensional Newtonian collision. The velocity and direction of ball 3 after being hit by ball 2 is:

(a) 7V, positive   (b) 7V, negative   (c) V/7, positive   (d) V/7, negative

96. Three balls, the first two having the same mass (m) and velocity (V), are traveling in the positive direction and along one line. Ball 3 is moving toward balls 1 and 2 in the negative direction at a velocity that is double the velocity of the first two. Balls 1 and 2 are traveling with a time delay t s. Ball 2 first hits ball 3, and ball 3 comes to a full stop. Ball 2 continues moving in the positive direction. In an instant ball 1 collides with balls 2 and 3 and the three balls join, moving in the negative direction at a velocity of V. Assume a one-dimensional Newtonian collision. The mass of ball 3 is:

(a) 4m   (b) m/4   (c) 2m/5   (d) 5m/2

97. Three balls, the first two having the same mass (m) and velocity (V), are traveling in the positive direction and along one line. Ball 3 is moving toward balls 1 and 2 in the negative direction at a velocity that is double the velocity of the first two. Balls 1 and 2 are traveling with a time delay t s. Ball 2 first hits ball 3, and ball 3 comes to a full stop. Ball 2 continues moving in an unknown direction. In an instant ball 1 collides with balls 2 and 3 and the three balls join, moving in the negative direction at a velocity of V. Assume a one-dimensional Newtonian collision. The velocity and direction of ball 2 after being hit by ball 3 is:

(a) 7V, negative   (b) 7V, positive   (c) V/7, negative   (d) V/7, positive

Answer Key									
1. F	**2.** (a)	**3.** (c)	**4.** (c)	**5.** (d)	**6.** (a)	**7.** (b)	**8.** (a)	**9.** (a)	**10.** (b)
11. (b)	**12.** (d)	**13.** (c)	**14.** (a)	**15.** (c)	**16.** (a)	**17.** (d)	**18.** (a)	**19.** (d)	**20.** (a)
21. (c)	**22.** (a)	**23.** (d)	**24.** (a)	**25.** (d)	**26.** (b)	**27.** (c)	**28.** (a)	**29.** (d)	**30.** (a)
31. (b)	**32.** (b)	**33.** (b)	**34.** (d)	**35.** (a)	**36.** (c)	**37.** (a)	**38.** (c)	**39.** (b)	**40.** (c)
41. (d)	**42.** (a)	**43.** (b)	**44.** (c)	**45.** (a)	46. (c)	**47.** (a)	**48.** (d)	**49.** (a)	**50.** (a)
51. (b)	**52.** (b)	**53.** (d)	**54.** (b)	**55.** (d)	**56.** (a)	**57.** (c)	**58.** (d)	**59.** (d)	**60.** (d)
61. (a)	**62.** (d)	**63.** (b)	**64.** (a)	**65.** (d)	**66.** (a)	**67.** (c)	**68.** (b)	**69.** (a)	**70.** (c)
71. (d)	**72.** (a)	**73.** (c)	**74.** (d)	**75.** (c)	**76.** (c)	**77.** (a)	**78.** (c)	**79.** (a)	**80.** (c)
81. (a)	**82.** (c)	**83.** T	**84.** (c)	**85.** (a)	**86.** (d)	**87.** (b)	**88.** (a)	**89.** (c)	**90.** (d)
91. (a)	**92.** (a)	**93.** (c)	**94.** (a)	**95.** (c)	**96.** (a)	**97.** (a)			

CHAPTER 2

PROBABILITY AND STATISTICAL ANALYSIS

1. The difference between an experiment and a calculation is that the outcome of the former is . . . while that of the latter is to be . . .
   (a) known / observed   (b) observed / known
   (c) known / known   (d) observed / observed

2. What is the probability of occurrence of heads if a fair coin is flipped once?
   (a) 0%   (b) 50%   (c) 100%   (d) 75%

3. What is the probability of getting tails every time if a fair coin is flipped 3 times?
   (a) 10%   (b) 25%   (c) 12.5%   (d) 15%

4. A statistician is asked to check whether a six-sided die is fair. She rolls the die 2 times and finds that 5 comes up on 9% of the occasions. She concludes that the die is unfair. What would be her boss's fair reaction? Analyze each of the following cases.
   1. Informs the statistician that her services are no longer required.
   2. Thanks the statistician for her smart observation and doubles her salary.
   3. Carries on with his regular work and ignores her findings.
   4. Reprimands her, stating that the observation was due yesterday.
   5. Forms a harassment team, following the statistician on her future job interviews.
   6. Investigates the supporting evidence.
   (a) (1) and (3)   (b) (2)   (c) (3) and (5)   (d) (4) and (5)

Solution Guide

It could be that the die is not fair, which makes any of the previous reactions obsolete, not to mention that some are illegal on their own. Investigation is always the first reasonable reaction. Having said that, the investigator is to be reliable and not bought by the company. Investigators are also fact finders in the majority of cases, not verdict providers, which constitutes conflict of interest.

5. COVID19 stats published by Canada on the cumulative number of confirmed cases from January 30, 2020 until April 08, 2020 are shown in Figure 17; data taken from the Government of Canada Public Health website [5]. Data is presented for the period of 68 days.

Select the correct plot showing the cumulative number of confirmed cases (Figure 18). Identify the applicable formula to the cumulative diagram.

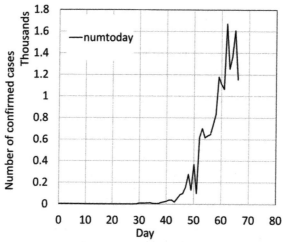

FIGURE 17. Number of daily confirmed cases over time.

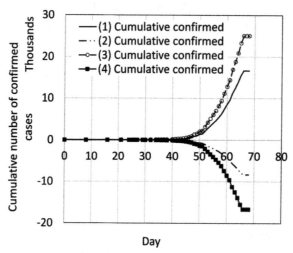

FIGURE 18. Number of cumulative daily confirmed cases over time.

(a) (1), $f(x) = \sum_{i=1}^{x} (\text{numtoday})_i$

(b) (3), $f(x) = \sum_{i=1}^{x} (\text{numtoday})_i/2$

(c) (2), $f(x) = \sum_{i=1}^{68} (\text{numtoday})_i$

(d) (4), $f(x) = \sum_{i=1}^{68} (\text{numtoday})_i/2$

6. COVID19 stats published by Canada on the cumulative number of confirmed cases from January 30, 2020 until April 08, 2020 are shown in Figure 19; cumulative data taken from the Government of Canada Public Health website [5]. Data is presented for the period of 68 days. Select the correct plot showing the cumulative number of deaths (Figure 20). Identify the applicable formula to the cumulative diagram.

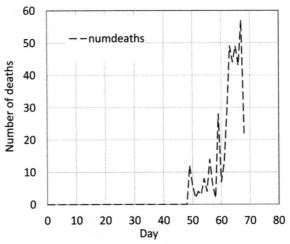

FIGURE 19. Number of daily deaths over time.

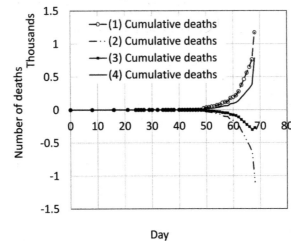

FIGURE 20. Number of cumulative deaths over time.

(a) (3), $f(x) = \sum_{i=1}^{x} (\text{numdeath})_i/2$

(b) (4), $f(x) = \sum_{i=1}^{x} (\text{numdeath})_i$

(c) (2), $f(x) = \sum_{i=1}^{68} (\text{numdeath})_i$

(d) (1), $f(x) = \sum_{i=1}^{68} (\text{numdeath})_i/2$

7. COVID19 stats published by Canada on the cumulative number of confirmed cases from January 30, 2020 until April 08, 2020 are shown in Figure 21; data taken from the Government of Canada Public Health website [5]. Data is presented for the period of 68 days. Select the correct plot showing the percentage of cumulative number of confirmed cases to number tested in Figure 22. Identify the applicable formula to the cumulative diagram.

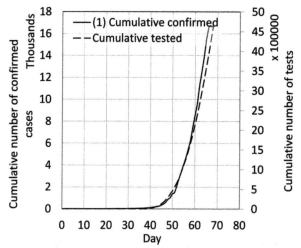

FIGURE 21. Number of cumulative confirmed cases and tests over time.

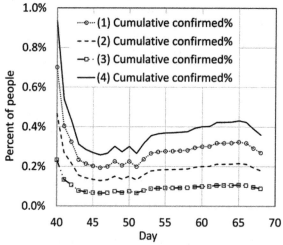

FIGURE 22. Percentage of cumulative confirmed cases to number tested over time.

(a) (1), $f(x) = \sum_{i=1}^{x}(\text{numconf})_i/2 \Big/ \sum_{i=1}^{x}(\text{numtested})_i$

(b) (3), $f(x) = \sum_{i=1}^{68}(\text{numconf})_i/2 \Big/ \sum_{i=1}^{68}(\text{numtested})_i$

(c) (2), $f(x) = \sum_{i=1}^{68}(\text{numconf})_i \Big/ \sum_{i=1}^{68}(\text{numtested})_i$

(d) (4) $f(x) = \sum_{i=1}^{x}(\text{numconf})_i \Big/ \sum_{i=1}^{x}(\text{numtested})_i$

8. COVID19 stats published by Canada on the cumulative number of deaths from January 30, 2020 until April 08, 2020 are shown in Figure 23 (data taken from the Government of Canada Public Health website [5]. Data is presented for the period of 68 days. Select the accurate percentage of cumulative number of deaths per number of confirmed cases in Figure 24. Identify the applicable formula to the cumulative diagram.

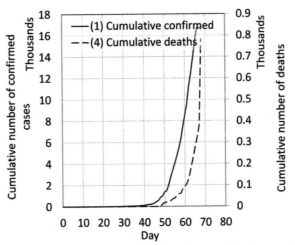

FIGURE 23. Number of cumulative confirmed cases and deaths over time.

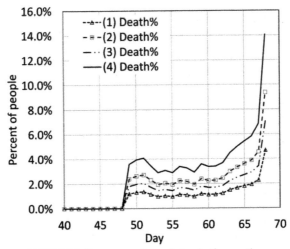

FIGURE 24. Percentage of cumulative deaths over time.

(a) $f(x) = \sum_{i=1}^{68} (\text{numdeaths})_i \bigg/ \sum_{i=1}^{68} (\text{numtoday})_i$

(b) $f(x) = \sum_{i=1}^{68} (\text{numdeaths})_i / 2 \bigg/ \sum_{i=1}^{68} (\text{numtoday})_i$

(c) $f(x) = \sum_{i=1}^{x} (\text{numdeaths})_i \bigg/ \sum_{i=1}^{x} (\text{numtoday})_i$

(d) $f(x) = \sum_{i=1}^{x} (\text{numdeaths})_i / 2 \bigg/ \sum_{i=1}^{x} (\text{numtoday})_i / 2$

9. The sample space is the list of . . . outcomes.

   (a) all possible   (b) impossible   (c) unrelated   (d) distinct

10. The sample space for an eight-sided die, numbered 4 to 11, is:

   (a) all eight sides of the die
   (b) all or none of the eight sides of the die
   (c) none of the eight sides of the die
   (d) any of the sides of the die

11. An eight-sided die numbered 4 to 11 is thrown once. The probability of observing number 10 is:

(a) 12.5%   (b) 25%   (c) 66.7%   (d) 8.3%

12. A nine-sided die numbered 4 to 12 is thrown once. The probability of observing numbers 8 or 12 is:

(a) 12.5%   (b) 22.2%   (c) 66.7%   (d) 8.3%

13. An eight-sided die numbered 4 to 11 is thrown 4 times. The probability of observing number 3 or 11 for all throws is:

(a) 0.02%   (b) 0.2%   (c) 0.05%   (d) 0.5%

14. A five-sided die numbered 5 to 9 is thrown twice. The probability of observing numbers 5 or 9 for all throws is:

(a) 16%   (b) 4%   (c) 1%   (d) 36%

15. A five-sided die numbered 5 to 9 is thrown 3 times. The probability of observing numbers 1, 3, or 5 for all throws is:

(a) 21.6%   (b) 0.8%   (c) 6.4%   (d) 0.9%

16. A six-sided die numbered 5 to 10 is thrown 5 times. The probability of observing only the even numbers for all throws is:

(a) 0.8%   (b) 0.03%   (c) 3.1%   (d) 1.0%

17. Among the following, which selection of statements best describes the conditions of a sample space?

1. Each single outcome should be possible.
2. All possible outcomes should be explored.
3. Outcomes cannot be identical.
4. Each single outcome should have been independently predetermined.

(a) 1, 2, and 4   (b) 1, 2, and 3
(c) 2, 3, and 4   (d) 1, 3, and 4

18. There are several possible outcomes for an event, each with a known probability. These outcomes are related as follows:

(a) The summation of their weighted values should be one.
(b) Their weighted sum divided by their sum should be one.
(c) Their product multiplied by their sum should be one.
(d) Their product divided by their sum should be one.

19. In statistics, frequency is:

(a) number of event occurrences   (b) number of time intervals
(c) frequency of losses   (d) frequency of wins

20. The event frequency can be predicted by means of historical happenings and evidence. (T/F)

21. Three balls labeled 2, 3, and 4 are placed in a bag. Balls are pulled randomly out of the bag one at a time and replaced afterward. The sample space in this case is:

(a) {2, 3, 4}   (b) {2, 2, 4}
(c) {1, 2, 3}   (d) {2, 4, 4}

22. A five-sided die, engraved with numbers 1, 5, 88, 7, and 10, is thrown 5 times. It is known that the probability of obtaining 88 is 16%. Would this die be considered *fair*? The probability of obtaining 10 on all throws is:
    (a) Yes, 0.041%   (b) No, 0.041%
    (c) Yes, 0.01%   (d) No, 0.01%

23. What does fairness in probability imply?
    (a) The probability of occurrences is equally distributed
    (b) The probability of occurrences is equally distributed, assuming similar frequencies
    (c) The probability of occurrences is not equally distributed, assuming dissimilar frequencies
    (d) The probability of occurrences is 1, assuming similar frequencies

24. Three people are pulling out one ball each from a bag containing balls numbered {1, 2, 3, 4, 5}. What is the probability that all balls have numbers less than 4?
    (a) 80%   (b) 40%   (c) 20%   (d) 60%

25. The difference between empirical probability and expectation is that the former is based on:
    (a) historical evidence while the latter is based on logic
    (b) logic while the latter is based on historical evidence
    (c) a guess while the latter is based on historical evidence
    (d) historical evidence while the latter is based on the person's intuition

26. A lottery is organized where they randomly pull out balls (numbered from 1 to 10) from a container. A prize is awarded for guessing correctly a sequence of 2 balls. What is your chance of winning the prize if you try once?
    (a) 11%   (b) 1.0%   (c) 0.1%   (d) 1.1%

27. ESPN has published statistical data for John Doe's basketball play for the past 50 years; the probability of his team winning when he was playing is 40.5%. Would you need empirical data or logic in order to calculate his losses? What is the probability for his team losing when he was playing for the given period? Note that in this scenario, him not being able to attend the game is not assumed a loss.
    (a) Empirical data, 40.5%
    (b) Logic, 29.8%
    (c) Both logic and empirical data, 59.5%
    (d) Both logic and empirical data, 20.3%

28. Empirical and logical evidence are always compatible. (T/F)

29. The empirical evidence can help to infer logical expectations. (T/F)

30. The law of large numbers says that with increasing the number of samples, the empirical evidence . . . logical expectations.
    (a) approaches   (b) diverges from
    (c) does not relate to   (d) becomes exactly equal to

31. Over the past 5 years, the team played in 300 matches. From these, they won 165 and tied 10 times. Estimate the probability of the team losing in their next match.
    (a) 41.7%   (b) 100%   (c) 61%   (d) 50%

32. Based on over 1,000 soccer matches, the following win probabilities can be estimated. The summation of probabilities of teams A, B, and C winning is 1.5. Team A is the strongest of all, with the probability of winning twice compared to team C and 3 times compared to team B. The probability that team B wins is:

(a) 81.8%   (b) 27.3%   (c) 40.9%   (d) 25.9%

33. You have been given a small box of colorful candies that contains three smaller compartments, with 15 pieces in each. After taking all the candies out of the boxes, you find there are 5 different colors present, with an equal number for each color. You place all the pieces in a fabric pouch and ask your friend to pick one piece at a time, without looking, from the pouch. You take note of the color picked and return the piece back in the pouch. What is the probability of your friend drawing a green piece on all 3 attempts?

(a) 0.8%   (b) 0.4%   (c) 0.2%   (d) 0.6%

34. You have been given a box containing 20 bubble gum pieces of 5 different colors, with an equal number for each color. You removed all the purple ones and put the rest in a small dish. You trained your cat, David, to pick randomly a piece of gum from the dish. David takes the gum each time and hides it somewhere in the house. Assuming green is one of the colors in the dish, what is the probability of David picking a green gum piece in 4 consecutive attempts, starting from a full dish.

(a) 0.041%   (b) 0.055%   (c) 0.032%   (d) 0.067%

35. You have been given a box containing 20 bubble gum pieces of 5 different colors, with an equal number for each color. You removed all the white, yellow, and orange ones and put the rest in a small dish. You trained your cat, Emma, to pick randomly a piece of gum from the dish. Emma takes the gum each time and hides it somewhere in the house. Assuming green is one of the colors in the dish, what is the probability of Emma picking a green gum piece in 4 consecutive attempts, starting from a full dish.

(a) 1.43%   (b) 0.72%   (c) 2.86%   (d) 2.3%

36. There are 10 prizes among which are an Apple computer, a walkie-talkie, a pair of shorts, a plush monkey, and a blonde doll. Five people (Jerry, Mani, Emily, Anna, and Tony) draw prize tickets from the box in order, without returning their ticket to the box. What is the probability that Jerry leaves with the walkie-talkie in his hand, Mani gets to wear the new pair of shorts, Emily ends up taking her new Apple computer back to the university, Anna leaves with the plush monkey, and Tony gets the blonde doll?

(a) 3%   (b) 0.3%   (c) 0.03%   (d) 0.003%

37. A fair eight-sided die produces one of the numbers 1 to 8, each with equal probability. If an eight-sided die is rolled many times, what is approximately the expected value of the average of the rolls with increasing number of dice rolled and why?

(a) 4.5, law of large numbers   (b) 3.5, law of small numbers
(c) 4, chaos of small numbers   (d) 3, chaos of large numbers

38. The chaos of small numbers is applicable when there are:

(a) a small number of data sets (about 25)
(b) too many data sets (more than 100)
(c) no data sets (under 1)
(d) over 1,000 data sets

39. The probability versus frequency (the number of occurrences) diagram represents the number of each:

(a) occurrence associated with the outcome probability
(b) probability associated with its frequency
(c) occurrence associated with its probability
(d) probability associated with its frequency over a specific period

40. The area under a probability-frequency plot is (note that frequency is the number of occurrences):

(a) 0   (b) 1   (c) 100   (d) −1

41. Among the probability-frequency diagrams presented in Figure 25, which one(s) are a fair distribution of events?

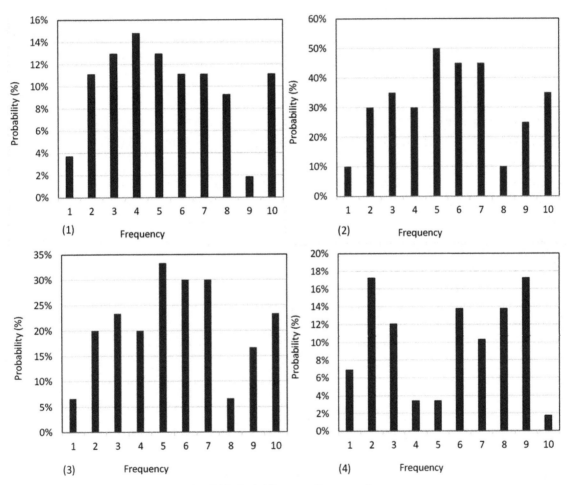

FIGURE 25. Probability versus frequency of events.

(a) (1) and (2)   (b) (1), (2), and (3)
(c) (1), (2), and (4)   (d) (1) and (4)

42. Table 1 presents 10 events; predict a, b, and c.

TABLE 1. Probability data sets.

Event	Average Frequency	Weight	Frequency Weight	Probability
1	1	0.05	0.05	5.43%
2	2	0.03	0.07	6.68%
3	3	0.02	0.06	5.91%
4	1	0.03	0.03	3.27%
5	2	0.06	0.13	12.68%
6	3	0.02	0.07	7.26%
7	1	0.09	0.09	8.82%
8	2	0.08	0.16	16.40%
9	3	0.04	0.13	13.35%
10	0.4	a	b	c

(a) $a = 0.56$, $b = 0.20$, $c = 20.20\%$ (b) $a = 0.20$, $b = 0.56$, $c = 20.20\%$
(c) $a = 0.56$, $b = 20.20$, $c = 0.20\%$ (d) $a = 20.20$, $b = 0.56$, $c = 0.20\%$

43. Table 2 presents 10 events; predict a and b.

TABLE 2. Probability data sets.

Event	Average Frequency	Weight	Frequency Weight	Probability
1	3	0.09	0.26	25.77%
2	2	0.05	0.10	9.50%
3	1	0.06	0.06	6.21%
4	3	0.02	0.07	7.14%
5	2	0.03	0.06	5.94%
6	1	0.08	0.08	7.59%
7	3	0.01	0.02	2.34%
8	2	0.07	0.14	14.18%
9	0.3	0.54	a	b
10	1	0.06	0.06	5.63%

(a) $a = 15.70$, $b = 0.16\%$ (b) $a = 0.46$, $b = 0.84\%$
(c) $a = 0.84$, $b = 0.46\%$ (d) $a = 0.16$, $b = 15.70\%$

44. If you flip a coin 3 times, the total number of possible outcomes is:
    (a) 2   (b) 8   (c) 4   (d) 16

45. The probability of team A winning a game is 45%, but it decreases by a factor of 1.5 with each subsequent game. The probability of losing after the third game is:
    (a) 20.1%   (b) 2.7%   (c) 86.7%   (d) 9.5%

46. The probability of team A winning a game is 34%, but it increases by a factor of 34% with the subsequent game. The probability of winning after the third subsequent game is:
  (a) 90.54%   (b) 9.46 %   (c) 4.73%   (d) 45.27%

47. You flip a fair coin 3 times. The probability of getting at least 2 tails is:
  (a) 19%   (b) 25%   (c) 38%   (d) 50%

48. Among the following sets of outcome probabilities, expressed as percentages, which ones represent unfair and hot hands?
  (a) {20, 25, 28, 26, 27}   (b) {17, 16, 17, 16, 17}
  (c) {20, 21, 20, 21, 20, 21}   (d) {8, 10, 12, 14, 16}

49. Hot hands are associated with:
  (a) unusually long runs of identical outcomes
  (b) long runs of similar events
  (c) short runs of identical events
  (d) very short runs of similar events

50. Among the following scenarios, which one contains an incorrect estimate of the associated event probability (assume 50% chance of the player scoring in a game)?
  (a) scoring in 15 consecutive games with the probability of it happening 0.2%
  (b) scoring in 10 consecutive games with the probability of it happening 0.1%
  (c) scoring in 12 consecutive games with the probability of it happening 0.024%
  (d) scoring in 3 consecutive games with the probability of it happening 12.5%

51. A physician ran 10 sets of experiments to investigate the effect of a family of drugs on a specific virus with successful outcomes. Each set used new patients and their number for each set decreased progressively from 5 to 1, in increments of 1, with the same sequence repeated after that point. The number of patients who have been tested for the new drug is:
  (a) 25   (b) 20   (c) 30   (d) 35

52. A physician ran 10 sets of experiments to investigate the effect of a family of drugs on a specific virus. There are 2 possible outcomes of equal probability (successful and not successful). The probability (P) of 9 consecutive trials being successful and the expected longest run of consecutive successful tests (n), if 2,000 test sets are run, are:
  (a) $P = 0.39\%, n = 10$ (b) $P = 0.19\%, n = 10$
  (c) $P = 0.19\%, n = 9$ (d) $P = 0.39\%, n = 9$

53. A fair coin is flipped repeatedly until n consecutive heads occurs. The expected number of coin flips is:
  (a) $2(2^n + 1)$ (b) 2^{n+1} (c) $2(2^n - 1)$ (d) 2^n

54. A fair coin is flipped repeatedly n times. The expected longest run of consecutive tails occurring is:
  (a) $\ln(0.5/n + 1)/\ln(0.5)$ (b) $\ln(0.5/n + 1)/\ln(2)$
  (c) $\ln(0.5n + 1)/\ln(0.5)$ (d) $\ln(0.5n + 1)/\ln(2)$

55. A physician ran 3,000 sets of experiments to investigate the effect of a family of drugs on a specific virus. There are 3 possible outcomes of equal probability (successful, not successful, and to be determined). The expected longest run of consecutive successful tests is:
  (a) 7   (b) 6   (c) 4   (d) 5

56. A fair coin is flipped repeatedly until 9 consecutive heads occurs. The expected number of coin flips is:

(a) 1,020   (b) 1,022   (c) 1,026   (d) 1,024

57. If a physician ran Y number of tests with 4 possible outcomes (success, failure, to be determined, and death) of equal probabilities, the expected longest run of consecutive successful tests is:

(a) $\ln(0.25Y + 1)/\ln(4)$ (b) $\ln(0.25Y)$
(c) $\log(Y)/\log(4)$ (d) $\log(0.25Y)$

58. If a basketball player has a 40% chance of scoring on a single shot, and she attempts 200 consecutive shots, the expected longest run of consecutive successful shots is?

(a) 7   (b) 5   (c) 9   (d) 3

59. If a basketball player has a 50% chance of scoring on a single shot, and she attempts 4 consecutive shots, identify the probability of runs of baskets with the minimum length of 2.

(a) 15.8%   (b) 31.5%   (c) 62.5%   (d) 25.7%

60. A subset of outcomes associated with sample space is called an:

(a) event   (b) outcome   (c) occurrence   (d) occasion

61. An empty set is a sample set in which:

(a) all events occur   (b) only failed events occur
(c) no events occur   (d) a single event occurs

62. An eight-sided die has numbers 1 to 8 printed on the sides. If the die is thrown once and the uppermost face is observed, a list of all possible events is:

(a) {1, 2, 3, 4, 5, 6, 7, 8}   (b) {1, 3, 5, 7}
(c) {2, 4, 6, 8, 10}   (d) {1, 2, 3, 6, 7, 8, 9}

63. A ten-sided die has numbers 1 to 6 printed on the sides. If the die is thrown 3 times and the uppermost face is observed, all possible data sets with even numbers that add up to 14 are:

(a) $A = \{2, 4, 8\}$, $B = \{8, 4, 6\}$, $C = \{6, 8\}$
(b) $A = \{8, 6\}$, $B = \{2, 4\}$, $C = \{6, 4\}$
(c) $A = \{2, 6, 6\}$, $B = \{4, 4, 6\}$, $C = \{4, 6, 4\}$, $D = \{6, 2, 6\}$, $E = \{6, 6, 2\}$
(d) $A = \{4, 6, 8\}$, $B = \{2, 4, 6\}$, $C = \{2, 6, 4\}$, $D = \{6, 8, 4\}$

64. A ten-sided die has numbers 1 to 10 printed on the sides. If the die is thrown once and the uppermost face is observed, the probability of observing an odd number less than 10 on its uppermost face is:

(a) 100%   (b) 25%   (c) 10%   (d) 50%

65. A ten-sided die has numbers 1 to 10 printed on the sides. If the die is thrown once and the uppermost face is observed, the probability of observing an even number less than 9 on its uppermost face is:

(a) 40%   (b) 50%   (c) 30%   (d) 60%

66. An eight-sided die has numbers 1 to 8 printed on the sides. If the die is thrown twice and the uppermost face is observed, the probability of observing an even number between 5 and 9 after each throw is:

(a) 6%   (b) 6.25%   (c) 25%   (d) 6.75%

67. A ten-sided die has numbers 1 to 10 printed on the sides. If the die is thrown 3 times and the uppermost face is observed, the probability of observing an odd number between 2 and 10 after each throw is:

(a) 40%   (b) 8%   (c) 16%   (d) 6.4%

68. An eight-sided die has numbers 1 to 8 printed on the sides. If the die is thrown twice and the uppermost face is observed, the probability of observing an odd number on each throw and having the sum of the two numbers observed being equal to 10 is:

(a) 5.8%   (b) 4.7%   (c) 3.2%   (d) 1.2%

69. A six-sided die has numbers 3 to 8 printed on the sides. If the die is thrown 3 times and the uppermost face is observed, the probability of observing number 5 after all throws is:

(a) 0.46%   (b) 2.78%   (c) 16.67%   (d) 32.3%

70. For an athlete, a 1-day rest will increase his performance by 12% (relative to no rest). If his average performance is 65% with no rest, his performance after 2 and a half-day rest is (assume a linear trend between rest and performance enhancement):

(a) 91.3%   (b) 86.4%   (c) 81.3%   (d) 73.4%

71. If the outcome of an occurrence depends on another event taking place, the outcome is associated with:

(a) unconditional probability   (b) qualified probability
(c) conditional probability   (d) restrictive probability

72. Assuming A and B are conditionally dependent events occurring in the S sample space, their probabilities are related by:

(a) $P(A|B) = \dfrac{P(A \cap B)}{P(B)}$ (b) $P(A|B) = P(A \cap B)P(B)$

(c) $P(A|B) = \dfrac{P(A)}{P(A \cap B)}$ (d) $P(A|B) = P(A)P(A \cap B)$

73. A and B are independent events. If we know the probability of both A and B occurring $P(A \cap B)$ and the probability of B occurring $P(B)$, the probability of A occuring $P(A)$ is:

(a) $P(A) = P(B)P(A \cap B)$ (b) $P(A) = P(B) / P(A \cap B)$

(c) $P(A) = \dfrac{P(A \cap B)}{P(B)}$ (d) $P(A) = P(B) - P(A \cap B)$

74. A, B, and C are independent events. If we know the probability of A, B, and C occurring $P(A \cap B \cap C)$ and the probability of B occurring $P(B)$ and C occuring $P(C)$, the probability of A occuring $P(A)$ is:

(a) $P(A) = P(B)P(C)P(A \cap B \cap C)$ (b) $P(A) = \dfrac{P(A \cap B \cap C)}{P(B)P(C)}$

(c) $P(A) = P(B|A|C)$ (d) $P(A) = P(B)P(C) / P(A \cap B \cap C)$

75. A, B, and C are independent events, with corresponding weight functions of $a(0.1)$, $b(0.2)$, and $c(0.7)$. The probability of event A is 3.5 times that of B. The probability of event C is 16 times that of A. $P(S) = 1$, where S is the sample space. The probabilities of A, B, and C occuring ($P(A)$, $P(B)$, and $P(C)$) are:

(a) $P(A) = 0.9\%$, $P(B) = 0.5\%$, $P(C) = 98.6\%$
(b) $P(A) = 5.1\%$, $P(B) = 84.8\%$, $P(C) = 10.1\%$

  (c) $P(A) = 5\%$, $P(B) = 10\%$, $P(C) = 85\%$
  (d) $P(A) = 8\%$, $P(B) = 75\%$, $P(C) = 17$

76. Assuming A, B, and C are independent events with the probabilities of 40%, 30%, and 20%, respectively, the probability of all 3 events occurring is:
  (a) 2.0%   (b) 2.4%   (c) 3.1%   (d) 3.5%

77. Bernoulli experiments focus on the experiments with . . . outcome(s).
  (a) two   (b) unknown   (c) one   (d) expected

78. In a Bernoulli experiment, if the probability of the failure event is 10% with a frequency of 2, the probability of an independent success with a frequency of 3 is:
  (a) 20%   (b) 15%   (c) 90%   (d) 27%

79. An unfair eight-sided die has numbers 1 to 8 printed on the sides. The probability for rolling any of the numbers smaller than 4 is 14%. Numbers 4 and greater all have equal probabilities which are equal to:
  (a) 11.6%   (b) 14.2%   (c) 12.5%   (d) 13.6%

80. An unfair eight-sided die has numbers 1 to 8 printed on the sides. The probability of each number under 6 is 10%. Numbers 6 and greater all have equal probabilities. If this die is thrown 4 times, the probability of obtaining {1, 5, 7, 8} is:
  (a) 0.019%   (b) 0.078%   (c) 0.028%   (d) 2.8%

81. If a soccer player makes 7 consecutive attempts to score a goal in independent Bernoulli trials, G representing scoring a goal and F failing to do so, and the probability of achieving {*GGFF*} is 5.5%, the probability (*P*) of him scoring a goal is (assume *P* is under 50%):
  (a) 35.4%   (b) 37.5%   (c) 31.5%   (d) 39.4%

82. A soccer player makes 9 consecutive attempts to score a goal in independent Bernoulli trials, G representing scoring a goal and F failing to do so. The probability of him scoring a goal is 40%. What is the probability of him achieving {*GFGGF*}?
  (a) 23.0%   (b) 6.13%   (c) 61.3%   (d) 2.3%

83. What is the expected longest run of consecutive goals if he attempts 12 times, assuming that the probability of him scoring a goal is 61%?
  (a) 3   (b) 5   (c) 2   (d) 4

84. What is the longest miss length of highest probability of a soccer player scoring consecutive goals if he attempts 20 times, assuming that the probability of him scoring a goal is 41%?
  (a) 5   (b) 2   (c) 4   (d) 3

85. A soccer player has a 10% chance to score a goal at each attempt. The probability of him scoring at least 1 goal during 25 attempts is:
  (a) 7.18%   (b) 71.79%   (c) 92.8%   (d) 9.28%

86. The number of the expected longest run of consecutive goals, if the probability of scoring a goal is *P*, and *k* attempts are made, is:

  (a) $-\dfrac{\ln\left((1-P)\times K\right)}{\ln(P)}$ (b) $-\dfrac{\ln(P\times K)}{\ln(1-P)}$

  (c) $-\dfrac{\ln(2\times(1-P)\times K)}{2\times\ln(P)}$ (d) $-\dfrac{\ln(2\times P\times K)}{2\times\ln(1-P)}$

87. The expected longest run of consecutive goals is 9. The probability of scoring a goal (P), if 2,500 attempts are made, is:

(a) 47.8%   (b) 44.8%   (c) 45.3%   (d) 50.1%

88. Analyze the probability diagrams presented in Figure 26 for the provided number of attempts (NOA) for the expected longest run of consecutive successful attempts. Identify the relationships used to obtain these diagrams. The probability of scoring a goal is P.

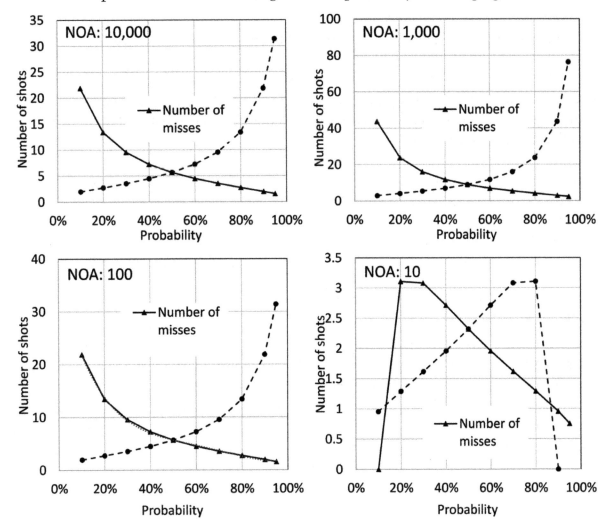

FIGURE 26. Number of shots versus probability.

(a) $LG_{goal} = -\dfrac{\ln(P \times NOA)}{\ln(P)}$, $LG_{miss} = -\dfrac{\ln(P \times NOA)}{\ln(1-P)}$

(b) $LG_{goal} = -\dfrac{\ln((1-P) \times NOA))}{\ln(P)}$, $G_{miss} = -\dfrac{\ln(P \times NOA)}{\ln(1-P)}$

(c) $LG_{goal} = -\dfrac{\ln(P \times NOA)}{\ln(P)}$, $LG_{miss} = -\dfrac{\ln((1-P) \times NOA)}{\ln(1-P)}$

(d) $LG_{goal} = -\dfrac{\ln((1-P) \times NOA)}{\ln(P)}$, $LG_{miss} = -\dfrac{\ln((1-P) \times NOA)}{\ln(1-P)}$

Solution Guide

The diagrams are based on the following formula for the expected longest run of consecutive goals: $LG_{goal} = -\dfrac{\ln((1-P) \times NOA)}{\ln(P)}$, where P is the probability of scoring a goal and NOA is the number of attempts. For the longest failure run length of highest probability, the formula is: $LG_{miss} = -\dfrac{\ln(P \times NOA)}{\ln(1-P)}$. Note that Bernoulli conditions apply to this scenario, where there are 2 independent possible outcomes: scoring a goal or failing to do so. With decreasing the number of attempts, the diagrams become less linear, although they can be fitted to a fourth-degree polynomial or an exponential function with a confidence level above 97% and 99% for the number of goals and misses, respectively.

89. In a data set, if discrete random variables have values $a_1 \cdots a_n$, and a probability is associated with each of the variables, $P(a_1) \ldots P(a_n)$, there is a probability distribution in which:

   (a) $0 \leq P(a_i) \leq 1$, $\sum\limits_{i=1}^{n} P(a_i) = 10$ (b) $0 \leq P(a_i) \leq 100$, $\sum\limits_{i=1}^{n} P(a_i) = 1$

   (c) $0 \leq P(a_i) \leq 1$, $\sum\limits_{i=1}^{n} P(a_i) = 1$ (d) $-1 \leq P(a_i) \leq 100$, $\sum\limits_{i=1}^{n} P(a_i) = 10$

90. For a fair Bernoulli experiment where a coin is flipped 5 times, identify the expected longest run of consecutive heads (n), and the probability (P) of obtaining the longest run of consecutive heads of 5, if the coin is flipped 10 times. Present the diagram and table for the sample sets.

   (a) 3, 50%   (b) 3, 23.5%   (c) 1, 50%   (d) 1, 31.8%

Solution Guide

$$S = \begin{cases} \text{HHHHH, HHHHT, HHHTH, HHHTT, HHTHH, HHTHT, HHTTH,} \\ \text{HHTTT, HTHHH, HTHHT, HTHTH, HTHTT, HTTHH, HTTHT,} \\ \text{HTTTH, HTTTT, THHHH, THHHT, THHTH, THHTT, THTHH,} \\ \text{THTHT, THTTH, THTTT, TTHHH, TTHHT, TTHTH, TTHTT,} \\ \text{TTTHH, TTTHT, TTTTH, TTTTT} \end{cases}$$

The data set is presented in Table 3. Associated probability diagrams for the number of heads and the expected longest run of consecutive heads are presented in Figure 27. Note that in order to count the number of runs and probabilities associated with the number of heads, any of the 32 possible scenarios above (2^5) are investigated for their number of heads (versus tails) and the expected longest run of consecutive heads. Similar ones are then added in a separate column (frequency: expected longest run of consecutive heads) and the results are divided by the total number of sample sets (32).

Associated probability diagrams are presented in Figure 27; they approach bell-shaped distributions with an increasing number of data sets. The probability-frequency diagram is already adopting a bell-shaped distribution.

You are encouraged to explore this observation by performing similar analyses for the case where a coin is thrown 6 times, resulting in 64 (2^6) data sets, and 4 times, resulting in 16 (2^4) data sets.

TABLE 3. Probability data sets.

Number of heads	Frequency (longest run)	Longest run probability (%)	Frequency (maximum number of head lengths)	Frequency probability (%)
0	1	3%	1	3%
1	12	38%	5	16%
2	11	34%	10	31%
3	5	16%	10	31%
4	2	6%	5	16%
5	1	3%	1	3%

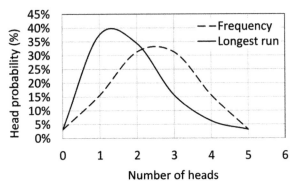

FIGURE 27. Heads probability versus number of heads.

91. A bar graph for a large set of discrete random data is most probably similar to a:
(a) bell-shaped diagram   (b) quadratic diagram
(c) elliptical diagram   (d) hyperbolic diagram

92. In a bar graph of a random data set of discrete numbers, the width of each bar is . . . the rest.
(a) independent of   (b) equal to
(c) different from   (d) not the same as

93. In a normalized probability bar graph for a large set of discrete random data, the total area under the bars is equal to:
(a) 0.5   (b) 0   (c) 1   (d) 100

94. A scattered probability distribution for a large data set, made of continuous large random numbers, forms a:
(a) quadratic diagram   (b) elliptical diagram
(c) hyperbolic diagram   (d) bell-shaped diagram

95. Identify the probability associated with frequency 4 on the bar graph presented in Figure 28, which shows a discrete number of data sets:

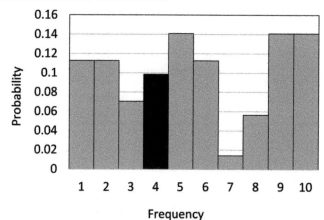

FIGURE 28. Probability versus frequency.

(a) 0.098   (b) 0.061   (c) 0.056   (d) 0.045

96. Identify the probability associated with frequency 3 in Table 4:

TABLE 4. Probability data sets.

Probability	Frequency
5.97%	1
11.94%	2
?	3
1.49%	4
13.43%	5
14.93%	6
5.97%	7
14.93%	8
10.45%	9
7.46%	10

(a) 11.94%   (b) 12.94%   (c) 13.44%   (d) 14.15%

97. Identify the most accurate representation of the normalized diagram presented in Figure 29, using the format $\mathrm{Norm.Dist}(X, \overline{X}, \sigma)$, where X is the frequency, $\overline{X}$ is the arithmetic mean of distribution, and s is the standard deviation ($0 \le x \le 9$):

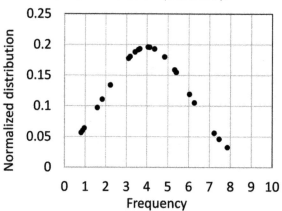

FIGURE 29. Normalized distribution versus frequency.

(a) Norm.Dist $(X, 2.4, 4.0)$ (b) Norm.Dist $(X, 3.1, 5.1)$
(c) Norm.Dist $(X, 5.1, 3.1)$ (d) Norm.Dist $(X, 4.0, 2.0)$

98. Assuming that the data were associated with the normalized probabilities, the area under the missing part of the curve in Figure 30 is approximately equal to $(0 \leq x \leq 9)$:

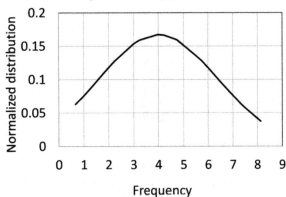

FIGURE 30. Normalized distribution versus frequency.

(a) 0.13   (b) 0.26   (c) 0.87   (d) 0.74

99. Figure 31 presents the function $f(x)$ diagram versus x. Identify the alternative name for the diagram if the data were associated with the normalized probabilities as well as the normalized relationship $(0 \leq x \leq 5)$:

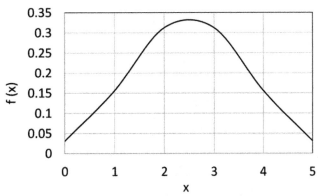

FIGURE 31. Normalized distribution versus frequency.

(a) $N_i = f(x_i) \Big/ \int_0^5 f(x_i)$, probability density curve

(b) $N_i = f(x_i) \Big/ \int_1^4 f(x_i)$, normalized function

(c) $N_i = \int_0^5 f(x_i) \Big/ f(x_i)$, probability normalized curve

(d) $N_i = \int_0^4 f(x_i) \Big/ f(x_i)$, density function

100. Table 5 presents the outcomes for given frequencies. The outcomes for the average $(\overline{X})$ and median (Med), respectively, are:

TABLE 5. Frequency and outcome data sets.

Outcome	Frequency
47	3
76	1
71	10
89	1
88	7
100	10
14	8
15	0
13	4
36	6
55	6

(a) $\overline{X} = 54.9, \text{Med} = 55.0$ (b) $\overline{X} = 55.0, \text{Med} = 54.9$

(c) $\overline{X} = 59.7, \text{Med} = 54.9$ (d) $\overline{X} = 54.9, \text{Med} = 59.7$

101. Table 6 presents the outcomes for given frequencies. The sum of the outcomes' median, average, minimum, and maximum is:

TABLE 6. Frequency and outcome data sets.

Outcome	Frequency
47	3
76	1
71	10
89	1
88	7
100	10
14	8
15	0
13	4
36	6
55	6

(a) 222.9   (b) 22.29   (c) 2229.1   (d) 2.229

102. Table 7 presents the outcomes for given frequencies. The expected outcome is:

TABLE 7. Frequency and outcome data sets.

Outcome	Frequency
47	3
76	1
71	10
89	1
88	7
100	10
14	8
15	0
13	4
36	6
55	6

(a) 60.7   (b) 59.0   (c) 59.7   (d) 60.0

103. Table 8 presents the outcomes for given frequencies. Parameters $a, b,$ and c are:

TABLE 8. Frequency and outcome data sets.

Outcome (O)	Frequency (F)	Probability (P)	O×P	F×O
47	3	5.4%	2.52	141
76	1	1.8%	1.36	76
71	10	17.9%	12.68	710
89	1	1.8%	1.59	89
88	7	12.5%	11.00	616
100	10	a	17.86	c
14	8	14.3%	2.00	112
15	0	0.0%	0.00	0
13	4	7.1%	b	52
36	6	10.7%	3.86	216
55	6	10.7%	5.89	330

(a) $a = 15.7\%, b = 0.84, c = 1,100$ (b) $a = 13.5\%, b = 0.75, c = 1,010$
(c) $a = 17.9\%, b = 0.93, c = 1,000$ (d) $a = 11.3\%, b = 0.66, c = 1,001$

104. Table 9 presents the outcomes for given frequencies. The expected outcome is:

TABLE 9. Frequency and outcome data sets.

Outcome (O)	Frequency (F)	Probability (P)	O×P	F×O
5	4	3.0%	0.15	20
15	8	6.0%	0.90	120
25	12	9.0%	2.24	300
35	16	11.9%	4.18	560
45	20	14.9%	6.72	900
55	24	17.9%	17.86	1320
10	2	1.5%	0.15	20
20	6	4.5%	0.90	120
30	10	7.5%	0.93	300
40	14	10.4%	4.18	560
50	18	13.4%	6.72	900

(a) 58.0   (b) 56.5   (c) 59.4   (d) 44.9

105. For the function presented in Figure 32, select the polynomial that best represents the diagram $(0 \leq x \leq 5)$:

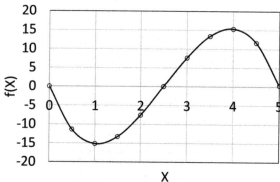

FIGURE 32. Polynomial function.

(a) $f(x) = 2.52X^3 - 18.93X^2 + 31.56X$ (b) $f(x) = -1.25X^3 + 9.47X^2 - 16.25X$

(c) $f(x) = -2.52X^3 + 18.93X^2 - 31.56X$ (d) $f(x) = 1.25X^3 - 9.47X^2 - 16.25X$

106. Using the polynomial fitted to the function presented in Figure 33, the area under the diagram $(0 \leq x \leq 5)$ is:

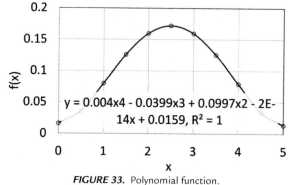

FIGURE 33. Polynomial function.

(a) 0.4   (b) 0.3   (c) 0.5   (d) 0.6

Solution Guide

$$A = \int_0^5 \left(0.004X^4 - 0.0399X^3 + 0.0997X^2 + 0.0159\right)dX$$

$$= \left[\frac{0.004}{5}X^5 - \frac{0.0399}{4}X^4 + \frac{0.0997}{3}X^3 + 0.0159X\right]_0^5$$

107. Figure 34 shows the integral of the distribution function $(f(x))$ presented in Figure 33. The x-interval, centered at 2.5, corresponding to 80% of the area under $f(x)$ is:

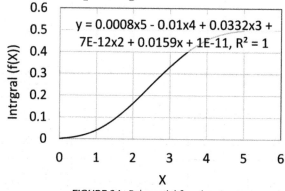

FIGURE 34. Polynomial function.

(a) $0 < x < 5$ (b) $0.8 < x < 4.2$ (c) $0.5 < x < 4.5$ (d) $1.1 < x < 3.9$

108. Figure 34 shows the integral of the distribution function $(f(x))$ presented in Figure 33. The percent of the area under $f(x)$ for the interval $0.7 < x < 2.5$ is:

(a) 67.2%   (b) 41.3%   (c) 45.4 %   (d) 39.8%

109. Figure 34 shows the integral of the distribution function $(f(x))$ presented in Figure 33. The percent of the area under $f(x)$ for the interval $2 < x < 3$ is:

(a) 27.1%   (b) 33.5%   (c) 25.3%   (d) 45.9%

110. For the data sets presented in Table 10, calculate the variance and standard deviation.

TABLE 10. Frequency and outcome data sets.

Outcome	Frequency
47	3
76	1
71	10
89	1
88	7
100	10
14	8
15	0
13	4
36	6
55	6

(a) $\sigma^2 = 958.33, \sigma = 30.96$

(b) $\sigma^2 = 479.17, \sigma = 21.89$

(c) $\sigma^2 = 30.96, \sigma = 958.33$

(d) $\sigma^2 = 21.89, \sigma = 479.17$

Solution Guide

Standard deviation (σ), is the square root of variance $\left(\sigma = \sqrt{\sigma^2}\right)$. Variance is the squared value of the summation of the deviation of the outcomes from the expected outcome $\sigma^2 = \sum_{i=1}^{n}(X_i - \mu)^2 P(x)$, given their probabilities, where (μ) is the expected value $\overline{X} = \sum_{i=1}^{n}X_i P(X_i)$. n is the number of data sets. The probabilities of individual outcomes are to be calculated as the first step $X_i P(X_i)$. The expected value is in fact the probability of each outcome, given its frequency, averaged over the entire data set. Data are presented in Table 11.

TABLE 11. Variance and standard deviation calculations.

Frequency	Outcome	Frequency× Outcome	Probability	Outcome× Probability	(Outcome-Expected)	(Outcome-Expected)^2	Probability×(Outcome-Expected)^2
3	47	141	0.054	2.52	-12.68	160.75	8.61
1	76	76	0.018	1.36	16.32	266.39	4.76
10	71	710	0.179	12.68	11.32	128.17	22.89
1	89	89	0.018	1.59	29.32	859.75	15.35
7	88	616	0.125	11.00	28.32	802.10	100.26
10	100	1000	0.179	17.86	40.32	1625.82	290.32
8	14	112	0.143	2.00	-45.68	2086.53	298.08
0	15	0	0.000	0.00	-44.68	1996.17	0.00
4	13	52	0.071	0.93	-46.68	2178.89	155.63
6	36	216	0.107	3.86	-23.68	560.67	60.07
6	55	330	0.107	5.89	-4.68	21.89	2.35
56.00 Sum		**59.68** Expected Outcome	**1.0** Sum	**59.68** Expected Outcome		Variance Standard Deviation	958.33 30.96

111. Data sets A, B, and C are three continuous random variables with the same mean and standard deviations related by ($\sigma_A < \sigma_C < \sigma_B$). The set that has the greatest variation in observations is data set:

(a) A   (b) C

(c) B   (d) Insufficient information to answer

112. For the Y data presented in Figure 35, standard deviation (σ) and expected value (μ) are equal to ($0 \leq x \leq 1$):

(a) $\sigma = 0.30, \mu = 0.49$ (b) $\sigma = 0.30, \mu = 0.55$

(c) $\sigma = 0.35, \mu = 0.55$ (d) $\sigma = 0.35, \mu = 0.49$

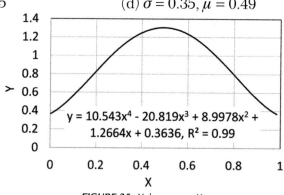

FIGURE 35. Y data versus X.

113. The plot in Figure 36 is an integral of function $Y(X)$ shown in Figure 35. This integral is shown as the polynomial function in Figure 36. The area under the curve $Y(X)$ between 0 and 0.5 is approximately:

(a) 0.91   (b) 0.25   (c) 0.125   (d) 0.46

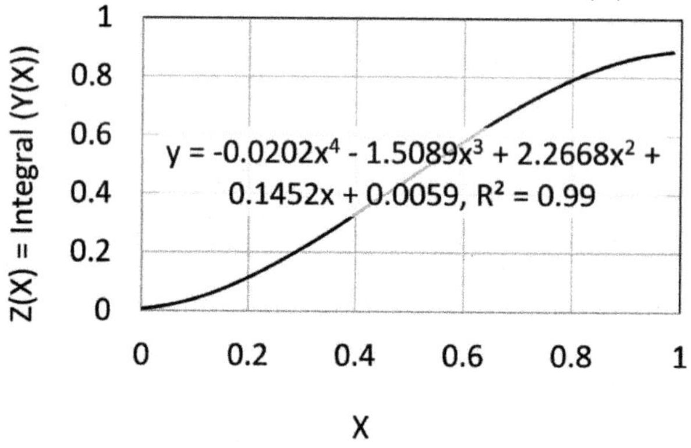

The $Z(X)$ = Integral (Y(X)) plot shows: $y = -0.0202x^4 - 1.5089x^3 + 2.2668x^2 + 0.1452x + 0.0059, R^2 = 0.99$

FIGURE 36. Integral of Y data versus X.

114. The plot in Figure 37 is an integral of function $Z(X)$ = Integral (Y) shown in Figure 36. This integral is shown as the polynomial function in Figure 37. Integral (Y) and Y are:

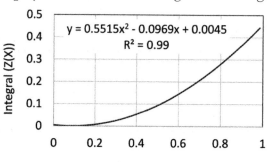

The Integral (Z(X)) plot shows: $y = 0.5515x^2 - 0.0969x + 0.0045, R^2 = 0.99$

FIGURE 37. Integral of Z data versus X.

(a) $\begin{cases} \text{Integral}(Y) = 1.103X - 0.0969 \\ \qquad Y = 1.103 \end{cases}$ (b) $\begin{cases} \text{Integral}(Y) = 0.5515X^2 - 0.0969X \\ \qquad Y = 1.103X - 0.0969 \end{cases}$

(c) $\begin{cases} \text{Integral}(Y) = 1.103X - 0.0969 \\ \text{Integral}(Y) = 0.5515X^2 - 0.0969X \end{cases}$ (d) $\begin{cases} \qquad Y = 1.103 \\ \text{Integral}(Y) = 1.103X - 0.0969 \end{cases}$

115. You have collected a large set of numerical data. If you are adding another outcome to this set, what is the probability of this outcome being within one standard deviation of this set's mean?

    (a) 68.28%   (b) 95.46%   (c) 99.73%   (d) 34.13%

116. The larger the standard deviation is, the further from the mean the outcomes are. (T/F)

117. You have collected a large set of numerical data. If you are adding another outcome to this set, what is the probability of this outcome being within two standard deviations of this set's mean?

    (a) 68.26%   (b) 95.46%   (c) 99.73%   (d) 34.13%

118. You have collected a large set of numerical data. If you are adding another outcome to this set, what is the probability of this outcome being within three standard deviations of this set's mean?

    (a) 68.26%   (b) 95.46%   (c) 99.73%   (d) 34.13%

119. Based on the discrete data presented in Table 12, what is the probability that an outcome falls within the 1.3 standard deviation?

TABLE 12. Probability calculations.

Outcome	Frequency	Probability	Probability×(Outcome-Expected)^2
47	3	0.054	8.61
76	1	0.018	4.76
71	10	0.179	22.89
89	1	0.018	15.35
88	7	0.125	100.26
100	10	0.179	290.32
14	8	0.143	298.08
15	0	0.000	0.00
13	4	0.071	155.63
36	6	0.107	60.07
55	6	0.107	2.35
59.68	56.00	1.000	958.33 Variance
Expected Outcome	SUM (Frequency)	SUM (Probability)	30.96 Standard Deviation

    (a) 60.71%   (b) 14.64%   (c) 99.73%   (d) 92.86%

120. For the data presented in Table 13, using the calculated standard deviation, the parameters representing probability (*a*), probability by outcome (*b*), variance (*c*), and standard deviation (*d*) are:

TABLE 13. Variance and standard deviation calculations.

Within Sigma	1.6	Outcome	Frequency	Probability	Probability×Outcome	Probability×(Outcome-Expected)^2
Mean-1.6 Sigma	1.56	37	3	0.054	1.982	24.19
Mean+1.6 Sigma	114.94	86	1	0.018	1.536	13.75
		61	10	0.179	10.893	1.35
		99	1	0.018	1.768	29.65
		78	7	0.125	9.750	48.76
		110	10	0.179	19.643	478.23
		4	8	0.143	0.571	420.44
		25	0	0.000	0.000	0.00
		3	4	0.071	0.214	218.04
		46	6	0.107	4.929	16.08
		65	6	0.107	6.964	4.88
				a SUM (Probability)	**b** SUM (Probability×Outcome)	**c** Variance
						d Standard Deviation

(a) $a = 60.71\%$, $b = 38.89$, $c = 214.29$, $d = 14.64$
(b) $a = 92.86\%$, $b = 58.75$, $c = 802.69$, $d = 28.33$
(c) $a = 100\%$, $b = 58.25$, $c = 1{,}255.37$, $d = 35.43$
(d) $a = 33.93\%$, $b = 21.09$, $c = 33.85$, $d = 5.82$

121. Table 14 filters out the outcomes that do not fall within one standard deviation interval $\sigma(-0.5\sigma \leq \mu \leq 0.5\sigma)$, and identifies those outcomes as *FALSE* in the table. Calculate the parameters representing probability (*a*), probability by outcome (*b*), variance (*c*), and standard deviation (*d*) based on the set shown in Table 14:

TABLE 14. Variance and standard deviation calculations.

Within Sigma	0.5	Outcome	Frequency	Probability	Probability×Outcome	Probability×(Outcome-Expected)^2
Mean-0.5*sigma	44.20	47	3	0.054	2.518	8.61
Mean+0.5*sigma	75.16	FALSE	-	-	-	-
		71	10	0.179	12.679	22.89
		FALSE	-	-	-	-
		FALSE	-	-	-	-
		FALSE	-	-	-	-
		FALSE	-	-	-	-
		FALSE	-	-	-	-
		FALSE	-	-	-	-
		FALSE	-	-	-	-
		55	6	0.107	5.893	2.35
				a SUM (Probability)	**b** SUM (Probability×Outcome)	**c** Variance
						d Standard Deviation

(a) $a = 60.71\%$, $b = 38.89$, $c = 214.29$, $d = 14.64$
(b) $a = 92.86\%$, $b = 58.75$, $c = 802.69$, $d = 28.33$
(c) $a = 99.99\%$, $b = 59.68$, $c = 958.33$, $d = 30.96$
(d) $a = 33.93\%$, $b = 21.09$, $c = 33.85$, $d = 5.82$

122. The higher the Z-score is for a particular outcome:
   (a) the closer the outcome is to the expected value
   (b) the further away the outcome is from the expected value
   (c) the better the accuracy of a certain outcome
   (d) the worse the accuracy of a certain outcome

123. The Z-score is a standardized function and may be used to decide if an outcome is hot handed or randomly generated. (T/F)

124. The average performance score to select NASA astronauts is 55, with a standard deviation of 5. Assuming that candidate A scored 50, her Z-score is:
   (a) −1   (b) −5   (c) 5   (d) 1

125. 1,000 candidates participated in a NASA competition's final round. Assume they have a fair chance for being selected for the entire process. Also assume they are to pass 5 tests, with equal chance for passing each test of 32%, with them all having equal chances as well. How many candidates would you expect to be selected (round off to the closest integer)?
   (a) 320   (b) 3   (c) 5   (d) 50

126. 52 candidates participated in a NASA competition's final round. Assume they have a fair chance for being selected for the entire process. Also assume they are to pass 4 tests, with equal chance for passing each test. If 5 people were selected at the end of this process, the probability of passing each test was:
   (a) 56%   (b) 25%   (c) 20%   (d) 5%

127. The Wald-Wolfowitz runs test is used to estimate the expected value (μ) and standard deviation (σ) of binomial data using the total number of samples (N) as well as the number of each population (N_0, N_1). The correct expressions are:

(a) $E(X) = \mu = \dfrac{2N_0N_1}{N} + 1, \sigma(X) = \sqrt{\dfrac{(\mu-1)(\mu-2)}{N-1}}$

(b) $E(X) = \mu = \dfrac{N_0N_1}{N} + 2, \sigma(X) = \sqrt{\dfrac{(\mu-2)(\mu-3)}{2N-1}}$

(c) $E(X) = \mu = \dfrac{3N_0N_1}{N} + 4, \sigma(X) = \sqrt{\dfrac{(\mu-1)(\mu-3)}{N-1}}$

(d) $E(X) = \mu = \dfrac{4N_0N_1}{N} + 8, \sigma(X) = \sqrt{\dfrac{(\mu-1)(\mu-2)}{2N-1}}$

128. Wald-Wolfowitz runs test formulae are used to estimate the expected value (μ) and standard deviation (σ) of binomial data. The probability of success is 58% and the total number of conducted tests is 115. The total number of failures (N_f), expected value, and standard deviation (σ) are:
   (a) $N_f = 46$, $\mu = 56.21$, $\sigma = 5.12$ (b) $N_f = 51$, $\mu = 57.77$, $\sigma = 5.27$
   (c) $N_f = 48$, $\mu = 56.93$, $\sigma = 5.19$ (d) $N_f = 44$, $\mu = 55.33$, $\sigma = 5.04$

129. An inspector has received test results showing that there have been 400 successful outcomes from a total of 1,000 tests conducted. The overall probability of success is 63%. Use the Wald-Wolfowitz runs test to: (1) report on whether the test results are hot handed; and

(2) estimate the total number of failures (N_f), the expected value (μ), and standard deviation (σ) for a large set of binomial data.
(a) Yes, $N_f = 370$, $\mu = 467.2$, $\sigma = 14.73$ (b) No, $N_f = 370$, $\mu = 467.2$, $\sigma = 14.73$
(c) Yes, $N_f = 470$, $\mu = 46.7$, $\sigma = 1.47$ (d) No, $N_f = 470$, $\mu = 46.7$, $\sigma = 1.47$

Solution Guide

Using the formulae, $E(X) = \mu = \dfrac{2N_0 N_1}{N} + 1$, $\sigma(X) = \sqrt{\dfrac{(\mu-1)(\mu-2)}{N-1}}$, one can calculate the expected value ($\mu = 467.20$), variance ($\sigma^2 = 217.09$), and standard deviation ($\sigma = 14.73$). Using the upper and lower criterion of the given data, over 95% of the data should be within the range ($\mu - 2\sigma < N_s < \mu + 2\sigma$), where N_s is the number of successful tests. Substituting the values in this equality results in $438 < N_s < 497$. Since the number of successes observed is outside the given range, it can be concluded that the data is hot handed.

130. An election was run with only two parties, A and B, on the ballot. Party B lost by a margin of 5,000 votes, with a total of 65,000 votes cast. The early polls suggested that the probability for Party B winning the race was about 65%. Use the Wald-Wolfowitz runs test to estimate the expected value (μ), and standard deviation (σ) for a large set of binomial data. Provide an analysis as to whether the election results are hot handed and identify the actual win probability outcome (P) as well as the standard deviation:
    (a) Definitely no, $P = 51.9\%$, $\sigma = 11.6$ (b) No, 51.9%, $P = 57.7\%$, $\sigma = 116$
    (c) It depends, $P = 51.9\%$, $\sigma = 11.6$ (d) Yes, $P = 57.7\%$, $\sigma = 116$

Solution Guide

Using the formulae, $E(X) = \mu = \dfrac{2N_0 N_1}{N} + 1$, $\sigma(X) = \sqrt{\dfrac{(\mu-1)(\mu-2)}{N-1}}$, one can calculate the expected value ($\mu = 29{,}576$), variance ($\sigma^2 = 13{,}456$), and standard deviation ($\sigma = 116$). Using the upper and lower criterion of the given data, over 95% of the data should be within the range ($\mu - 2\sigma < N_s < \mu + 2\sigma$), where N_s is the number of positive votes. Substituting the values in this equality results in $29{,}343 < N_s < 29{,}808$. Since the number of votes for Party B is outside the predicted range ($N_s = 27{,}500$), it can be concluded that the election results are hot handed with a 95% confidence level. The win probability based on the win outcome is the ratio of the number of actual wins to the total number of votes $\left(P = \dfrac{37{,}500}{65{,}000} = 57.7\% \right)$.

131. Allen enters a conference room where his 2 colleagues are already sitting. The probability that Allen shares his birthday with at least 1 of them is:
    (a) 0.4%   (b) 0.8%   (c) 1.6%   (d) 2.4%

132. Allen enters a conference room and finds there are 10 of his coworkers already there. The probability that Allen shares his birthday with at least 1 of them is:
    (a) 14.1%   (b) 85.9%   (c) 62.1%   (d) 37.9%

133. Allen enters a conference room and finds there are 18 of his coworkers already there. The probability that Allen shares his birthday with at least 1 of them is:
    (a) 14.1%   (b) 85.9%   (c) 62.1%   (d) 37.9%

134. If 14 integer numbers are randomly selected from a sample set {1 to 160}, the probability of selecting at least 2 identical numbers is:
    (a) 44.32%   (b) 85.2%   (c) 4.32%   (d) 8.52%

135. When selecting random integer numbers from a sample set {1 to 100}, how many would one expect to select to have a probability of 50% of having 2 identical numbers in the selection?

(a) 6 (b) 12 (c) 36 (d) 48

136. If n is the number of integers taken randomly from the sample set {1 to m}, with probability P of pulling at least 2 identical numbers, n can be predicted by:

(a) $n(P;m) = \sqrt{2 \times m \times \ln\left(\dfrac{1}{1-P}\right)}$

(b) $n(P;m) = \sqrt{m \times \ln\left(\dfrac{2}{1-P}\right)}$

(c) $n(P;m) = \sqrt{3 \times m \times \ln\left(\dfrac{1}{1-2 \times P}\right)}$

(d) $n(P;m) = \sqrt{4 \times m \times \ln\left(\dfrac{3}{1-P}\right)}$

137. Anecdotal evidence is a carefully recorded testimony collected as part of research studies. (T/F)

138. Category ranking systems are often used to predict the best performer within a particular category. (T/F)

139. Contestants play in a game where they receive points for correct answers; for incorrect answers, points are deducted. Over many years of playing the game, it was found that the average earned points is zero with a 100-point standard deviation. The probability of a contestant scoring between 50 and 70 points is:

(a) 6.7% (b) 8.4% (c) 5.5% (d) 15.1%

140. Assume a basketball player's mean score and standard deviation are 22 and 10. His Z-score if he scores 30 for the next game is:

(a) −0.8 (b) −1.1 (c) 0.8 (d) 1.1

141. Contestants play in a game where they receive points for correct answers; for incorrect answers, points are deducted. Over many years of playing the game, it was found that the average earned points is zero with 1-point standard deviation. A contestant scored −1.25. The probability of him scoring above −1.25 points for another game is 30%. His Z-score for another game is:

(a) 0.5 (b) −0.5 (c) 1.2 (d) −1.2

142. Probability . . . ; probability density function . . .
(a) can be between −1 and 1 / must be from 0 to 1
(b) must be less than one / can be any positive number
(c) must be a positive number / can be a negative number
(d) must be greater than 1 / must be between 0 and 1

143. The standard normal distribution probability density function is:

(a) $f(x) = \dfrac{1}{\sqrt{2\pi}} e^{-\frac{x^2}{2}}$

(b) $f(x) = \dfrac{2}{\sqrt{\pi}} e^{-(2x)^2/2}$

(c) $f(x) = \dfrac{1}{\sqrt{4\pi}} e^{-2x^2/4}$

(d) $f(x) = \dfrac{4}{\sqrt{\pi}} e^{-x^2/4}$

144. Figure 38 presents the probability density function (*PDF*) plotted versus *X*. Estimate area under this curve.

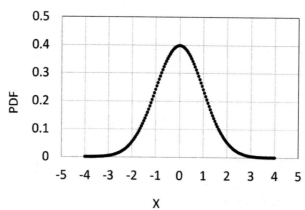

FIGURE 38. Probability distribution function versus X.

(a) 1   (b) 2   (c) 1.5   (d) 0.4

145. The area under the curve presented in Figure 39 shows the cumulative probability distribution. The probabilities of *X* being between $0 < X < 0.5$ and $1 < X < 4$ in Figure 39, respectively, are:

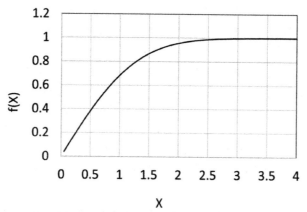

FIGURE 39. Cumulative probability distribution.

(a) 32%, 38%   (b) 16%, 19%
(c) 19%, 16%   (d) 38%, 32%

146. The area under the normal distribution probability density function (*PDF*) for the range $(-\infty, x)$ is equal to:

(a) $\dfrac{1}{2}\mathrm{erf}(\sqrt{0.5}x)$ (b) $\dfrac{1}{2}\mathrm{erfc}(\sqrt{0.5}x)$

(c) $\dfrac{1}{4}\mathrm{erf}(\sqrt{0.25}x)$ (d) $\dfrac{1}{4}\mathrm{erfc}(\sqrt{0.25}x)$

147. For a standard normal $\mathrm{PDF}(x) = \dfrac{1}{\sqrt{2\pi}}e^{-\frac{x^2}{2}}$, the probability of *x* being in the range $\{2, 5\}$ is:

(a) 0.023   (b) 0.015   (c) 0.032   (d) 0

148. Tree diagrams are useful for developing selection strategy. (T/F)

149. Using the probability tree presented in Figure 40, the probabilities for four events A, B, C, and D, are:

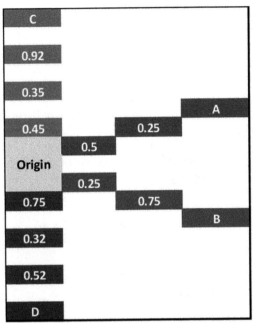

FIGURE 40. Probability tree for four events.

(a) $P(A) = 10.00\%$, $P(B) = 15.31\%$, $P(C) = 11.94\%$, $P(D) = 62.75\%$
(b) $P(A) = 25.00\%$, $P(B) = 10.57\%$, $P(C) = 21.37\%$, $P(D) = 43.06\%$
(c) $P(A) = 12.50\%$, $P(B) = 18.75\%$, $P(C) = 14.49\%$, $P(D) = 12.48\%$
(d) $P(A) = 15.68\%$, $P(B) = 25.37\%$, $P(C) = 16.25\%$, $P(D) = 42.70\%$

150. There are 6 blue and 8 red balls inside a pouch. The experimenter pulls 3 balls from the pouch, one ball at a time. Present a probability tree for the selected events, assuming that the experimenter does not return balls back to the pouch after each draw. Identify the probabilities for drawing balls with the following order {RBB, BRB, RBR, BRR}:

Solution Guide

Figure 41 shows the probability tree for the color of balls pulled out and associated probabilities.

				Probability
			R (6/12)	15.38%
		R (7/13)		
	R (8/14)		B (6/12)	15.38%
			B (5/12)	10.99%
		B (6/13)		
			R (7/12)	15.38%
Origin				
			R (7/12)	15.38%
		R (8/13)		
	B (6/14)		B (5/12)	10.99%
			B (4/12)	5.49%
		B (5/13)		
			R (8/12)	10.99%

FIGURE 41. Probability tree for the color of balls pulled out.

(a) $(RBB) = 10.99\%$, $P(BRB) = 10.99\%$, $P(RBR) = 15.38\%$, $P(BRR) = 15.38\%$
(b) $(RBB) = 15.38\%$, $P(BRB) = 10.99\%$, $P(RBR) = 15.38\%$, $P(BRR) = 14.20\%$
(c) $(RBB) = 15.38\%$, $P(BRB) = 14.20\%$, $P(RBR) = 5.49\%$, $P(BRR) = 10.99\%$
(d) $(RBB) = 5.49\%$, $P(BRB) = 15.38\%$, $P(RBR) = 10.99\%$, $P(BRR) = 14.20\%$

151. There are 6 blue (B) and 8 red (R) balls inside a pouch. The experimenter pulls 3 balls from the pouch, one ball at a time. Present a probability tree for the selected events, assuming that the experimenter does not return balls back to the pouch after each draw. Identify the probability of drawing 2 reds in a row followed by either R or B:

Solution Guide

Figure 42 shows the probability tree for the color of balls pulled out and associated probabilities.

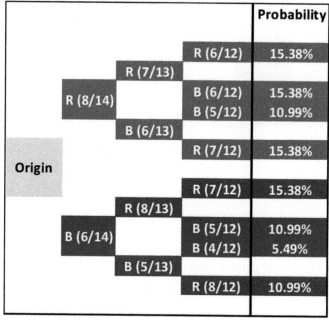

FIGURE 42. Probability tree for the color of balls pulled out.

(a) $P(RR \cap R) \cup (RR \cap B) = 30.77\%$ (b) $P(RR \cup R) \cap (RR \cup B) = 30.76\%$
(c) $P(RR \cap R) \cup (RR \cap B) = 15.38\%$ (d) $P(RR \cup R) \cap (RR \cup B) = 15.38\%$

152. Several sets of tests are conducted to find a vaccine for protection against a virus, with each set resulting either in success or failure. Three distinct attempts are made. If the test is successful in each attempt, the possibility for success in the next attempt increases by 5%. If it fails any attempt, the possibility for failure increases by 5% in the next attempt. Present a probability tree for the 3 attempts, assuming that the probability of the initial attempt being successful is 60%. These attempts result in introducing a number of strategies. Identify and rank these strategies based on the predicted outcomes. The highest probability of having a successful attempt from any of the selected strategies is:

(a) 12.6%   (b) 27.3%   (c) 13.2%   (d) 9.0%

Solution Guide

Figure 43 shows the probability tree for the 3 attempts. Note that for the strategies to be selected, the final attempt should produce positive test results. Table 15 presents the

4 selected strategies. Note that there are a total of 8 strategies that are divided into 4 distinct categories (1 to 4). There are 2 options associated with each strategy, *a* presenting the higher path and *b* presenting the lower path. Therefore, the best strategy (*1a*) is the higher path of strategy 1, which results in all-successful attempts {*WWW*}, while the weakest strategy (*4b*) is the lower path of strategy 4, which results in 2 failed attempts followed by 1 successful attempt {*LLW*}.

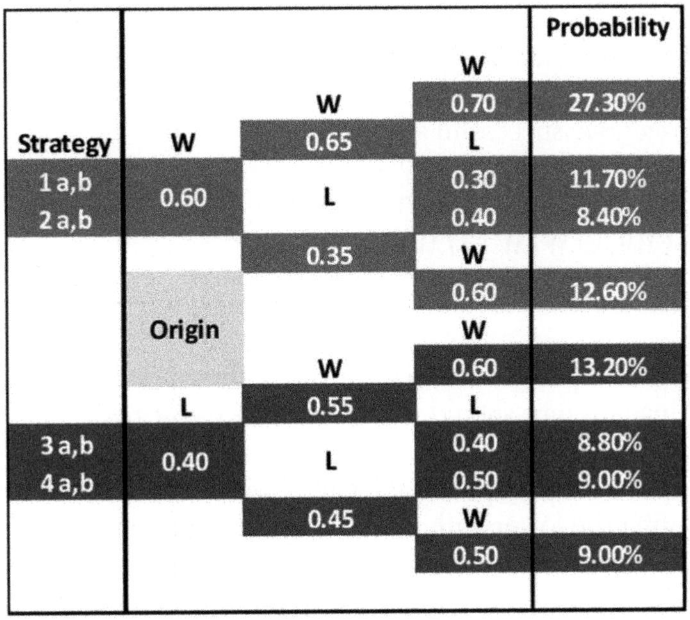

Strategy				Probability
			W	
		W	0.70	27.30%
Strategy	W	0.65	L	
1 a,b	0.60	L	0.30	11.70%
2 a,b			0.40	8.40%
		0.35	W	
			0.60	12.60%
	Origin		W	
		W	0.60	13.20%
	L	0.55	L	
3 a,b	0.40	L	0.40	8.80%
4 a,b			0.50	9.00%
		0.45	W	
			0.50	9.00%

FIGURE 43. Probability tree.

TABLE 15. Success probability tree for the introduced strategies.

Rank	Strategy	P (Win)
1	1a	27.30%
2	3a	13.20%
3	2b	12.60%
4	1b	11.70%

153. Several sets of tests are conducted to find a cure for a virus, with each set resulting either in success or failure. Three distinct attempts are made. If the test is successful in each attempt, the possibility for success in the next attempt increases by 5%. If it fails any attempt, the possibility for failure increases by 5% in the next attempt. Present a probability tree for the 3 attempts, assuming that the probability of the initial attempt being successful is 60%. These attempts result in introducing a number of strategies. The probability of having at least 2 successful attempts is:

(a) $P(WWW \cup WLW \cup LWW \cup LLW) = 27.3\%$
(b) $P(WWW \cup WWL \cup WLW \cup LWW) = 64.8\%$
(c) $P(WWW \cup WLW \cup LWW \cup LLW) = 12.6\%$
(d) $P(WWW \cup WLW \cup LWW \cup LLW) = 36.3\%$

Solution Guide

Figure 43 shows the probability tree for the 3 attempts. Note that for the strategies to be selected, the final attempt should produce positive test results. Table 15 presents the 4 selected strategies. Note that there are a total of 8 strategies that are divided into 4 distinct categories (1 to 4). There are 2 options associated with each strategy, *a* presenting the higher path and *b* presenting the lower path. Two successful attempts are required with any combination or order.

154. There are 2 scenarios of success and failure associated with tests carried out to find a vaccine for protection against a virus. Three distinct attempts are made. If the test is successful in each attempt, the possibility for success in the next attempt increases by 5%. If it fails any attempt, the possibility for failure increases by 5% in the next attempt. Present a probability tree for the 3 attempts, assuming that the probability that the initial attempt is successful is 60%. These attempts result in introducing a number of strategies. The probability of having at least 2 consecutive successful attempts is:

(a) $P(WWW \cup WWL \cup LWW) = 52.2\%$
(b) $P(WWW \cup WWL \cup WLW \cup LWW) = 27.3\%$
(c) $P(WWW \cup WWL \cup WLW \cup LWW) = 24.9\%$
(d) $P(WWW \cup WWL \cup WLW \cup LWW) = 37.5\%$

Solution Guide

Figure 43 shows the probability tree for the 3 attempts. Note that for the strategies to be selected, the final attempt should produce positive test results. Table 15 presents the 4 selected strategies. Note that there are a total of 8 strategies that are divided into 4 distinct categories (1 to 4). There are 2 options associated with each strategy, *a* presenting the higher path and *b* presenting the lower path. Two successful (and failed) consecutive attempts are required with any combination or order.

155. There is a company interested in selling 2 different products in multiple regions. Each has 2 preferred cities to sell the products. The first city has a population of about 2 million while the second one has a population of about 5.8 million. People like to purchase the products that are more compatible with their values; therefore, if each of the products is sold to the more compatible city, they would obtain 85% of the entire market of that city. The early surveys have shown that if the 2 products are sold in the same city, product A has a 65% chance of success. Suggest the payoff matrix for the products being sold. The best-case scenario (dominant strategy) for Product 1 to be sold is:

(a) Product 1 for City 1 and Product 2 for City 1
(b) Product 1 for City 1 and Product 2 for City 2
(c) Product 1 for City 2 and Product 2 for City 1
(d) Product 1 for City 2 and Product 2 for City 2

Solution Guide

Table 16 presents the payoff results for the two products. If Product 1 is sold in City 1, and Product 2 is sold in City 1, Product 2 wins. If Product 1 is sold in City 2, and Product 2 is sold in City 2, Product 2 wins. If Product 1 is sold in City 2, and Product 2 is sold in City 1, Product 1 wins. If Product 1 is sold in City 1, and Product 2 is sold in City 2, Product 2 wins.

TABLE 16. Payoff matrix (to be multiplied by 1,000).

Product 1	City	Product 2	
		1	2
	1	(700, 1300)	(2570, 5230)
	2	(4930, 2570)	(2030, 3770)

156. Constant-sum games have . . . while zero-sum games have constant . . .
(a) zero-total gains / total gains
(b) zero losses for all teams / equal losses for all teams
(c) zero losses for any team / losses for any team
(d) zero total gains for all teams / total gains for all teams

157. An equilibrium point when playing sports is the strategy that players would:
(a) change if the other players changed their strategies.
(b) not change if the other players changed their strategies.
(c) change if the other players did not change their strategies.
(d) not change if the other players did not change their strategies.

158. A saddle point is a(an) . . . in a zero-sum or a constant-sum game.
(a) high point   (b) equilibrium point
(c) low point   (d) point

159. A saddle point is where the players would:
(a) change their own strategies if the other players did not change theirs.
(b) select their strategies whichever way they wish despite other players.
(c) not select their strategies without considering other players.
(d) not change their own strategies if the other players did not change theirs.

160. Find the saddle point for the scenarios (Table 17) in which there are two players working independently, who follow five distinct strategies. They are to decide when to change their strategies. Analyze your decision.

TABLE 17. Success probabilities matrix for two players.

Method	Player 2				
	1	2	3	4	5
1	5.5%	13.0%	10.0%	8.0%	7.0%
2	7.0%	10.5%	9.5%	9.0%	6.0%
3	8.0%	8.0%	9.0%	11.0%	5.0%
4	9.0%	5.5%	8.5%	5.0%	4.0%
5	10.0%	11.0%	12.0%	11.0%	14.0%

(a) Player 1 (Method 5) and Player 2 (Method 1)
(b) Player 1 (Method 1 or 5) and Player 2 (Method 5 or 1)
(c) Player 1 (Method 1) and Player 2 (Method 5)
(d) Player 1 (Method 1 and 5) and Player 2 (Method 5 and 1)

Solution Guide

Table 18 presents the completed table in which the minimum success probability of one player (Player 1) is compared to the maximum success probability of another player (Player 2). The saddle point is selected by equating the minimum of one with the maximum of another one. In other words, assuming that the strategy of one does not vary, the strategy of the other one should not vary either, producing fair chance of success for both players. There is one saddle point for these scenarios.

TABLE 18. Success probabilities matrix for two players.

| | Method | \multicolumn{5}{c}{Player 2} | Minimum |
		1	2	3	4	5	
Player 1	1	5.5%	13.0%	10.0%	8.0%	7.0%	5.5%
	2	7.0%	10.5%	9.5%	9.0%	6.0%	6.0%
	3	8.0%	8.0%	9.0%	11.0%	5.0%	5.0%
	4	9.0%	5.5%	8.5%	5.0%	4.0%	4.0%
	5	10.0%	11.0%	12.0%	11.0%	14.0%	10.0%
Maximum		10.0%	13.0%	12.0%	11.0%	14.0%	

161. In a zero-sum game, to reach the saddle point, the players attempt to:

(a) minimize their payoff column by minimizing the competitor's payoff row
(b) minimize their payoff column by maximizing the competitor's payoff row
(c) maximize their payoff column by maximizing the competitor's payoff row
(d) maximize their payoff column by minimizing the competitor's payoff row

162. In a mixed-strategy game, competitors have a variety of winning options available to them whose probabilities sum up to:

(a) 1　　　　(b) 0　　　　(c) −1　　　　(d) c (c is a constant)

163. Two candidates (Willy and Nilly) are running for the same senate seat. They each have 3 strategies to follow with their associated probabilities to win. The early polls have suggested that Willy and Nilly select the strategies with the following probabilities: $\{P_1, P_2, P_3\}$ and $\{Q_1, Q_2, Q_3\}$. Their payoff matrix is presented in Table 19. To obtain the saddle points, one may:

TABLE 19. Payoff strategies matrix.

| | Strategy | \multicolumn{3}{c}{Nilly} |
		1	2	3
Willy	1	$\alpha11$	$\alpha12$	$\alpha13$
	2	$\alpha21$	$\alpha22$	$\alpha23$
	3	$\alpha31$	$\alpha32$	$\alpha33$

(a) minimize their payoff column by minimizing the competitor's payoff row
(b) minimize their payoff column by maximizing the competitor's payoff row
(c) maximize their payoff column by maximizing the competitor's payoff row
(d) maximize their payoff column by minimizing the competitor's payoff row

164. Two candidates (Willy and Nilly) are running for the same senate seat. They each have 3 strategies to follow with their associated probabilities to win. The early polls have suggested that Willy and Nilly select the strategies with the following probabilities: $\{P_1, P_2, P_3\}$ and $\{Q_1, Q_2, Q_3\}$. Their payoff matrix is presented in Table 20. Analyze the expected success probabilities matrix data.

TABLE 20. Payoff strategies matrix.

| | Strategy | \multicolumn{3}{c}{Nilly} |
		1	2	3
Willy	1	$\alpha11$	$\alpha12$	$\alpha13$
	2	$\alpha21$	$\alpha22$	$\alpha23$
	3	$\alpha31$	$\alpha32$	$\alpha33$

(a) $E(x) = \sum_{i=1}^{3}\sum_{j=1}^{3}\alpha_{ij} \times P_i \times Q_j$ (b) $E(x) = \sum_{i=1}^{3}\alpha_{ij} \times \sum_{j=1}^{3}P_i \times Q_j$

(c) $E(x) = \sum_{i,j\ 1} {}_{ij} \times P_i \times Q_j$ (d) $E(x) = \sum_{i=1}^{3}\alpha_i \times \sum_{j=1}^{3}\alpha_j \times P_i \times Q_j$

Solution Guide

The success probabilities matrix data are presented in Table 21.

TABLE 21. Success probabilities matrix for two candidates.

Selection	Payoff (α_{ij})	Probability (WiPi)	XiP(Xi)
W1N1	$\alpha 11$	P1×Q1	P1×Q1×$\alpha 11$
W1N2	$\alpha 12$	P1×Q2	P1×Q2×$\alpha 12$
W1N3	$\alpha 13$	P1×Q3	P1×Q3×$\alpha 13$
W2N1	$\alpha 21$	P2×Q1	P2×Q1×$\alpha 14$
W2N2	$\alpha 22$	P2×Q2	P2×Q2×$\alpha 15$
W2N3	$\alpha 23$	P2×Q3	P2×Q3×$\alpha 16$
W3N1	$\alpha 31$	P3×Q1	P3×Q1×$\alpha 17$
W3N2	$\alpha 32$	P3×Q2	P3×Q2×$\alpha 18$
W3N3	αij	P3×Q3	P3×Q3×$\alpha 19$
			E(X)=ΣXiP(Xi)

165. Two candidates (Willy and Nilly) are running for the same senate seat. They each have 3 strategies to follow with their associated probabilities to win. The early polls have suggested that Willy and Nilly select the strategies with the following probabilities: {0.2, 0.5, 0.3} and {0.6, 0.3, 01}. Their payoff matrix is presented in Table 22. Present the expected success probability for Willy.

TABLE 22. Payoff strategies matrix.

		Nilly		
	Strategy	1	2	3
Willy	1	0.4	0.3	-0.2
	2	0.2	-0.3	0.1
	3	0.1	-0.1	-0.5

(a) 0.076%   (b) 0.76%   (c) 7.6%   (d) 76%

Solution Guide

The success probabilities matrix data are presented in Table 23.

TABLE 23. Success probabilities matrix for two candidates.

Selection	Payoff (α_{ij})	Probability (WiPi)	XP(X)
W1N1	0.4	12%	4.8%
W1N2	0.3	6%	1.8%
W1N3	−0.2	2%	-0.4%
W2N1	0.2	30%	6.0%
W2N2	−0.3	15%	-4.5%
W2N3	0.1	5%	0.5%
W3N1	0.1	18%	1.8%
W3N2	−0.1	9%	-0.9%
W3N3	−0.5	3%	-1.5%
		E(X)=ΣXiP(Xi)	7.6%

166. Al and Dao are two competitors who are playing against each other in a tournament. Each of them has their own style of play, with Al opting for Strategies 1 and 2 (with the selection probabilities of P and $(1 - P)$ and Dao opting for Strategies 1 and 2 (with the selection probabilities of Q and $(1 - Q)$. The payoff matrix for these strategies is presented in Table 24. Analyze the expected success probabilities matrix data.

TABLE 24. Payoff strategies matrix.

		Dao	
	Strategy	**1**	**2**
Al	**1**	$\alpha 1$	$\alpha 2$
	2	$\alpha 3$	$\alpha 4$

(a) $E(x) = P \times Q \times (\alpha 1 - \alpha 2 - \alpha 3 + \alpha 4) + P \times (\alpha 2 - \alpha 4) + Q \times (\alpha 3 - \alpha 4) + \alpha 4$
(b) $E(x) = P \times Q \times (\alpha 1 - \alpha 2 + \alpha 3 + \alpha 4) + Q \times (\alpha 2 - \alpha 4) + P \times (\alpha 3 - \alpha 4) - \alpha 4$
(c) $E(x) = P \times Q \times (\alpha 1 - \alpha 2 - \alpha 3 - \alpha 4) - P \times (\alpha 2 - \alpha 4) - Q \times (\alpha 3 - \alpha 4) + \alpha 4$
(d) $E(x) = P \times Q \times (\alpha 1 + \alpha 2 - \alpha 3 + \alpha 4) - Q \times (\alpha 2 - \alpha 4) - P \times (\alpha 3 - \alpha 4) - \alpha 4$

Solution Guide

Probabilities for the introduced strategies for both players are presented in Table 25.

TABLE 25. Success probabilities matrix for two players.

Selection	Payoff	Probability	XP(X)
A1D1	$\alpha 1$	$P \times Q$	$P \times Q \times \alpha 1$
A1D2	$\alpha 2$	$P \times (1-Q)$	$P \times (1-Q) \times \alpha 2$
A2D1	$\alpha 3$	$(1-P) \times Q$	$(1-P) \times Q \times \alpha 3$
A2D2	$\alpha 4$	$(1-P) \times (1-Q)$	$(1-P) \times (1-Q) \times \alpha 4$
			$E(X) = \Sigma XiP(Xi)$

167. Al and Dao are two competitors who are playing against each other in a tournament. Each of them has their own style of play, with Al opting for Strategies 1 and 2 (with the selection probabilities of P and $(1 - P)$ and Dao opting for Strategies 1 and 2 (with the selection probabilities of Q and $(1 - Q)$. The payoff matrix for these strategies is presented in Table 26. Present the probabilities associated with the optimum decisions for both players.

TABLE 26. Payoff strategies matrix.

		Dao	
	Strategy	**1**	**2**
Al	**1**	$\alpha 1$	$\alpha 2$
	2	$\alpha 3$	$\alpha 4$

(a) $P = \left(\dfrac{\alpha 3 - \alpha 4}{\alpha 1 - \alpha 2 - \alpha 3 + \alpha 4} \right), Q = \left(\dfrac{\alpha 2 - \alpha 4}{\alpha 1 - \alpha 2 - \alpha 3 + \alpha 4} \right)$

(b) $P = \left(-\dfrac{\alpha 3 - \alpha 4}{\alpha 1 - \alpha 2 - \alpha 3 + \alpha 4} \right), Q = \left(-\dfrac{\alpha 2 - \alpha 4}{\alpha 1 - \alpha 2 - \alpha 3 + \alpha 4} \right)$

(c) $Q = \left(-\dfrac{\alpha 3 - \alpha 4}{\alpha 1 - \alpha 2 - \alpha 3 + \alpha 4} \right), P = \left(-\dfrac{\alpha 2 - \alpha 4}{\alpha 1 - \alpha 2 - \alpha 3 + \alpha 4} \right)$

(d) $Q = \left(\dfrac{\alpha 3 - \alpha 4}{\alpha 1 - \alpha 2 - \alpha 3 + \alpha 4} \right), P = \left(\dfrac{2\alpha - 4\alpha}{\alpha 1 - 2\alpha - 3\alpha + 4\alpha} \right)$

Solution Guide

Probabilities for the introduced strategies for both players are presented in Table 27.

TABLE 27. Success probabilities matrix for two players.

Selection	Payoff	Probability	XP(X)
A1D1	$\alpha 1$	P×Q	P×Q×$\alpha 1$
A1D2	$\alpha 2$	P×(1−Q)	P×(1−Q)×$\alpha 2$
A2D1	$\alpha 3$	(1-P)×Q	(1-P)×Q×$\alpha 3$
A2D2	$\alpha 4$	(1-P)×(1−Q)	(1-P)×(1−Q)×$\alpha 4$
			E(X)=ΣXiP(Xi)

$$E(x) = P \times Q \times (\alpha 1 - \alpha 2 - \alpha 3 + \alpha 4) + P \times (2\alpha - 4\alpha) + Q \times (3\alpha - 4\alpha) + 4\alpha$$

$$\frac{dE(x)}{dP} = 0$$

$$\therefore Q \times (\alpha 1 - \alpha 2 - \alpha 3 + \alpha 4) + (\alpha 2 - \alpha 4) = 0$$

$$\therefore Q = \left(-\frac{\alpha 2 - \alpha 4}{\alpha 1 - \alpha 2 - \alpha 3 + 4\alpha} \right)$$

$$\frac{dE(x)}{dQ} = 0$$

$$\therefore P \times (\alpha 1 - \alpha 2 - \alpha 3 + \alpha 4) + (\alpha 3 - \alpha 4) = 0$$

$$\therefore P = \left(-\frac{\alpha 3 - \alpha 4}{\alpha 1 - \alpha 2 - \alpha 3 + \alpha 4} \right)$$

168. Two basketball teams (Alpha and Beta) are competing in a playoff (zero-sum game). Each of them has their own style of play, with Team Alpha opting for Strategies 1 and 2 (with the selection probabilities of 0.2 and 0.8) and Team Beta opting for Strategies 1 and 2 (with the selection probabilities of 0.45 and 0.55). The payoff matrix for these strategies is presented in Table 28. Present the probabilities associated with the optimum decisions for both teams.

TABLE 28. Payoff strategies matrix.

		Beta	
	Strategy	1	2
Alpha	1	1	−1.5
	2	−0.5	1

(a) $E(x) = 81.5\%$, $P = 37.5\%$, $Q = 62.5\%$
(b) $E(x) = 18.5\%$, $P = 62.5\%$, $Q = 37.5\%$
(c) $E(x) = 81.5\%$, $P = 37.5\%$, $Q = 62.5\%$
(d) $E(x) = 18.5\%$, $P = 37.5\%$, $Q = 62.5\%$

Solution Guide

Probabilities for the introduced strategies for both teams are presented in Table 29.

TABLE 29. Success probabilities matrix for two teams.

Selection	Payoff	Probability	XP(X)
A1B1	1	9.00%	9.0%
A1B2	−1.5	11.00%	-16.5%
A2B1	−0.5	36.00%	-18.0%
A2B2	1	44.00%	44.0%
		E(X)=Σ XiP(Xi)	18.5%

169. Two basketball teams (Alpha and Beta) are competing in a playoff (a constant-sum game with the total points of 1.2). Each of them has their own style of play, with Team Alpha opting for Strategies 1 and 2 (with the selection probabilities of 0.7 and 0.3) and Team Beta opting for Strategies 1 and 2 (with the selection probabilities of 0.4 and 0.6). The payoff matrix for these strategies is presented in Table 30. Calculate the missing payoff information X. Present the probabilities associated with the optimum decisions for both players.

TABLE 30. Payoff strategies matrix.

		Beta	
	Strategy	**1**	**2**
Alpha	1	0.1	0.3
	2	0.6	X

(a) $X = 0.2$, $E(x) = 26.2\%$, $P = 66.7\%$, $Q = 16.7\%$
(b) $X = 1.2$, $E(x) = 26.2\%$, $P = 16.7\%$, $Q = 66.7\%$
(c) $X = 0.2$, $E(x) = 73.8\%$, $P = 33.3\%$, $Q = 83.3\%$
(d) $X = 1.2$, $E(x) = 73.8\%$, $P = 83.3\%$, $Q = 33.3\%$

Solution Guide

Probabilities for the introduced strategies for both teams are presented in Table 31.

TABLE 31. Success probabilities matrix for two teams.

Selection	Payoff	Probability	XP(X)
A1B1	0.1	28.00%	2.8%
A1B2	0.3	42.00%	12.6%
A2B1	0.6	12.00%	7.2%
A2B2	0.2	18.00%	3.6%
		E(X)=Σ XiP(Xi)	26.2%

170. Player rankings are presented in Table 32. Rankings are made in 5 different criteria numbered 1 to 5, equally weighted. Identify the top three players.

TABLE 32. Player rankings.

	Criteria				
Player	**1**	**2**	**3**	**4**	**5**
1	1	1	3	2	1
2	2	1	2	1	3
3	3	2	1	3	2
4	4	3	5	4	3
5	5	4	4	5	4

(a) Players 1, 2, and 3
(b) Players 1, 3, and 5
(c) Players 2, 3, and 5
(d) Players 3, 4, and 5

171. Player rankings are presented in Table 33. Rankings are made in 5 different criteria numbered 1 to 5. Identify the top two players.

TABLE 33. Player rankings.

Player	Criteria				
	1	**2**	**3**	**4**	**5**
1	1	1	3	2	1
2	2	1	2	1	3
3	3	2	1	3	2
4	4	3	5	4	3
5	5	4	4	5	4

(a) Players 1 and 3   (b) Players 1 and 2
(c) Players 2 and 3   (d) Players 3 and 4

172. Player rankings are presented in Table 34. Rankings are made in 5 different criteria numbered 1 to 5. Identify the top player using the instant runoff method.

TABLE 34. Player rankings.

Player	Criteria				
	1	**2**	**3**	**4**	**5**
1	1	1	5	5	1
2	2	1	2	1	3
3	3	2	1	3	2
4	4	3	3	4	3
5	5	4	4	2	4

(a) Player 1   (b) Player 2   (c) Player 3   (d) Player 4

173. A coach is picking players for his team. Rankings for 5 players in 5 different criteria (1 to 5) are listed in Table 35. The coach associates an importance weight function of {0.3, 0.2, 0.4, 0.1, 0} to the average outcome. If the coach uses the instant runoff method for the selection of the best player, who is the best player and what is his lead percent?

TABLE 35. Player rankings.

Player	Criteria				
	1	**2**	**3**	**4**	**5**
1	1	1	5	5	1
2	2	1	2	1	3
3	3	2	1	3	2
4	4	3	3	4	3
5	5	4	4	2	4

(a) Player 1 by 15% lead   (b) Player 1 by 1.15% lead
(c) Player 2 by 1.15% lead   (d) Player 2 by 15% lead

174. Player rankings are presented in Table 36. Rankings are made in 5 different criteria numbered 1 to 5. Identify the worst 2 players using a head-to-head count.

TABLE 36. Player rankings.

Player	Criteria				
	1	2	3	4	5
1	1	1	5	5	1
2	2	1	2	1	3
3	3	2	1	3	2
4	4	3	3	4	3
5	5	4	4	2	4

(a) Players 3 and 4   (b) Players 4 and 5
(c) Players 3 and 5   (d) Players 2 and 5

175. Player rankings are presented in Table 37. Rankings are made in 5 different criteria numbered 1 to 5. Identify the Condorcet winner, if any. Player 2 would be a Condorcet winner if (s)he:

TABLE 37. Player rankings.

Player	Criteria				
	1	2	3	4	5
1	1	1	5	5	1
2	2	1	2	1	3
3	3	2	1	3	2
4	4	3	3	4	3
5	5	4	4	2	4

(a) player 1, won criterion 1 overall
(b) player 2, won criterion 5 overall
(c) there is no Condorcet winner, won criterion 1 overall
(d) no Condorcet winners, won criterion 5 overall

176. Player rankings are presented in Table 38. Rankings are made in 5 different criteria numbered 1 to 5. Identify the best player using the Copeland (head-to-head) method.

TABLE 38. Player rankings.

Player	Criteria				
	1	2	3	4	5
1	1	1	5	5	1
2	2	1	2	1	3
3	3	2	1	3	2
4	4	3	3	4	3
5	5	4	4	2	4

(a) Player 3   (b) Player 2
(c) Player 1   (d) Player 4

Solution Guide

Table 39 presents the Copeland analysis for the presented data. It is concluded that Player 2 has a larger number of wins over other players, followed by Players 3, 1, 4, and 5.

TABLE 39. Copeland analysis.

Player	Player 1	2	3	4	5	Head-to-Head
1	0	2,2	3,2	3,2	3,2	3
2	2,2	0	3,2	4,0	5,0	10
3	2,3	2,3	0	5,0	4,1	6
4	2,3	0,4	0,5	0	4,1	-7
5	2,3	0,5	1,4	1,4	0	-12

177. In a good voting system, there are no restrictions as to the order in which candidates are to be ranked; the individual voter does not identify the outcome of the election; and if all or a majority of the voters prefer Candidate A over B, Candidate B is not elected by a group choice of voters. (T/F)

178. A party is using polls to select the best candidate to nominate for running in an election. They asked participants to rank each of the 3 candidates in 5 criteria (Table 40). Based on the Pareto optimality, the party's choice should be:

TABLE 40. Poll participants' choices.

Candidate	Criteria 1	2	3	4	5
1	1	1	1	2	1
2	1	3	2	1	1
3	3	1	5	1	2

(a) Candidate 2   (b) Candidate 3
(c) Candidates 2 or 3   (d) Candidate 1

179. A party is using polls to select the best candidate to nominate for running in an election. They asked participants to rank each of the 4 candidates in 5 criteria (Table 41). Note that Candidate 4 (Table 40) is added to the poll participants' choices presented in Table 41. The party's choice should be . . . and the selection process is . . . of the irrelevant alternatives:

TABLE 41. Poll participants' choices.

Candidate	Criteria 1	2	3	4	5
1	1	1	1	2	1
2	1	3	2	1	1
3	3	1	3	1	2
4	1	1	1	1	1

(a) Candidate 4 / not independent   (b) Candidate 4 / independent
(c) Candidate 1 / not independent   (d) Candidate 1 / independent

180. A party is using polls to select the best candidate to nominate for running in an election. They asked participants to rank each of the 5 candidates in 5 criteria (Table 42). Note that Candidate 5 (Table 41) is added to the poll's participants choice presented in Table 42. The party's choice using the Copeland (head-to-head) method should be . . . and the selection process is . . . the irrelevant alternatives. The selection . . . Pareto optimality and . . . principles.

TABLE 42. Poll participants' choices.

Candidate	Criteria				
	1	2	3	4	5
1	1	1	1	2	1
2	1	3	2	1	1
3	3	1	3	1	2
4	1	1	1	1	1
5	2	2	1	1	3

(a) Candidate 4 / dependent upon / meets / non-dictatorship
(b) Candidate 4 / independent of / meets / non-dictatorship
(c) Candidate 5 / independent of / does not meet / non-dictatorship
(d) Candidate 5 / dependent upon / does not meet / dictatorship

Solution Guide

The Copeland analysis is presented in Table 43.

TABLE 43. Copeland analysis.

Candidate	Candidate					Head-to-Head
	1	2	3	4	5	
1	0	2,1	3,1	0,1	3,1	4
2	1,2	0	3,1	0,2	2,2	-1
3	1,3	1,3	0	0,3	2,2	-7
4	1,0	2,0	3,0	0	3,0	9
5	1,3	2,2	2,2	0,3	0	-5

181. A party is using polls to select the best candidate to nominate for running in an election. They asked participants to rank each of 5 candidates in 5 criteria (Table 42). Candidate 3 received disapproval from the voters for suggesting a foreign policy promoting friendship with underprivileged states and assisting the rest of the world to eliminate a viral pandemic. The election follows the universal domain rules. (T/F)

182. A party is using polls to select the best candidate to nominate for running in an election. They asked participants to rank each of the 5 candidates in 5 criteria (Table 42). Rank the candidates before and after eliminating Candidate 5 using the averaging method.
    (a) Candidates (4, 1, 2, 5, and 3) versus Candidates (4, 1, 2, and 3)
    (b) Candidates (3, 1, 2, 5, and 4) versus Candidates (3, 1, 2, and 4)
    (c) Candidates (1, 4, 2, 5, and 3) versus Candidates (1, 4, 2, and 3)
    (d) Candidates (3, 4, 2, 5, and 1) versus Candidates (3, 4, 2, and 1)

183. Empirical analysis:
    (a) means analyzing the experimentally collected data
    (b) does not necessarily require the experimental data
    (c) is used to derive new relationships based on fundamental theories
    (d) is a kind of numerical analysis

184. The Pearson correlation coefficient identifies:
    (a) degree of association between two continuous variables
    (b) covariance of the continuous numbers
    (c) standard deviation of the continuous numbers
    (d) non-data related information

185. The Pearson correlation coefficient is calculated by:

(a) $PCC = \dfrac{\sum_{i=0}^{n}(X_i - \bar{Y})(Y_i - \bar{X})}{\sqrt{\sum_{i=0}^{n}(X_i - \bar{Y})^2}\sqrt{\sum_{i=0}^{n}(Y_i - \bar{X})^2}}$

(b) $PCC = \dfrac{\sum_{i=0}^{n}(X_i - \bar{X})^2(Y_i - \bar{Y})^2}{\sum_{i=0}^{n}(X_i - \bar{X})^2 \sum_{i=0}^{n}(Y_i - \bar{Y})^2}$

(c) $PCC = \dfrac{\sum_{i=0}^{n}(X_i - \bar{X})(Y_i - \bar{Y})}{\sum_{i=0}^{n}(X_i - \bar{X})^2 \sum_{i=0}^{n}(Y_i - \bar{Y})^2}$

(d) $PCC = \dfrac{\sum_{i=0}^{n}(X_i - \bar{X})(Y_i - \bar{Y})}{\sqrt{\sum_{i=0}^{n}(X_i - \bar{X})^2}\sqrt{\sum_{i=0}^{n}(Y_i - \bar{Y})^2}}$

186. The Pearson correlation coefficient for the data presented in Table 44 is:

TABLE 44. Statistical data.

X	Y
1	-8
2	-6
3	-4
4	-2
5	0
6	2
7	4
8	6
9	8
10	10

(a) 0.75 (b) 0.5 (c) 1 (d) 0.25

Solution Guide

Calculations are presented in Table 45:

TABLE 45. Pearson correlation coefficient calculations.

X	Y	X-X_avg	Y-Y_avg	(X-X_avg)^2	(Y-Y_avg)^2	(X-X_avg)(Y-Y_avg)
1	-8	-4.5	-9	20.25	81	40.5
2	-6	-3.5	-7	12.25	49	24.5
3	-4	-2.5	-5	6.25	25	12.5
4	-2	-1.5	-3	2.25	9	4.5
5	0	-0.5	-1	0.25	1	0.5
6	2	0.5	1	0.25	1	0.5
7	4	1.5	3	2.25	9	4.5
8	6	2.5	5	6.25	25	12.5
9	8	3.5	7	12.25	49	24.5
10	10	4.5	9	20.25	81	40.5
5.5	1	0	0	82.5	330	165
Average X and Y		Column Summation				

187. Further analysis is conducted for the data presented in Table 45. Variable Y is drawn versus X (Figure 44). Which relationship describes $Y = f(X)$?

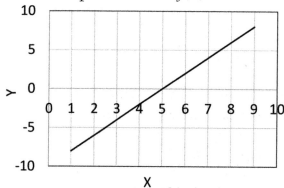

FIGURE 44. Plot of Y versus X.

(a) $Y = 2X - 10$ (b) $Y = X - 5$ (c) $Y = 0.5X + 10$ (d) $Y = 2X + 10$

188. For the diagram presented in Figure 45, select the relationship that describes $Y = f(X)$ most accurately.

FIGURE 45. Plot of Y versus X.

(a) $Y = -0.537X + 11.947$ (b) $Y = -2.024\ln(X) + 12.051$
(c) $Y = -0.026X^2 - 0.251X + 11.374$ (d) $Y = 0.003X^3 - 0.075X^2 - 0.024X$

189. For the data presented in the form of $Y = f(X)$, to fit a linear relationship using the least squares method, the expressions ... are used, where $(\overline{X}, \overline{Y})$ are the average values of X and Y and (S_{XX}, S_{YY}, S_{XY}) are their variances.

(a) $a = \dfrac{S_{XY}}{S_{XX}}, b = \overline{Y} - a\overline{X}, Y = aX + b$ (b) $a = \dfrac{S_{YY}}{S_{XX}}, b = \overline{X} - a\overline{Y}, X = aY - b$

(c) $a = \dfrac{S_{XY}}{S_{YY}}, b = \overline{X} - a\overline{Y}, Y = aX - b$ (d) $a = \dfrac{S_{XX}}{S_{XY}}, b = \overline{Y} - a\overline{X}, Y = -aX + b$

190. For a relationship with two degrees of freedom Y and using the average value $(\overline{Y})$, chi-square (χ^2) and standard deviation (σ) are given by:

(a) $\chi^2 = \displaystyle\sum_{i=1}^{n} \left(\frac{Y_i - \overline{Y}}{\sigma_i} \right)^2, \sigma_i = \sqrt{\displaystyle\sum_{i=1}^{n} \frac{1}{n-2}\left(Y_i - \overline{Y} \right)^2}$

(b) $\chi^2 = \displaystyle\sum_{i=1}^{n} \left(\frac{X_i - X(Y_i)}{\sigma_i} \right)^2, \sigma_i = \sqrt{\displaystyle\sum_{i=1}^{n} \frac{1}{n-2}\left(Y_i - Y(X_i) \right)}$

(c) $\chi^2 = \sum_{i=1}^{n}\left(\frac{X_i - X(Y_i)}{2\sigma_i}\right)^2, \sigma_i = \sqrt{\sum_{i=1}^{n}\frac{1}{n-1}(Y_i - Y(X_i))}$

(d) $\chi^2 = \sum_{i=1}^{n}\left(\frac{Y_i - \overline{Y}}{2\sigma_i}\right)^2, \sigma_i = \sqrt{\sum_{i=1}^{n}\frac{1}{n-1}(Y_i - \overline{Y})^2}$

191. For a relationship with one degree of freedom X and using the average value $(\overline{X})$, chi-square (χ^2) and standard deviation (σ) are given by:

(a) $\chi^2 = \sum_{i=1}^{n}\left(\frac{X_i - \overline{X}}{\sigma_i}\right), \sigma_i = \sqrt{\sum_{i=1}^{n}\frac{1}{n}(X_i - \overline{X})^2}$

(b) $\chi^2 = \sum_{i=1}^{n}\left(\frac{X_i - \overline{X}}{\sigma_i}\right)^2, \sigma_i = \sqrt{\sum_{i=1}^{n}\frac{1}{2n-1}(X_i - \overline{X})}$

(c) $\chi^2 = \sum_{i=1}^{n}\left(\frac{X_i - \overline{X}}{\sigma_i}\right), \sigma_i = \sqrt{\sum_{i=1}^{n}\frac{1}{n}(X_i - \overline{X})}$

(d) $\chi^2 = \sum_{i=1}^{n}\left(\frac{X_i - \overline{X}}{\sigma_i}\right)^2, \sigma_i = \sqrt{\sum_{i=1}^{n}\frac{1}{n-1}(X_i - \overline{X})^2}$

192. The smaller the chi-square (χ^2) test static is, the better the observed data fits the expected data. (T/F)

Answer Key									
1. (b)	2. (b)	3. (c)	4. (b)	5. (a)	6. (b)	7. (d)	8. (d)	9. (a)	10. (d)
11. (a)	12. (b)	13. (a)	14. (a)	15. (b)	16. (c)	17. (b)	18. (a)	19. (a)	20. T
21. (a)	22. (b)	23. (b)	24. (d)	25. (a)	26. (d)	27. (c)	28. F	29. T	30. (a)
31. (a)	32. (b)	33. (a)	34. (b)	35. (a)	36. (d)	37. (a)	38. (a)	39. (c)	40. (b)
41. (d)	42. (a)	43. (d)	44. (b)	45. (c)	46. (b)	47. (d)	48. (c)	49. (a)	50. (a)
51. (c)	52. (a)	53. (c)	54. (d)	55. (b)	56. (b)	57. (a)	58. (b)	59. (c)	60. (a)
61. (c)	62. (a)	63. (c)	64. (d)	65. (a)	66. (b)	67. (d)	68. (b)	69. (a)	70. (b)
71. (c)	72. (a)	73. (c)	74. (b)	75. (a)	76. (b)	77. (a)	78. (d)	79. (a)	80. (c)
81. (b)	82. (d)	83. (a)	84. (c)	85. (c)	86. (a)	87. (b)	88. (b)	89. (c)	90. (d)
91. (a)	92. (b)	93. (c)	94. (d)	95. (a)	96. (c)	97. (d)	98. (a)	99. (a)	100. (b)
101. (a)	102. (c)	103. (c)	104. (d)	105. (c)	106. (c)	107. (d)	108. (c)	109. (b)	110. (a)
111. (c)	112. (a)	113. (d)	114. (a)	115. (a)	116. T	117. (b)	118. (c)	119. (a)	120. (c)
121. (d)	122. (b)	123. T	124. (a)	125. (b)	126. (a)	127. (a)	128. (c)	129. (a)	130. (d)
131. (b)	132. (a)	133. (d)	134. (a)	135. (b)	136. (a)	137. F	138. T	139. (a)	140. (c)
141. (b)	142. (b)	143. (a)	144. (a)	145. (d)	146. (a)	147. (a)	148. T	149. (c)	150. (a)
151. (a)	152. (b)	153. (b)	154. (a)	155. (c)	156. (a)	157. (d)	158. (b)	159. (d)	160. (a)
161. (d)	162. (a)	163. (d)	164. (a)	165. (c)	166. (a)	167. (b)	168. (d)	169. (a)	170. (a)
171. (b)	172. (a)	173. (a)	174. (b)	175. (c)	176. (b)	177. T	178. (d)	179. (a)	180. (b)
181. F	182. (a)	183. (a)	184. (a)	185. (d)	186. (c)	187. (a)	188. (c)	189. (a)	190. (a)
191. (d)	192. T								

CHAPTER 3

ENGINEERING MECHANICS

1. The second Newton's law relates force, body mass, and:
   - (a) velocity
   - (b) acceleration
   - (c) torque
   - (d) moment of inertia

2. The field of study that assumes the resultant force on a body is zero is:
   - (a) dynamics
   - (b) hydrostatic
   - (c) statics
   - (d) mechanics

3. A free body diagram shows . . . applied to the body in . . .
   - (a) torques / one direction
   - (b) forces / one direction
   - (c) torques / all directions
   - (d) forces / all directions

4. Identify the forces applied to a car moving on a surface:
   - (a) weight, friction, and thrust
   - (b) weight, reaction, and surface friction
   - (c) thrust, surface friction, and reaction force
   - (d) weight, thrust, rolling friction, and reaction force

5. Trusses are designed to carry:
   - (a) axial load
   - (b) torsional load
   - (c) torque
   - (d) shear load

6. To analyze the truss, which approach(es) is(are) mainly employed?
   - (a) joint method, where forces in joined parts are calculated
   - (b) equilibrium method, where forces are calculated for a section of the body
   - (c) multiple parts are assessed simultaneously
   - (d) both (a) and (b)

7. If a free-hanging cable/chain is under its own weight, then forces in the cable/chain are:
   - (a) different for tension and compression
   - (b) often not in balance
   - (c) perpendicular to the chain/cable
   - (d) uniform with respect to the length of the chain/cable

8. Mass m is attached to a massless cable with length L, forming a pendulum. At the top of the pendulum swing, the cable makes the angle of θ with respect to the vertical line (Figure 46). The force component on the mass along the cable direction is:

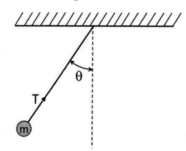

FIGURE 46. Mass attached to a cable, forming a pendulum.

(a) mg sin θ

(b) $\dfrac{d}{dx}(mg\cos\theta)$

(c) $\dfrac{d}{dy}(mg\sin\theta)$

(d) mg cos θ

9. Mass m is attached to a massless cable with length L, forming a pendulum. At the top of the pendulum swing, the cable makes the angle of θ with respect to the vertical line (Figure 46). The force component on the mass perpendicular to the cable direction is:

(a) mg sin θ

(b) $\dfrac{d}{dx}(mg\cos\theta)$

(c) $\dfrac{d}{dy}(mg\sin\theta)$

(d) mg cos θ

10. Mass m is attached to a massless cable with length L, forming a pendulum. If the pendulum is held at the angle of θ and released (Figure 46), speed of mass at the bottom of the swing is:

(a) $\sin\left(\dfrac{\theta}{2}\right)\sqrt{gL}$

(b) $2\sin\left(\dfrac{\theta}{2}\right)\sqrt{gL}$

(c) $\cos\left(\dfrac{\theta}{2}\right)\sqrt{gL}$

(d) $2\cos\left(\dfrac{\theta}{2}\right)\sqrt{gL}$

11. Dynamics is related to the application of force, causing motion. (T/F)

12. Kinetics is related to displacement and its derivatives. (T/F)

13. A truss element is loaded by:
    (a) shear forces and moments
    (b) two axial loads equal and in opposite directions
    (c) two axial loads carrying moments
    (d) external loads applied to its end points

14. A mechanical system is a group of particles in a continuum mode interacting with one another. (T/F)

15. A force is completely defined by a positive or negative scalar, while its sign depends on the force's direction. (T/F)

16. The magnitude of a force with components of F_x, F_y, and F_z in the $x - y - z$ Cartesian coordinates is given by:

    (a) $\sqrt{F_x^2 + F_y^2 + F_z^2}$

    (b) $\sqrt[3]{F_x^2 + F_y^2 + F_z^2}$

    (c) $F_x^2 + F_y^2 + F_z^2$

    (d) $\left(F_x^2 + F_y^2 + F_z^2\right)^2$

17. Forces experienced by rigid bodies are mechanical forces if the bodies are at rest or moving. (T/F)

18. The forces applied to a flying aircraft are:
    (a) drag, thrust, lift, and weight
    (b) drag, lift, and weight
    (c) lift, thrust, and friction
    (d) friction, lift, and weight

19. The difference between a point force and distributed force is:
    (a) distributed force is applied to a line, surface, or volume; point force is applied to a single point on a line, surface, or volume
    (b) distributed force is applied to a point; point force is applied to the surface
    (c) distributed force is applied to two points; point force is applied to points on a surface
    (d) distributed force is applied to points on a surface; point force is the summation of the distributed forces

20. The balance of all forces applied to a stationary rigid body is:
    (a) 0   (b) > 0   (c) either (a) or (b)   (d) unknown

21. The distances between all particles belonging to a system remain constant, when moving or at rest, for:
    (a) liquids   (b) solids   (c) rigid bodies   (d) gases

22. A system of forces consists of:
    (a) multiple forces acting on the body
    (b) two forces applied in opposite directions to a rigid body
    (c) forces from multiple fields (e.g., gravity) applied in different directions to a body
    (d) none of the above

23. The equivalent of all forces applied to the system is:
    (a) a single force that results in the same reaction as the system of forces
    (b) a single force that results in a greater or equal reaction as the system of forces
    (c) a set of two forces that are applied in opposite directions
    (d) all of the above

24. When a balanced set of forces is acting on a system:
    (a) system accelerates
    (b) system moves at the variable speed
    (c) system becomes stationary
    (d) system is at equilibrium

25. Mechanical properties, also known as axiomatic properties:
    (a) are verifiable by tests or obtained from experiments
    (b) can be measured or estimated by simple observations
    (c) can be represented by mathematical formulae
    (d) all of the above

26. If two forces are applied to a rigid body, which have the same magnitude (F) and the same direction, the resultant force is:

    (a) $2F$ (b) F (c) 0   (d) $-F$

27. Two forces (F_1 and F_2) are applied to a rigid body (Figure 47). The forces lie in the same plane, have the same magnitude, but point in opposite directions, as shown in the figure. The resultant force is:

FIGURE 47. Forces applied to a rigid body.

    (a) $2F$ (b) F (c) 0   (d) $-F$

28. If you add forces in balanced condition to a balanced system, the system:

    (a) decelerates
    (b) accelerates
    (c) becomes stationary
    (d) remains at mechanical equilibrium state

29. The balance of forces with different magnitudes (F_x and F_y), with the resultant force (F) making the angle θ with one another, is:

    (a) $\sqrt{F_x + F_y + 2F_x F_y \cos\theta}$ (b) $\sqrt{F_x^2 + F_y^2 - 2F_x F_y \cos\theta}$

    (c) $\sqrt{F_x^2 + F_y^2 + 2F_x F_y \sin\theta}$ (d) $\sqrt{F_x^2 + F_y^2 - 2F_x F_y \sin\theta}$

30. If forces are applied to a deformable body at equilibrium:

    (a) the body loses its equilibrium state under equal pressures from opposite sides
    (b) the body remains at equilibrium state if with increasing forces the body resists elastically to deformation
    (c) the body is at equilibrium in all conditions
    (d) none of the above

31. The force applied to a rigid body can:

    (a) not be translated along its line of action
    (b) not be translated in all conditions
    (c) be translated along its line of action
    (d) be translated in all directions

32. The resultant force is applied to the coinciding point, and its magnitude is equal to the magnitude of the vector sum of all applied forces. (T/F)

33. A body can have a set of applied forces which result in equilibrium (no acceleration) or a net force with acceleration. (T/F)

34. A constraint restricts all movement of a body in space. (T/F)

35. The friction forces on a smooth surface can be . . . and the reaction force is applied . . . to the surface.

    (a) significant / at the 45-degree angle

(b) ignored / normal

(c) dependent on an object mass / parallel

(d) significant / parallel

36. A mechanical support point is located . . . and is . . . to the support surface.

(a) within the contact area / normal

(b) at the end of the joints / parallel

(c) on the upper surface of an object / normal

(d) on the vertical surface of an object / parallel

37. For an ideal thread, the weight of the load on the screw significantly affects the friction forces. (T/F)

38. For an ideal thread, the thread angle increases the friction and wear of a screw. (T/F)

39. In an ideal thread, reaction forces are . . . and mechanical advantage of the screw depends on the screw . . .

(a) along the thread / distance between the threads

(b) normal to the thread / diameter

(c) parallel to the thread / length

(d) normal to the thread / pitch angle

40. Two bodies connected by a cylindrical joint . . . rotate about and . . . translate along the joint axis.

(a) cannot / cannot   (b) cannot / can

(c) can / cannot   (d) can / can

41. In a hinged movable support, one body can only . . . relative to another body . . . and reaction forces are . . . to the . . .

(a) translate / about three axes / parallel / contact surface

(b) translate / about one axis / parallel / support surface

(c) rotate / about three axes / normal / contact surface

(d) rotate / about one axis / normal / support surface

42. In a spherical linkage, translational motion is possible in all directions. (T/F)

43. In cylindrical and spherical joints, the resultant force on the joints is the . . . of the applied force vectors.

(a) difference   (b) dot product

(c) cross product   (d) vector summation

44. In an ideal bar, weight is important; (T/F) loads are applied to the end joints only; (T/F) reaction forces can apply in any direction relative to the bar. (T/F)

(a) F / T / F   (b) F / F / F   (c) T / T / T   (d) T / T / F

45. A step bearing:

(a) supports only axial loads   (b) supports loads in all directions

(c) allows for axial translation   (d) allows for translation in one plane

46. A spherical plain bearing permits:

(a) angular rotation about a central point in two orthogonal directions

(b) angular rotation about a central point and axial translation in two orthogonal directions

(c) axial translation in two orthogonal directions

(d) angular rotation in two orthogonal directions and translation in one direction

47. An anchorage restricts translation and rotation of a body in:
(a) one direction where load is applied
(b) one direction where moment is applied
(c) all directions
(d) two directions where load and moments are applied

48. A cantilever beam subjected to a structural load carries the load to the support, applying a shear stress and a bending moment to the supported end. (T/F)

49. A constrained body can be treated as a free body if:
(a) active forces are present
(b) constraints are only partial
(c) constraints are flexible
(d) constraints can be substituted with the reaction forces

50. A body is unconstrained if its translation and rotation are . . . in a given space.
(a) restricted (b) free
(c) both (a) and (b) (d) none of the above

51. Forces exerted on the body by constraints are:
(a) reaction or constraint forces (b) reactive or constraint forces
(c) active or constraint forces (d) active or reaction forces

52. Active forces are applied to the . . . body and are . . . the reaction forces:
(a) rigid / independent of (b) free / independent of
(c) rigid / dependent on (d) free / dependent on

53. The moment of forces about the center of gravity is:
(a) > 0
(b) < 0
(c) equal to mass multiplied by body length
(d) 0

54. For a volume subdivided into n sub-volumes V_i with centroids located at (x_i, y_i, z_i), the centroid in a 3D Cartesian space is defined as:

(a) $x_c = \dfrac{\sum_{i=1}^{n} x_i V_i}{\sum_{i=1}^{n} V_i}, y_c = \dfrac{\sum_{i=1}^{n} y_i V_i}{\sum_{i=1}^{n} V_i}, z_c = 0$ (b) $x_c = \dfrac{\sum_{i=1}^{n} x_i V_i}{\sum_{i=1}^{n} V_i}, y_c = \dfrac{\sum_{i=1}^{n} y_i V_i}{\sum_{i=1}^{n} V_i}, z_c = \dfrac{\sum_{i=1}^{n} z_i V_i}{\sum_{i=1}^{n} V_i}$

(c) $x_c = \dfrac{\sum_{i=1}^{n} x_i V_i}{\sum_{i=1}^{n} V_i}, y_c = 0, z_c = \dfrac{\sum_{i=1}^{n} z_i V_i}{\sum_{i=1}^{n} V_i}$ (d) $x_c = 0, y_c = \dfrac{\sum_{i=1}^{n} y_i V_i}{\sum_{i=1}^{n} V_i}, z_c = \dfrac{\sum_{i=1}^{n} z_i V_i}{\sum_{i=1}^{n} V_i}$

55. A planar shape is in a 3D Cartesian space. It is oriented so as to be non-parallel to any of the principal planes. If this planar shape is subdivided into n sub-areas A_i with centroids located at (x_i, y_i, z_i), the centroid in a 3D Cartesian space is defined as:

(a) $x_c = \dfrac{\sum_{i=1}^{n} x_i A_i}{\sum_{i=1}^{n} A_i}, y_c = \dfrac{\sum_{i=1}^{n} y_i A_i}{\sum_{i=1}^{n} A_i}, z_c = 0$ (b) $x_c = \dfrac{\sum_{i=1}^{n} x_i A_i}{\sum_{i=1}^{n} A_i}, y_c = \dfrac{\sum_{i=1}^{n} y_i A_i}{\sum_{i=1}^{n} A_i}, z_c = \dfrac{\sum_{i=1}^{n} z_i A_i}{\sum_{i=1}^{n} A_i}$

(c) $x_c = \dfrac{\sum\limits_{i=1}^{n} x_i A_i}{\sum\limits_{i=1}^{n} A_i}, y_c = 0, z_c = \dfrac{\sum\limits_{i=1}^{n} z_i A_i}{\sum\limits_{i=1}^{n} A_i}$ (d) $x_c = 0, y_c = \dfrac{\sum\limits_{i=1}^{n} y_i A_i}{\sum\limits_{i=1}^{n} A_i}, z_c = \dfrac{\sum\limits_{i=1}^{n} z_i A_i}{\sum\limits_{i=1}^{n} A_i}$

56. An arc-shaped curve is in a 3D Cartesian space. It is oriented so as to be non-parallel to any of the principal planes. If this curve is subdivided into n sub-lengths L_i with centroids located at (x_i, y_i, z_i), the centroid in a 3D Cartesian space is defined as:

(a) $x_c = \dfrac{\sum\limits_{i=1}^{n} x_i L_i}{\sum\limits_{i=1}^{n} L_i}, y_c = \dfrac{\sum\limits_{i=1}^{n} y_i L_i}{\sum\limits_{i=1}^{n} L_i}, z_c = 0$ (b) $x_c = \dfrac{\sum\limits_{i=1}^{n} x_i L_i}{\sum\limits_{i=1}^{n} L_i}, y_c = \dfrac{\sum\limits_{i=1}^{n} y_i L_i}{\sum\limits_{i=1}^{n} L_i}, z_c = \dfrac{\sum\limits_{i=1}^{n} z_i L_i}{\sum\limits_{i=1}^{n} L_i}$

(c) $x_c = \dfrac{\sum\limits_{i=1}^{n} x_i L_i}{\sum\limits_{i=1}^{n} L_i}, y_c = 0, z_c = \dfrac{\sum\limits_{i=1}^{n} z_i L_i}{\sum\limits_{i=1}^{n} L_i}$ (d) $x_c = 0, y_c = \dfrac{\sum\limits_{i=1}^{n} y_i L_i}{\sum\limits_{i=1}^{n} L_i}, z_c = \dfrac{\sum\limits_{i=1}^{n} z_i L_i}{\sum\limits_{i=1}^{n} L_i}$

57. On a semi-circular curve with radius R (Figure 48), the centroid is located at:

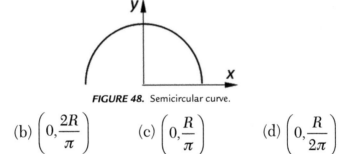

FIGURE 48. Semicircular curve.

(a) $(0, 0)$ (b) $\left(0, \dfrac{2R}{\pi}\right)$ (c) $\left(0, \dfrac{R}{\pi}\right)$ (d) $\left(0, \dfrac{R}{2\pi}\right)$

58. Moment of force is related to:
    (a) rotational effects of the forces applied to a fixed or moving body
    (b) translational effects of the forces applied to a fixed or moving body
    (c) rotational and translational effects of the forces applied to a fixed or moving body
    (d) none of the above

59. Tightening torque has a similar magnitude for all mechanical devices. (T/F)

60. Moments are related to physical quantities located at some distance with respect to a fixed reference point. (T/F)

61. The moment is the . . . of the . . .; additionally, it is . . . the . . . plane.
    (a) cross product / force and its position vector from the center of gravity / normal to / force-distance
    (b) dot product / force and distance from the center / parallel to / force-distance
    (c) dot product / force and distance from the center / on / force-distance
    (d) cross product / force and distance from the center / parallel to / force-distance

62. $\vec{r}$ is the position vector of the force vector $\vec{F}$ relative to the body's center of gravity, and θ is the angle between the two vectors. The moment calculated relative to the center of gravity on a plane is obtained from:

(a) $M = \vec{r} \times \vec{F} = |\vec{r}||\vec{F}|\cos\theta$

(b) $M = \vec{r} \times \vec{F} = -|\vec{r}||\vec{F}|\sin\theta$

(c) $M = \vec{r} \times \vec{F} = |\vec{r}||\vec{F}|\sin\theta$

(d) $M = \vec{r} \times \vec{F} = -|\vec{r}||\vec{F}|\cos\theta$

63. In order to analytically calculate the moment, one can use the following formula:

(a) $\begin{vmatrix} \vec{i} & \vec{j} & \vec{k} \\ x & y & z \\ F_x & F_y & F_z \end{vmatrix} = -(-yF_z + zF_y)\vec{i} - (-zF_x + xF_z)\vec{j} - (-xF_y + yF_x)\vec{k}$

(b) $\begin{vmatrix} \vec{j} & \vec{i} & \vec{k} \\ y & x & z \\ F_y & F_x & F_z \end{vmatrix} = -(yF_z - zF_y)\vec{i} + (zF_x - xF_z)\vec{j} - (xF_y - yF_x)\vec{k}$

(c) $\begin{vmatrix} \vec{i} & \vec{j} & \vec{k} \\ x & y & z \\ F_x & F_y & F_z \end{vmatrix} = (-yF_z + zF_y)\vec{i} + (-zF_x + xF_z)\vec{j} + (-xF_y + yF_x)\vec{k}$

(d) $\begin{vmatrix} \vec{j} & \vec{k} & \vec{i} \\ y & z & x \\ F_y & F_z & F_x \end{vmatrix} = -(yF_z - zF_y)\vec{i} - (zF_x - xF_z)\vec{j} + (xF_y - yF_x)\vec{k}$

64. If the moment of a force with respect to an axis is desired:

(a) projection of the force on the plane of the axis is multiplied by the distance between the projected force and the axis

(b) force is multiplied by the distance from the axis

(c) force is multiplied by the distance from the projected force

(d) projection of the force on the plane of the axis is multiplied by the distance from the axis

65. A force couple is a system of two forces that:

(a) cannot rotate an object about any plane parallel to the plane of action

(b) are equal in magnitude and parallel but in opposite directions

(c) both (a) and (b)

(d) none of the above

66. A system of multiple force couples can be replaced by one equivalent couple. (T/F)

67. The arm of the couple is equal to . . . and, at equilibrium, the sum of the couple's forces is . . .

(a) half of the distance between the forces / greater than zero

(b) the distance between the forces / greater than zero

(c) half of the distance between the forces / zero

(d) the perpendicular distance between the forces / zero

68. A truss is supported by a movable joint and a fixed joint (Figure 49); identify joint types and R_x and R_y at each joint:

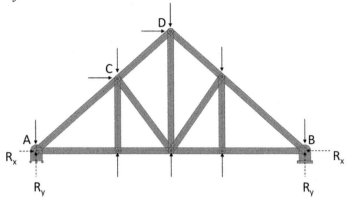

FIGURE 49. Truss bars free body diagram.

(a) A: movable joint $(R_x = R_{xA} = R_C, R_y = R_{yA})$, B: fixed joint $(R_x = R_{xB}, R_y = 0)$

(b) A: movable joint $(R_x = 0, R_y = R_{yA})$, B: fixed joint $(R_x = R_{xB}, R_y = R_{yB})$

(c) A: fixed joint $(R_x = R_{xA}, R_y = 0)$, B: movable joint $(R_x = R_{xB}, R_y = R_{yB} = R_D)$

(d) A: fixed joint $(R_x = 0 = R_C, R_y = R_{yA})$, B: movable joint $(R_x = 0, R_y = R_{yB})$

69. Which diagram best describes a complete set of loads applied to the mechanical system presented in two positions in Figure 50? Note that the bars connecting the object to the fixed (left hinge) and sliding (right hinge) hinges are weightless while the moving body is weighted.

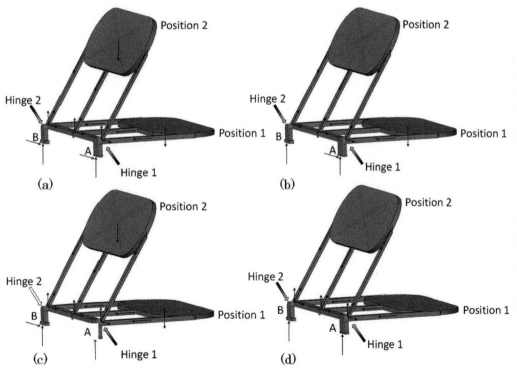

FIGURE 50. Force diagram of the object attached by hinges.

70. Forces applied along different lines of action that intersect at one point are called . . .; the resultant force is the . . . of the forces.
(a) concurrent / vector summation
(b) non-concurrent / cross product
(c) intersecting / vector summation
(d) non-intersecting / cross product

71. The . . . of . . . forces from one vector to another in space determines the force component magnitudes.
(a) projection / non-parallel
(b) normal / parallel
(c) projection / parallel
(d) normalization / non-parallel

72. The equilibrium condition for multiple forces passing through a common point is achieved by:
(a) maximizing the sum of the force components
(b) minimizing the force components
(c) zeroing the sum of the force components
(d) maximizing the force components

73. If the number of unknowns in force balance relations is larger than the number of constraints, the problem is:
(a) statically determined
(b) statically undetermined
(c) statically significant
(d) statically insignificant

74. If the number of unknowns in force balance relations is larger than that of the constraints, the problem is solved by . . . and . . . relations.
(a) force / torsion
(b) moment / displacement
(c) moment / torsion
(d) force / displacement

75. Displacement equations are based on the . . . compatibility condition, also known as . . .
(a) strain / Hooke's law
(b) stress / Hooke's law
(c) strain / Newton's law
(d) stress / Newton's law

76. For the balance of forces to happen, the lines of action of forces should:
(a) intersect at multiple points
(b) intersect at a common point
(c) not intersect at a common point
(d) none of the above

77. The principal moment is the:
(a) cross products of forces multiplied by their distances from any point in the space
(b) dot products of forces multiplied by their distances from any point in the space
(c) sum of cross products of the forces and their associated perpendicular distances from a reference point to the force vector
(d) summation of forces multiplied by their distances from any point in the space

78. The point of reduction is the point at which the . . . applied to a rigid body are transferred.
(a) principal forces
(b) projected forces
(c) normal forces
(d) force components

79. When incorporating point of reduction in the moment balance, the . . . should be included in the equilibrium calculations; this is equivalent to a . . . system.
(a) moment force and a new force / moment and force couple
(b) original moment force and a force component / force and force couple
(c) moment force and a new moment force / moment and force couple
(d) original moment force and a new moment force / force and force couple

80. A catenary line is a:

(a) trajectory that a weightless cable with rotating ends takes
(b) curve that a weighted cable with moving ends takes
(c) curve that a weighted cable with supported ends assumes
(d) trajectory that a weightless cable with sliding ends assumes

81. The center of gravity lies in the center of the:

(a) suspension
(b) object
(c) largest compartment volume
(d) smallest compartment volume

82. Calculate the centroid location (X_{COG}, Y_{COG}) for the 2D geometry with respect to the coordinate system shown in Figure 51 (O is the center of origin).

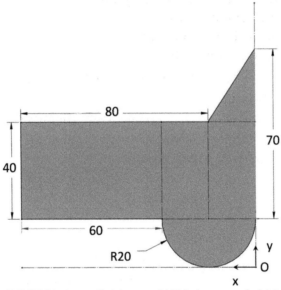

FIGURE 51. Area to find the centroid (O is the center of origin).

(a) $X_{COG} = 38.19$, $Y_{COG} = 43.54$ (b) $X_{COG} = 43.54$, $Y_{COG} = 38.19$
(c) $X_{COG} = 22.46$, $Y_{COG} = 45.12$ (d) $X_{COG} = 43.54$, $Y_{COG} = 38.19$

Solution Guide

One technique to calculate the centroid for complex geometries is to subdivide the geometry into shapes with known centers of gravity. The resultant centroid is then the summation of the products of shape areas and their centroid coordinates divided by the total area.

$$X_{COG} = \frac{1}{\sum_{k=1}^{n} A_{k-COG}} \left(\sum_{k=1}^{n} X_{k-COG} A_{k-COG} \right) \quad Y_{COG} = \frac{1}{\sum_{k=1}^{n} A_{k-COG}} \left(\sum_{k=1}^{n} Y_{k-COG} A_{k-COG} \right)$$

$$Z_{COG} = \frac{1}{\sum_{k=1}^{n} A_{k-COG}} \left(\sum_{k=1}^{n} Z_{k-COG} A_{k-COG} \right)$$

83. Calculate the centroid (X_{COG}, Y_{COG}, Z_{COG}) for the 3D geometry with respect to the point O presented in Figure 51. The frame origin is at point O, the X-axis points to the left; the Z-axis is pointing up.

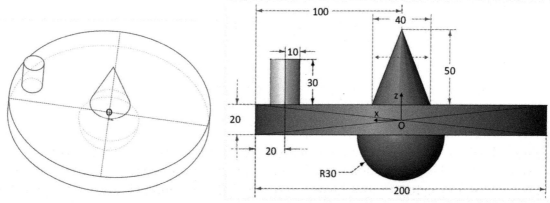

FIGURE 52. Volume to find the centroid (O is the center of origin).

(a) $X_{COG} = 7.46$, $Y_{COG} = 2.87$, $Z_{COG} = 4.17$
(b) $X_{COG} = 4.87$, $Y_{COG} = 4.17$, $Z_{COG} = 7.46$
(c) $X_{COG} = 1.05$, $Y_{COG} = 0$, $Z_{COG} = 2.67$
(d) $X_{COG} = 4.17$, $Y_{COG} = 2.87$, $Z_{COG} = 7.46$

Solution Guide

One technique to calculate the centroid for complex geometries is to subdivide the geometry into the sub-volumes with known centers of gravity. In this example, the resultant centroid is the summation of the products of each sub-volume by its centroid divided by the total volume.

$$X_{COG} = \frac{1}{\sum_{k=1}^{n} V_{k-COG}} \left(\sum_{k=1}^{n} X_{k-COG} V_{k-COG} \right) \quad Y_{COG} = \frac{1}{\sum_{k=1}^{n} V_{k-COG}} \left(\sum_{k=1}^{n} Y_{k-COG} V_{k-COG} \right)$$

$$Z_{COG} = \frac{1}{\sum_{k=1}^{n} V_{k-COG}} \left(\sum_{k=1}^{n} Z_{k-COG} V_{k-COG} \right)$$

84. The center of gravity in an unstable mechanical system is below the centroid. (T/F)

85. The centroid is the . . . of an area/volume.

(a) geometric center   (b) center of mass
(c) center of gravity   (d) none of the above

86. Both analytical and empirical methods can be used to calculate the centroid location. (T/F)

87. Derive the centroid for the analytical function plotted in Figure 53 ($0 < x < 3$).

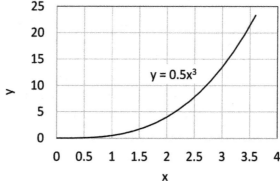

FIGURE 53. Analytical function to find the centroid.

(a) 1.9, 3.9   (b) 2.5, 1.9   (c) 2.4, 3.8   (d) 2.4, 3.9

Solution Guide

$$\overline{x} = \frac{\int_0^a x dA}{A} = \frac{\int_0^3 x(0.5x^3)dx}{\int_0^3 (0.5x^3)dx}$$

$$\overline{y} = \frac{\int_0^b y dA}{A} = \frac{\int_0^3 (0.25x^3)(0.5x^3)dx}{\int_0^3 (0.5x^3)dx}$$

88. Derive the forces in the weighted chain presented in Figure 54 and Figure 55 as a function of the position along the chain. T_0 is the chain's lowest part tension, S is the chain length, W is the chain weight per unit length, h is the chain sag, and L is the chain span. If β is 60 degrees, the cable weight is 100 N/m, and L is 2 m, the chain tension at the midpoint is:

FIGURE 54. Two posts with a hanging weighted chain.

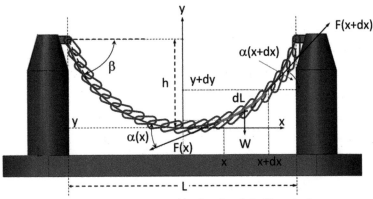

FIGURE 55. Two posts with a hanging chain (front view).

(a) 57.7 N   (b) 5.8 N   (c) 0.58 N   (d) 577.4 N

Solution Guide

The equilibrium shown is reached when the balance of forces is applied. This means that the forces applied to both sides of the selected link segment in addition to the weight should be balanced (T_0 is the tension of the cable at the lowest point $x = L/2$).

$$\overline{R} = \sum_{k=1}^{n} \overline{F_k} = 0$$

$$\sum_{k=1}^{n} F_x = 0$$

$$\therefore \ F(x+dx)\cos(\alpha(x+dx)) - F(x)\cos(\alpha x) = 0$$

$$\sum_{k=1}^{n} F_y = 0$$

$$\therefore \ F(x+dx)\sin(\alpha(x+dx)) - F(x)\sin(\alpha x) - Wdx = 0$$

at $x = 0$

$$\therefore \ F(dx)\cos(\alpha(dx)) = T_0, F(dx)\sin(\alpha(dx)) = Wdx$$

$$\frac{dy}{dx} = \tan\alpha(x) = \frac{W}{T_0}x$$

$$\therefore \ y = \frac{W}{2T_0}x^2$$

$$T = F(x) = \sqrt{T_0^2 + (Wx)^2}$$

$$\therefore \ T\Big|_{x=\frac{L}{2}} = F(x)\Big|_{x=\frac{L}{2}} = \sqrt{T_0^2 + \left(\frac{WL}{2}\right)^2}$$

89. A 5-kg planter, weighing W, is suspended by two weightless steel cables as shown in Figure 56. The cables have equal lengths (1 m) and the angles between the planter and each cable are equal ($\alpha = \alpha_1 = \alpha_2$). Derive the forces in the planter presented in Figure 56 and Figure 57. If the angle α is equal to 30 degrees, cable tension is:

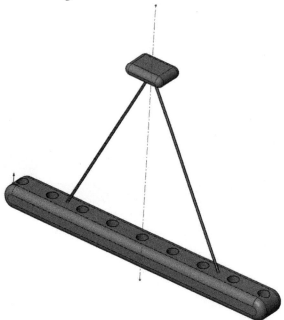

FIGURE 56. A planter suspended by weightless steel cables.

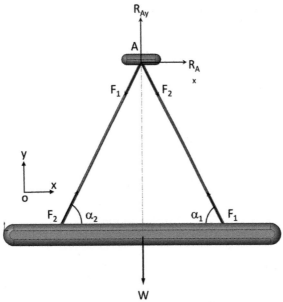

FIGURE 57. A planter suspended by weightless steel cables (front view).

(a) 29 N   (b) 98 N   (c) 49 N   (d) 12 N

Solution Guide

$$\vec{R} = \sum_{k=1}^{n} \overline{F_k} = 0$$

$$\sum_{k=1}^{n} F_x = 0$$

$$\therefore F_1 \cos \alpha_1 - F_2 \cos \alpha_2 = 0$$

$$\sum_{k=1}^{n} F_y = 0$$

$$\therefore F_1 \sin \alpha_1 + F_2 \sin \alpha_2 - W = 0$$

$$\sum_{k=1}^{n} F_x = 0$$

$$\therefore -F_1 \cos \alpha_1 + F_2 \cos \alpha_2 + R_{Ax} = 0$$

$$\sum_{k=1}^{n} F_y = 0$$

$$\therefore -F_1 \sin \alpha_1 - F_2 \sin \alpha_2 + R_{Ay} = 0$$

$$\alpha = \alpha_1 = \alpha_2$$

$$\therefore F = F_1 = F_2$$

$$\therefore F = \frac{W}{2 \sin \alpha}, R_{Ay} = 2F \sin \alpha, R_{Ax} = 0$$

90. Derive forces on the oil tanker pushed by three tugboats (Figure 58) if the ship is accelerating from velocity zero to V in the positive x-direction (mass M) over time Δt. Explain the moment balance, given the center of gravity at G (Figure 59). If $\alpha_1 = \alpha_2 = \alpha_3 = 45$ degrees, $b = 25$ m, and $c = 19$ m, d and acceleration (a) are (ignore water resistance, assume $F_1 = F_2 = F_3$):

(a) $d = 6$ m, $a = -F_1/M$ (b) $d = 25$ m, $a = F_1/M$
(c) $d = 125$ m, $a = F_1/2M$ (d) $d = 51$ m, $a = -F_1/2M$

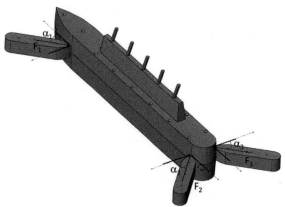

FIGURE 58. An oil tanker pushed by three tugboats.

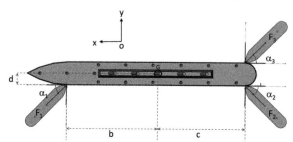

FIGURE 59. An oil tanker pushed by three tugboats (top view).

Solution Guide

$$\vec{R} = \sum_{k=1}^{n} \overrightarrow{F_k} = 0$$

$$\sum_{k=1}^{n} F_x = ma$$

$$\therefore -F_1 \cos\alpha_1 + F_2 \cos\alpha_2 + F_3 \cos\alpha_3 = ma = m\ddot{x} = m\dot{V} = m\frac{\Delta V}{\Delta t}$$

$$\sum_{k=1}^{n} F_y = 0$$

$$\therefore F_1 \sin\alpha_1 + F_2 \sin\alpha_2 - F_3 \sin\alpha_3 = 0$$

$$M|_G = 0$$

$$\therefore d(F_1 \cos\alpha_1 - F_2 \cos\alpha_2 + F_3 \cos\alpha_3) - bF_1 \sin\alpha_1 + c(F_2 \sin\alpha_2 - F_3 \sin\alpha_3) = 0$$

91. Derive the relation between forces in the truss structure presented in Figure 60. Find reaction forces ($\overrightarrow{R_A}$ and $\overrightarrow{R_B}$). Repeat the same analysis for the broken view (Figure 61) and find the interconnecting forces (F_{31}, F_{32}, F_{33}). If $\overrightarrow{F_1} = \overrightarrow{F_2} = \overrightarrow{F} = ... = \overrightarrow{F}$, $\overrightarrow{R_A} + \overrightarrow{R_B}$ is (bars have equal lengths):

(a) $-7F(\vec{i} + \vec{j})$ (b) $-7F\vec{j}$ (c) $14F(\vec{i} + \vec{j})$ (d) $14F\vec{j}$

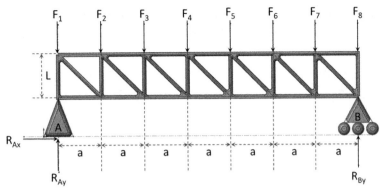

FIGURE 60. Truss on fixed and sliding bases.

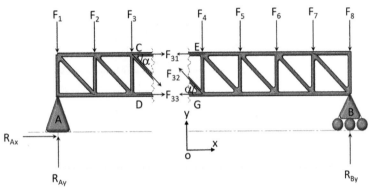

FIGURE 61. Truss on fixed and sliding bases (broken view).

Solution Guide

$$\vec{R} = \sum_{k=1}^{n} \overrightarrow{F_k} = 0$$

$$\sum_{i=1}^{n} F_{xi} = 0$$

$$\therefore R_{Bx} = 0$$

$$\therefore \sum_{i=1}^{n} F_{ix} = R_{Ax} + R_{Bx} = 0$$

$$\sum_{j=1}^{n} F_{yi} = 0$$

$$\therefore \sum_{j=1}^{n} F_{jy} = R_{Ay} + R_{By}$$

$$M\big|_A = 0$$

$$\therefore -a(F_2 + 2F_3 + 3F_4 + 4F_5 + 5F_6 + 6F_7 + 7F_8) + 7aR_{By} = 0$$

$$M\big|_B = 0$$

$$\therefore -a(F_7 + 2F_6 + 3F_5 + 4F_4 + 5F_3 + 6F_2 + 7F_1) + 7aR_{Ay} = 0$$

$$M\big|_A = 0$$

$$\therefore -a(F_2 + 2F_3 + 2F_{32}\sin\alpha) - L(F_{31} + F_{32}\cos\alpha) = 0$$

$$M|_B = 0$$

$$\therefore \ a(4F_4 - 4F_{32}\sin\alpha + 3F_5 + 2F_6 + F_7) + L(F_{31}) = 0$$

92. Derive the forces in the ring connected rigidly to the wall (Figure 62). Find reaction force components $(\overrightarrow{R_A})$. If $\overrightarrow{F_1} = \overrightarrow{F_2} = 116\,\text{N}$ and $\alpha = 60$ degrees, $\overrightarrow{R}$ is (reaction moment at point A is zero):

FIGURE 62. Wall-connected ring; disc shape represents the wall.

(a) 115 N   (b) 157 N   (c) 99 N   (d) 214 N

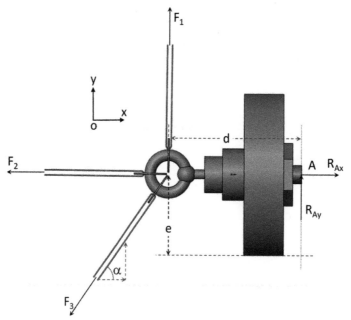

FIGURE 63. Wall-connected ring (top view).

Solution Guide

$$\overrightarrow{R} = \sum_{k=1}^{n}\overrightarrow{F_k} = 0$$

$$\sum_{i=1}^{n}F_{xi} = -F_2 - F_3\cos\alpha + R_{Ax} = 0$$

$$\sum_{j=1}^{n} F_{yj} = F_1 - F_3 \sin \alpha + R_{Ay} = 0$$

$$M\big|_A = 0$$

$$\therefore d\left(F_3 \sin \alpha - F_1\right) = 0$$

93. In a coplanar system, forces are applied to:

(a) different planes

(b) an object and produce an equilibrium state with a zero sum of the moments of the forces about any point in the plane

(c) different objects and produce a stationary state, and the sum of the moments of the forces about any point in the plane is zero

(d) an object and produce a stationary state, and the sum of the moments of the forces about any point in the plane is nonzero

94. An airliner is flying with only engine 1 operating (Figure 64 and Figure 65). The engine generates 325 kN of thrust. The plane is flying straight-and-level and drag force is F_D. A rudder applies force F_R along the negative y-direction (a is 5 m, b is 20 m). The magnitude of this force is:

FIGURE 64. Aircraft.

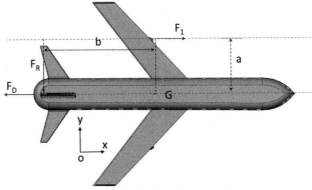

FIGURE 65. Aircraft (top view).

(a) 162.5 kN   (b) 81.2 kN   (c) 40.5 kN   (d) 325 kN

95. When transferring forces from their origins to new reference points, the:

(a) force system is transferred to the center

(b) coupled force system is transferred to the center

(c) forces' moments along the new reference point are transferred

(d) force system is transferred to the center plus its moment along the new reference point

96. The rolling friction force is the distribution of forces over:

    (a) the contact surface area   (b) the entire body

    (c) the contact point   (d) none of the above

97. The rolling and sliding friction forces are described by the same friction law. (T/F)

98. Joints:

    (a) connect rigid bodies that structurally do not change

    (b) introduce internal forces along the connecting bars/trusses

    (c) are exposed to the external forces at the end of the connecting bars/trusses

    (d) all of the above

99. A system of bars and joints is statically determinate if the number of joints (j), bars (s), and support reactions (r) are related by:

    (a) $s = 2j + r$ (b) $s = 2j - r$ (c) $s = j + r$ (d) $s = j - r$

100. If there are 5 bars and 3 support reactions, the number of joints required to make the problem statically determinate is:

    (a) 3   (b) 5   (c) 4   (d) 7

101. There are 12 bars, connected by 7 joints, with 3 support reactions. Is this a statically determinate truss mechanism? The number of excess members is:

    (a) No, 1   (b) No, 0   (c) Yes, 1   (d) Yes, 0

102. In order to find forces within the bars (or trusses), . . . is employed.

    (a) Newton's force balance   (b) Raphson's iteration method

    (c) Ritter's sectioning method   (d) all of the above

103. When analyzing forces at the joints, it is assumed that bars (or trusses) are under:

    (a) tension

    (b) compression

    (c) either tension or compression

    (d) none of the above

104. When analyzing forces applied to the bars at the joints:

    (a) compression means forces are toward the joint

    (b) tension means forces are directed away from the joints

    (c) zero load means the bar is not under tension or compression

    (d) all of the above

105. The rubber static friction coefficient is larger than that of the . . . against steel for dry contact:

    (a) steel   (b) ice   (c) wood   (d) aluminum

106. The aluminum sliding friction coefficient is higher than that of:

    (a) bronze   (b) copper   (c) nickel   (d) rubber

107. Sliding friction follows the model proposed by:

    (a) Amontons-Coulomb   (b) Ohm

    (c) Lavoisier   (d) Newton

108. Select the most accurate statement about the maximum sliding friction force:

    (a) It is independent of the contact area.

(b) It is determined by the value of the normal reaction.

(c) It is related to the material and the condition of the contacting surfaces.

(d) All of the above

109. The difference between the coefficient of rolling friction and sliding is that:

(a) the former has the units of of length-force; the latter has the units of length

(b) the former has the units of force-area; the latter is dimensionless

(c) both are dimensionless

(d) the former is dimensionless; the latter has the units of length

110. Which is applicable to the contact surface between the wall and a step ladder, when the ladder leaning against the wall is falling?

(a) Frictionless   (b) Rolling friction

(c) Static friction   (d) Sliding friction

111. A four-wheel-drive car is accelerating on a smooth, horizontal, dry road coated with standard asphalt (Figure 66 and Figure 67). The friction coefficient (μ), vehicle mass (m), thrust (F), car's tire radius (r), and acceleration (a) are provided. The center of gravity is in the middle of the car, at the intersection of the wheel's symmetry planes. Show the expression for the maximum drag force friction coefficient by analyzing the maximum drag force (F_D).

FIGURE 66. Tire.

(a) $\mu = \dfrac{(F - ma)}{mg}$ (b) $\mu = \dfrac{(4F - ma)}{m\sigma}$ (c) $\mu = \dfrac{(F - 4ma)}{4m\sigma}$ (d) $\mu = \dfrac{4(F - ma)}{mg}$

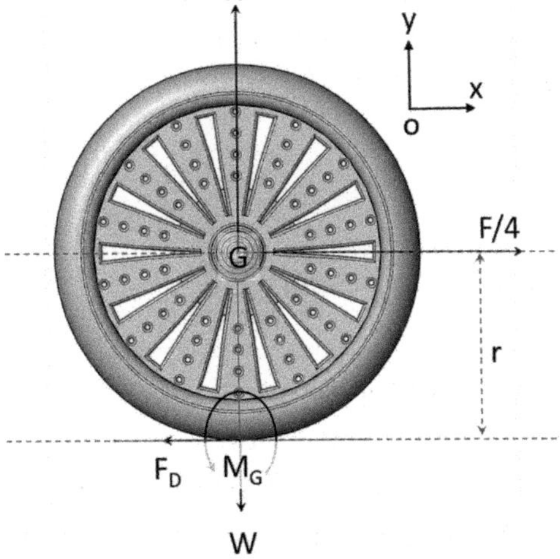

FIGURE 67. Tire (front view).

Solution Guide

$$\vec{R} = \sum_{k=1}^{n} \vec{F_k} = ma$$

$$\sum_{i=1}^{n} F_{xi} = \frac{F}{4} - F_D - \frac{m}{4}a$$

$$\therefore F_D = \frac{1}{4}(F - ma)$$

$$\sum_{j=1}^{n} F_{yj} = N - W = 0$$

$$\therefore N = W = \frac{m}{4}g$$

$$F_{D-\max} = \mu N = \mu \frac{m}{4}g$$

$$M\big|_G = \frac{rF}{4}$$

112. A wheelbarrow is shown in Figure 68. It holds three masses (m_1, m_2, and m_3) with their centers of gravity located as shown in Figure 69. W_1, W_2, and W_3 are the weights of each of the masses. The weight of the body (supported by the two wheels and the axel) has mass m and its weight is W. The location of the center of gravity is shown. (1) find the parametric relation for the horizontal position of center of gravity (X_{COG}); (2) derive the forces when the wheelbarrow is moving at a constant speed (V); and (3) derive the forces when the wheelbarrow is stationary. Complete the free body diagram presented in Figure 69. If $a = 2b$, $b = 2c$, $c = d$, $e = 2a$, and $W_1 = W_2 = W_3 = W/4$, find the center of gravity.

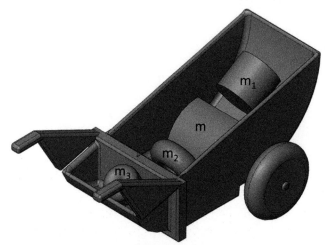

FIGURE 68. Wheelbarrow with three masses.

(a) $X_{COG} = \dfrac{2}{7}a$

(b) $X_{COG} = \dfrac{3}{14}a$

(c) $X_{COG} = \dfrac{7}{28}a$

(d) $X_{COG} = \dfrac{9}{28}a$

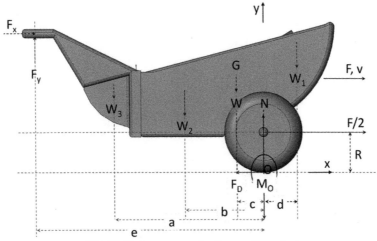

FIGURE 69. Forces on a wheelbarrow (front view).

Solution Guide

$$\vec{R} = \sum \vec{F} =$$

$$M\big|_O = 0$$

$$\therefore aW_3 + bW_2 + cW - dW_1 - eF_y = 0$$

$$N = \sum_{k=1}^{n} W_k$$

$$\sum_{i=1}^{n} F_{xi} = \frac{F}{2} - F_D + F_x = ma \therefore F_x = \mu N - \frac{F}{2} + ma$$

$$\sum_{j=1}^{n} F_{yj} = N + F_y = W + W_1 + W_2 + W_3 + F_y$$

$$X_{COG} = \frac{\sum_{k=1}^{n} \overrightarrow{W_k} \times \overrightarrow{r_k}\big|_O}{\sum_{k=1}^{n} \overrightarrow{W_k}}$$

$$\therefore aW_3 \quad bW_2 + cW - dW_1 = \ W + W_1 + W_2 \quad W_3 \quad _{COG}$$

113. A step ladder is leaning against the wall in Figure 70 and Figure 71. A person, weighing W, is climbing the ladder ($W = 100$ N). The friction coefficient between the ground and the ladder is μ. The wall is smooth. If $b = 5a$ and the weight of the ladder can be ignored, complete the free body diagram presented in Figure 70 and Figure 71. The contact force and its angle are:

FIGURE 70. Step ladder leaning against the wall.

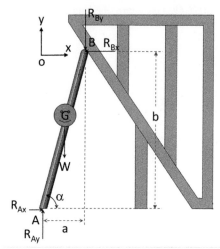

FIGURE 71. Step ladder (front view).

(a) $R = 102$ N, $\alpha = 42.1$ degrees   (b) $R = 204$ N, $\alpha = 84.3$ degrees

(c) $R = 102$ N, $\alpha = 84.3$ degrees   (d) $R = 204$ N, $\alpha = 42.1$ degrees

Solution Guide

$$\vec{R} = \sum_{k=1}^{n} \vec{F_k} = 0$$

$$R_{By} = 0, F_{Ax} = \mu N$$

$$M\big|_A = 0$$

$$\therefore a\left(-R_{By} - \frac{W}{2}\right) + R_{Bx}b = 0$$

$$\therefore R_{Bx} = \frac{aW}{2b}$$

$$\sum_{i=1}^{n} F_{xi} = R_{Ax} - R_{Bx} = 0$$

$$\therefore R_{Ax} = R_{Bx}$$

$$\sum_{j=1}^{n} F_{yj} = R_{Ay} - R_{By} - W = 0$$

$$\therefore R_{Ay} = W$$

$$R = \sqrt{R_{Ax}^2 + R_{Ay}^2} = \sqrt{\left(\frac{aW}{2b}\right)^2 + W^2} = W\sqrt{\left(\frac{a}{2b}\right)^2 + 1}$$

$$\tan\alpha = \frac{R_{Ay}}{R_{Ax}} = \frac{2b}{a}$$

$$\therefore \alpha = \arctan\left(\frac{2b}{a}\right)$$

114. A rope with mass m is suspended between two points which are in the same horizontal plane and separated by distance L. The rope sags due to its own weight (Figure 72). The variation of the rope sag (h) along the rope length (S), and the rope tension at the lowest point (T_0) are:

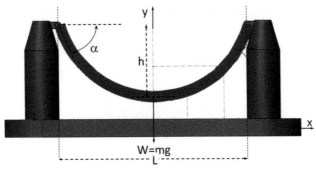

FIGURE 72. Two posts with a hanging weighted rope.

(a) $h = \sqrt{\frac{3}{8}L(S-L)}$, $T_0 = \frac{WL}{2\tan\alpha} = \frac{WL^2}{8h}$

(b) $h = \sqrt{\frac{8}{3}L(S-L)}$, $T_0 = \frac{2WL}{\tan\alpha} = \frac{8WL^2}{h}$

(c) $h = \sqrt{\frac{3}{8}L(L-S)}$, $T_0 = \frac{2\tan\alpha}{WL} = \frac{8h}{WL^2}$

(d) $h = \sqrt{\frac{8}{3}L(L-S)}$, $T_0 = \frac{\tan\alpha}{2WL} = \frac{h}{8WL^2}$

Solution Guide

$$F(x)\cos(\alpha(x)) = T_0$$

$$F(x)\sin(\alpha(x)) = Wx$$

$$\frac{dy}{dx} = \tan\alpha(x) = \frac{Wx}{T_0}$$

$$\therefore y = \frac{W}{2T_0}x^2$$

$$\frac{dS'}{dx} = \sqrt{1 + \left(\frac{dy}{dx}\right)^2}$$

$$\therefore \frac{dS'}{dx} = \sqrt{1 + \left(\frac{Wx}{T_0}\right)^2}$$

$$\therefore S' = \int_0^x \sqrt{1 + \left(\frac{Wx}{T_0}\right)^2}\, dx$$

$$\therefore S' = \left| x + \frac{1}{6}\left(\frac{W}{T_0}\right)^2 x^3 - \frac{1}{40}\left(\frac{W}{T_0}\right)^4 x^5 \right|_0^{\frac{L}{2}}$$

$$\therefore S = 2S' = L + 8h^2/(3L)$$

115. The system of the gears and pulleys presented in Figure 73 and Figure 74 consists of 4 gears with diameters d_G and 2 pulleys (A and B) with diameters d_P, and a disc C with diameter d_D. Force F_1 is applied to the pulley A and force F_2 is applied to the pulley B. Force F is applied to the edge of the disc C. The magnitude of F needed to balance the forces applied to the pulleys is:

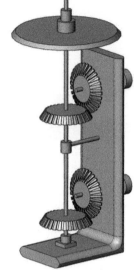

FIGURE 73. Gear system.

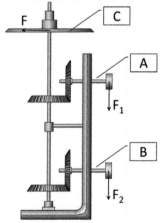

FIGURE 74. Gear system (front view).

(a) $F = \dfrac{2d_P}{d_D}(F_1 + F_2)$

(b) $F = \dfrac{d_P}{2d_D}(F_1 + F_2)$

(c) $F = \dfrac{d_D}{d_P}(F_1 + F_2)$

(d) $F = \dfrac{d_P}{d_D}(F_1 + F_2)$

Answer Key

1. (b)	**2.** (c)	**3.** (d)	**4.** (d)	**5.** (a)	**6.** (d)	**7.** (d)	**8.** (d)	**9.** (a)	**10.** (b)
11. T	**12.** F	**13.** (b)	**14.** T	**15.** F	**16.** (a)	**17.** T	**18.** (a)	**19.** (a)	**20.** (a)
21. (c)	**22.** (a)	**23.** (a)	**24.** (d)	**25.** (d)	**26.** (a)	**27.** (c)	**28.** (d)	**29.** (b)	**30.** (b)
31. (c)	**32.** T	**33.** T	**34.** F	**35.** (b)	**36.** (a)	**37.** F	**38.** T	**39.** (a)	**40.** (c)
41. (d)	**42.** F	**43.** (d)	**44.** (a)	**45.** (a)	**46.** (a)	**47.** (c)	**48.** T	**49.** (d)	**50.** (b)
51. (a)	**52.** (a)	**53.** (d)	**54.** (b)	**55.** (b)	**56.** (b)	**57.** (b)	**58.** (a)	**59.** F	**60.** T
61. (a)	**62.** (c)	**63.** (a)	**64.** (a)	**65.** (b)	**66.** T	**67.** (d)	**68.** (b)	**69.** (c)	**70.** (a)
71. (a)	**72.** (c)	**73.** (b)	**74.** (d)	**75.** (a)	**76.** (b)	**77.** (c)	**78.** (a)	**79.** (d)	**80.** (c)
81. (a)	**82.** (b)	**83.** (c)	**84.** F	**85.** (a)	**86.** (c)	**87.** (c)	**88.** (a)	**89.** (c)	**90.** (b)
91. (d)	**92.** (b)	**93.** (b)	**94.** (b)	**95.** (d)	**96.** (a)	**97.** T	**98.** (d)	**99.** (b)	**100.** (c)
101. (a)	**102.** (d)	**103.** (c)	**104.** (d)	**105.** (d)	**106.** (a)	**107.** (a)	**108.** (d)	**109.** (c)	**110.** (d)
111. (a)	**112.** (d)	**113.** (c)	**114.** (a)	**115.** (d)					

CHAPTER 4

STRUCTURE OF MATERIALS

1. A material's imperfections and defects may result in:
(a) dislocations
(b) loss of structural strength
(c) diffusion
(d) all of the above

2. Solid materials can be divided into the following major groups:
(a) Crystalline and amorphous
(b) Semi-crystalline and reinforced
(c) Amorphous and reinforced
(d) Semi-crystalline and amorphous

3. The main difference between crystalline and amorphous materials is:
(a) crosslinks are missing in crystalline materials
(b) crystalline solids do not have a definite melting point contrary to the amorphous solids
(c) crystalline solids have a long-range order contrary to the amorphous solids
(d) there are no sudden changes in volume when crystalline solids melt

4. A material's characterization is done based on its:
(a) properties
(b) strength
(c) electrical conduction
(d) thermal conduction

5. For the following statement, select the correct combination of True/False answers. In general, metals are:
(1) conductors (electrical and thermal); (T/F) (2) ductile (deform before failure); (T/F) and (3) shiny and opaque in appearance. (T/F)
(a) T / F / T   (b) T / T / T   (c) F / F / T   (d) F / F / F

6. For the following statement, select the correct combination of True/False answers. In general, polymers are:
(1) conductors (electrical and thermal); (T/F) (2) often transparent; (T/F) and (3) high tensile strength. (T/F)
(a) T / F / T   (b) T / T / T   (c) F / F / T   (d) F / T / T

7. For the following statement, select the correct combination of True/False answers. In general, ceramics (compounds of metals and nonmetals) are:

(1) conductors (electrical and thermal); (T/F) (2) ductile; (T/F) and (3) always transparent (T/F)

    (a) T / T / F   (b) F / F / T   (c) F / F / F   (d) T / T / F

8. If you know the position of an atom within a crystal, the position of other atoms in the crystal unit structure can be identified. (T/F)

9. If a material has a defect, the regular arrangement of atoms is disrupted. (T/F)

10. If the arrangement of atoms is disrupted, either the same or different bonds may be formed. (T/F) This happens in many biological processes; molecules break down into their component atoms and are reassembled into a more useful molecule. (T/F)

    (a) T / T   (b) T / F   (c) F / T   (d) F / F

11. Atoms consist of the following particles (m is the particle's mass):

    (a) Protons ($2\,m$), electrons ($>> m$), and neutrons (m)
    (b) Protons (m), electrons ($<< m$), and neutrons (m)
    (c) Protons (m), electrons ($<< 2m$), and nucleus ($2m$)
    (d) Protons ($2m$), electrons ($<< 2m$), and nucleus ($2m$)

12. An element that has a face-centered cubic unit cell has 8 atoms at the corners of the cube and 6 atoms on the faces. The atoms on a face are shared by . . . unit cell(s), each counts as . . . atom(s) per unit cell, resulting in a total of . . . atoms per unit cell. Atoms on the corner are shared by . . . unit cells; therefore, contribute only . . . atom(s) per unit cell, resulting in a total of . . . atom per unit cell. The total number of atoms in each unit cell is then . . .

    (a) 2 / 0.5 / 3 / 8 / 0.125 / 1/4   (b) 1 / 1 / 2 / 6 / 0.5 / 1 / 3
    (c) 2 / 0.5 / 3 / 8 / 0.75 / 1 / 4   (d) 1 / 1 / 2 / 6 / 1 / 1 / 3

13. Crystalline material properties can be predicted if the material's . . . are known.

    (a) symmetry, periodicity, and defects
    (b) configuration and temperature
    (c) temperature and pressure
    (d) Gibbs free energy and enthalpy

14. Quartz glass is generally . . . and . . . with . . . properties compared to silica glass.

    (a) lighter / less organized / anisotropic
    (b) denser / more organized / anisotropic
    (c) denser / more organized / isotropic
    (d) lighter / less organized / isotropic

15. Isotropic properties mean that there are:

    (a) different properties in all directions
    (b) identical properties in two directions but not the third one
    (c) identical properties in all directions
    (d) different properties in some directions

16. Silica and quartz are both . . . and are composed of . . . elements.

    (a) glasses / mostly the same
    (b) semi-metallic / different

(c) ceramics / mostly the same

(d) semi-glasses / different

17. Which of the following lists the three symmetry types in crystalline structures?

(a) Reflection, translation, and mirror

(b) Translation, reflection, and glide

(c) Mirror, rotation, and translation

(d) Translation, glide, and screw

18. Screw symmetry involves . . . ; glide symmetry involves . . .

(a) rotation and reflection / reflection and translation

(b) rotation and translation / reflection and translation

(c) rotation and translation / reflection and rotation

(d) reflection and translation / rotation and translation

19. Point, line, and dislocation are all types of crystal structure defects. (T/F)

20. Amorphous materials are composed of randomly arranged small crystals. (T/F)

21. All matter is in either solid or liquid form. (T/F)

22. The main difference between gases and liquids is the:

(a) atom's travel time

(b) atom's Knudsen number

(c) defects

(d) distance between atoms

23. The average distance between atoms in solids is . . . that in gases.

(a) shorter than

(b) greater than

(c) about the same as

(d) the same or greater than

24. Gases:

(a) contract and occupy space partially

(b) expand and occupy the entire available container space

(c) fill a defined volume by self-leveling

(d) have a defined volume

25. Solids:

(a) contract and occupy space partially

(b) expand and occupy the entire available container space

(c) fill a defined volume by self-leveling

(d) have a defined volume

26. Liquids:

(a) contract and occupy space partially

(b) expand and occupy the entire available container space

(c) fill a defined volume by self-leveling

(d) have a defined volume

27. Which statement best describes the equilibrium state in atoms?

(a) Atoms neither attract nor repel

(b) Atoms attract and repel with maximum internal energy

(c) Atoms attract and repel slightly with minimum internal energy

(d) Atoms attract and repel significantly with maximum internal energy

28. Melting temperature depends on the:
   (a) van der Waals energy and bond energy
   (b) bond energy and glass transition temperature
   (c) van der Waals energy and glass transition temperature
   (d) Gibbs free energy

29. If a material in a liquid state is cooled rapidly, the material:
   (a) reaches an equilibrium state and forms a semi-crystalline structure
   (b) does not reach an equilibrium state and forms an amorphous structure
   (c) reaches an equilibrium state and forms an amorphous structure
   (d) does not reach an equilibrium state and forms a semi-crystalline structure

30. Amorphous to crystalline is like:
   (a) melting temperature to glass transition temperature
   (b) glass transition temperature to melting temperature
   (c) glass transition temperature to critical temperature
   (d) critical temperature to melting temperature

31. Two identical molten materials starting from the same initial state can reach different final states after a cooling process. (T/F)

32. Amorphous materials have a . . . and crystalline materials have a . . .
   (a) long-range order structure / short-range order structure
   (b) short-range order structure / short-range order structure
   (c) long-range order structure / long-range order structure
   (d) short-range order structure / long-range order structure

33. Glasses in high-pressure form:
   (a) disordered structures, forming short-range orders
   (b) ordered structures, forming long-range orders
   (c) disordered structures, forming long-range orders
   (d) ordered structures, forming short-range orders

34. Short-range order is:
   (a) hundred-length bonds expanding to many
   (b) the regular and predictable arrangement of the atoms over a short distance (one to two atom spacings)
   (c) ten-length bonds
   (d) a statistically significant bond length (> 100)

35. How many atoms are contained in one mole?
   (a) the Knudsen number
   (b) depends on the material's density
   (c) depends on the molar weight
   (d) the Avogadro number

36. In order to characterize the material's structure quantitatively, . . . is(are) used:
   (a) descriptors   (b) indices
   (c) symmetry properties   (d) melting temperature

37. Atomistic descriptors predict . . . ; statistical descriptors predict . . .
   (a) the average atom properties / exact atom properties

(b) exact atom properties / approximate atom properties

(c) exact atom properties / the average atom properties

(d) the average atom properties / typical atom properties

38. To characterize crystalline materials, what characteristics can be used?

(a) symmetry, periodicity, and reflection

(b) averaging, translation, and rotation

(c) rotation, periodicity, and process parameters

(d) all of the above

39. Descriptors are:

(a) random and can be measured

(b) measured experimentally

(c) nonrandom and cannot be measured

(d) none of the above

40. The main two descriptors used for non-crystalline and amorphous materials are:

(a) pair distribution function and free volume

(b) free distribution function and pair volume

(c) pair and free distribution function

(d) pair and free volume

41. What are decimal numbers 12, 122, and 1222 expressed in binary?

(a) 1100 / 1111010 / 10011000110

(b) 1010 / 1101010 / 10011000110

(c) 1100 / 1011110 / 10011001111

(d) 1010 / 1010110 / 10011010111

42. What are binary numbers 1000, 1110, and 10100 expressed in decimal?

(a) 6 / 12 / 18   (b) 8 / 10 / 16   (c) 6 / 1 0 / 14   (d) 8 / 14 / 20

43. For a solid, free volume is:

(a) space not occupied by the molecules

(b) space between the fixed atoms

(c) volume occupied by the molecules

(d) volumes not occupied by the free atoms

44. Free volume increases linearly with . . . in amorphous materials when . . .

(a) melting temperature / temperature is higher than the glass transition temperature

(b) pressure / temperature is lower than the glass transition pressure

(c) temperature / temperature is higher than the glass transition temperature

(d) Gibbs free energy / temperature is lower than the glass transition pressure

45. Solid material's free volume:

(a) depends on its rate of cooling

(b) depends on its rate of heating after cooling

(c) is independent of heating or cooling rates

(d) depends on the atmospheric pressure

46. The smallest repeat units in polymers are called:

(a) monomers   (b) molecules

(c) mers   (d) crosslinks

47. For solid materials, viscosity and glass transition temperature depend on the free volume. (T/F)

48. As a system's size increases, its intensive properties increase proportionately in their magnitude. (T/F)

49. For any system, its extensive properties remain independent of the system's size. (T/F)

50. Pair distribution function is equal to zero at the large distance r from a reference atom. (T/F)

51. To find the pair distribution function, one needs to imagine a . . . and . . .
    (a) sphere around the atom and expand it by any radius / normalize it with respect to the total volume
    (b) trapezoid around the atom and expand it by any distance / equalize it with respect to the total volume
    (c) sphere around the atom and expand it by any radius / subtract the created volume from the total volume
    (d) trapezoid around the atom and expand it by any distance / add the created volume to the total volume

52. When defining the pair distribution function, the average distance from the first shell of the nearest neighbors is (where R is the assumed sphere radius surrounding the atoms):
    (a) $2R$ (b) R (c) $3R$ (d) $4R$

53. The distance measured from a reference atom, at which there is a 100% probability of finding another atom, is called the:
    (a) average distance   (b) sphere radius
    (c) correlation distance   (d) sphere diameter

54. Crystalline materials crystallize at:
    (a) melting temperature with a sudden change in volume
    (b) glass transition temperature with a sudden change in volume
    (c) melting temperature with no change in volume
    (d) glass transition temperature with no change in volume

55. Glass materials crystallize at:
    (a) melting temperature with a sudden change in volume
    (b) glass transition temperature with a sudden change in volume
    (c) melting temperature with no change in volume
    (d) glass transition temperature with no change in volume

56. The approximate cooling rate for pure-element metals when changing to amorphous states is:
    (a) 10^{12} K/s   (b) 10^2 K/s   (c) 10^6 K/s   (d) 10 K/s

57. For compounded plastics, impact modifiers are the additives that:
    (a) change the crystalline structure
    (b) improve the durability and toughness of a variety of plastic resins
    (c) are needed to adjust the level of impact resistance for custom applications
    (d) all of the above

58. Modifiers are added so that the resultant material has:
    (a) workable viscosity-temperature range
    (b) large network connectivity

(c) nonworkable viscosity-temperature range

(d) small network connectivity

59. Sp^3 bonds (e.g., silicon atoms in silica glass SiO_2 and baron atoms in boron trifluoride BF_3) are:

(a) individual units that are randomly distributed

(b) weaker than sp^2 bonds, forming 109.5-degree bond angles

(c) van der Waals and covalent, very weak, and with very high viscosity

(d) individual units that are distributed in an organized fashion

60. Sp^2 bonds (e.g., oxygen atoms in boron trioxide B_2O_3 and carbon atoms in benzene ring C_6H_6) are:

(a) shorter and stronger than sp^3 bonds with 120-degree bond angles

(b) individual units that are distributed in an organized fashion

(c) van der Waals with very high viscosity

(d) covalent and very weak

61. For silicate, with adding modifiers, the free volume . . . ; moreover, the correlation distance . . .

(a) decreases / increases   (b) increases / decreases

(c) decreases / decreases   (d) increase / increases

62. For borate, with adding modifiers, the free volume . . . ; moreover, the correlation distance . . .

(a) decreases / decreases   (b) increases / increases

(c) decreases / increases   (d) increases / decreases

63. Which are polymers in the following list?

(1) Wood; (2) Nylon; (3) Glass; (4) Cotton; and (5) Salt

(a) 1 / 2 / 3   (b) 2 / 3 / 5   (c) 2 / 4 / 5   (d) 1 / 2 / 4

64. Polymers are called organics because they contain molecular structures made of:

(a) hydrogen and oxygen atoms   (b) carbon and nitrogen atoms

(c) nitrogen and silicon atoms   (d) carbon and hydrogen atoms

65. A random walk is the:

(a) random number of monomers within each polymer chain

(b) random distribution of monomers in space

(c) sequence of n steps where the successive distances are not correlated

(d) random variation in polymer chain directional distribution

66. For polymers, the random walk models describe the:

(a) expansion of polymers in space in all directions

(b) number of monomers within each polymer chain

(c) distribution of monomers in space

(d) variation in polymer chain directional distribution

67. Random walk models assume the:

(a) expansion of the polymers with fixed lengths and angles

(b) path of monomers moving in an environment with unknown steps

(c) number of monomers within each chain

(d) behavior of monomers in the space

68. Random walk models may have the same lengths but different angles. (T/F)

69. The average distance traveled by the random walk monomers take in a molecular structure is:

(a) 0   (b) > 0   (c) < 0   (d) 1

70. The average spatial dimension traveled by monomers in random walk models:

(a) is equal to the root mean square of the end-to-end distance
(b) shows the average spatial dimension monomers travel
(c) is the standard deviation of the end-to-end distance
(d) both (a) and (b)

71. The space that a polymer chain occupies is given by (where n is the number of steps, L is the length of each random walk step, and R_0 is the end-to-end displacement that is equal to the radius of the sphere surrounding the monomer):

(a) $R_0 = n^{0.5}L$ (b) $R_0 = nL$ (c) $R_0 = n^{1.5}L$ (d) $R_0 = n^2 L$

72. The scaling exponent of polymers obtained from random walk models is:

(a) 1.5   (b) 1   (c) 0.5   (d) 2

73. If there are 10,000 monomer units in a polymer, and the length of each unit is 1 μm, the average volume the polymer chain occupies is:

(a) 0.0021 mm^3 (b) 0.0084 mm^3 (c) 0.0064 mm^3 (d) 0.0042 mm^3

74. Viscosity decreases with decreasing temperature. (T/F)

75. For the same temperature, with increasing bond strength, the viscosity increases. (T/F)

76. When in solutions, polymers form coils, which can adopt coil shapes that can be described:

(a) by self-avoiding walk models
(b) as a sequence of vertices, where each vertex is the nearest neighbor of its predecessor
(c) as a random path which visits a lattice site not more than once
(d) all of the above

77. If 10,000 steps are assumed, the scaling multipliers of polymers estimated by using random walk, collapsed, and self-avoiding walk coil models are, respectively:

(a) 100 / 21 / 251
(b) 21 / 100 / 251
(c) 251 / 21 / 100
(d) All of the above, depends on the chain orientation

78. Among the following, which one is the most accurate description of the polymer-solvent mixture conformation:

(a) Coil shapes occupying different volumes depending on their interaction.
(b) Coil shapes occupying different volumes depending on the monomer-solvent interaction.
(c) Coil shapes occupying the same volumes depending on the monomer-monomer interaction.
(d) Coil shapes occupying different volumes depending on the monomer-monomer interaction.

79. Polymer properties can change by varying:

(a) coil shapes and volume

(b) coil shapes and void characteristics

(c) free volume and void characteristics

(d) all of the above

80. It is easier for amorphous polymers to come to existence than the crystalline ones. (T/F) This depends on the:

(a) T / fast cooling rate and unorganized long-range chains

(b) T / slow heating rate and organized long-range chains

(c) F / formation method and unorganized short-range chains

(d) F / slow cooling rate and organized short-range chains

81. If a polymer is made up of groups arranged along the same side of the chain, it follows:

(a) atactic tacticity

(b) syndiotactic tacticity

(c) isotactic tacticity

(d) eutectic tacticity

82. If a polymer is made up of groups arranged so that they alternate between opposite sides of the polymer chain, it follows:

(a) atactic tacticity

(b) syndiotactic tacticity

(c) isotactic tacticity

(d) eutectic tacticity

83. If a polymer is made up of groups that are arranged randomly relative to the direction of the polymer chain, it follows:

(a) atactic tacticity

(b) syndiotactic tacticity

(c) isotactic tacticity

(d) eutectic tacticity

84. If a polymer is made up of groups that can be arranged in a specific but complex sequence of the positions along the chain, it follows:

(a) atactic tacticity

(b) syndiotactic tacticity

(c) isotactic tacticity

(d) eutectic tacticity

85. Thermoplastic bonds are . . . versus those of thermosets which are . . . and therefore thermoplastic bonds are . . .

(a) van der Waals / covalent / stronger

(b) covalent / Van der Waals / weaker

(c) van der Waals / covalent / weaker

(d) covalent / Van der Waals / stronger

86. Thermoset plastics, when heated sufficiently, will melt and will return to their original state when they cool down. (T/F)

87. Thermoplastics transition to viscous liquid or a rubbery state when they are heated to . . . temperature.

(a) glass transition

(b) melting

(c) critical

(d) vaporization

88. Thermosets do not soften when heated and can become irreversibly damaged after being exposed to a certain temperature for a certain time duration. (T/F)

89. To characterize amorphous and crystalline materials, generic descriptors can be used. (T/F) These descriptors are:

(a) T / free volume and pair distribution function

(b) F / free volume and pair distribution function

(c) T / free space and individual distribution method

(d) F / free space and individual distribution method

90. Like polymers, glasses can be described by random walk models. (T/F)

91. In a crystalline material, knowing the position of one atom can be used to predict the position of other atoms. (T/F)

92. Assuming atoms and molecules are arranged periodically in all three dimensions, periodicity:
    (a) should be defined in three dimensions
    (b) should be defined in two dimensions
    (c) should be defined in one dimension
    (d) does not need to be defined in any dimensions

93. A crystal consists of atoms packed together in a regular pattern. For filling space without holes, a unit cell of a pattern must be either . . . or . . .
    (a) a lattice / a net
    (b) a parallelogram (in 2D) / a parallelepiped (in 3D)
    (c) two non-coincident vectors in 3D / three non-coplanar vectors in 2D
    (d) (a) or (b) depending on the process parameters

94. In addition to the crystalline structure (lattice or net), there may also be . . . with . . . structure(s).
    (a) motifs sitting on lattice sides / symmetrical
    (b) particles filling up the nets / unsymmetrical
    (c) particles sitting on the neighboring nets / triangular
    (d) motifs sitting on lattice vertices / any

95. Periodicity of the motif is defined by:
    (a) identical lattice arrangements   (b) different motif arrangements
    (c) identical net arrangements   (d) different lattice arrangements

96. All crystals have . . . symmetry.
    (a) translational   (b) rotational   (c) reflectional   (c) periodic

97. Motifs can be identical, but the lattices can vary. (T/F)

98. Which type of symmetry applies when an equilateral triangle is rotated about its centroid by 120 degrees?
    (a) Translational   (b) Reflectional   (c) Rotational   (d) Periodic

99. The difference between primitive and multiple cells is that the former contains . . . lattice point(s), while the latter contains . . . lattice point(s).
    (a) a single / a single   (b) multiple / a single
    (c) a single / multiple   (d) multiple / multiple

100. The volume occupied by a single crystalline lattice is obtained from one of the following. The lattice angles should be 90 degrees. (T/F)

(a) $V = \left| \vec{a_1} \cdot \left(\vec{a_2} \times \vec{a_3} \right) \right| = \left| \vec{a_2} \cdot \left(\vec{a_3} \times \vec{a_1} \right) \right| = \left| \vec{a_3} \cdot \left(\vec{a_1} \times \vec{a_2} \right) \right|$ / F

(b) $V = \left| \vec{a_1} \cdot \left(\vec{a_2} \times \vec{a_3} \right) \right| = \left| \vec{a_1} \cdot \left(\vec{a_3} \times \vec{a_2} \right) \right| = \left| \vec{a_3} \cdot \left(\vec{a_2} \times \vec{a_1} \right) \right|$ / T

(c) $V = \left| \vec{a_1} \times \left(\vec{a_3} \times \vec{a_2} \right) \right| = \left| \vec{a_2} \times \left(\vec{a_3} \times \vec{a_1} \right) \right| = \left| \vec{a_3} \times \left(\vec{a_2} \times \vec{a_1} \right) \right|$ / F

(d) $V = \left| \vec{a_1} \times \left(\vec{a_2} \times \vec{a_3} \right) \right| = \left| \vec{a_2} \times \left(\vec{a_3} \times \vec{a_1} \right) \right| = \left| \vec{a_3} \times \left(\vec{a_1} \times \vec{a_2} \right) \right|$ / T

101. Glide symmetry consists of:

(a) translation and reflection   (b) translation and rotation

(c) rotation and reflection   (d) translation only

102. Screw symmetry consists of:

(a) translation and reflection   (b) translation and rotation

(c) rotation and reflection   (d) rotation only

103. Among the following, which ones are crystal systems?

(a) Cubic, hexagonal, and tetragonal

(b) Trigonal, orthorhombic, and triclinic

(c) Orthorhombic, monoclinic, and triclinic

(d) All of the above

104. Identify the shape of the crystal systems presented in Figure 75 associated with the diagrams (from top left to bottom right):

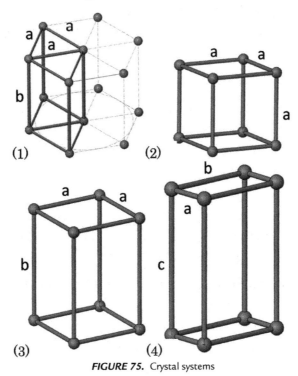

FIGURE 75. Crystal systems

(a) (1) Hexagonal, (2) cubic, (3) tetragonal, (4) orthorhombic

(b) (1) Orthorhombic, (2) tetragonal, (3) cubic, (4) rhombohedral

(c) (1) Monoclinic, (2) cubic, (3) tetragonal, (4) hexagonal

(d) (1) Triclinic, (2) tetragonal, (3) cubic, (4) orthorhombic

105. A crystallographic direction is the . . . between two set points, starting from the . . . to the . . . located on the . . .

(a) line / point of origin / point / cell

(b) area / cell / point of origin / cell

(c) volume / point of origin / point / cell

(d) area / point of origin / point / cell

106. For a point located at (1, 0, 1/6), identifying (x, y, z) coordinates, the crystallographic direction is defined as:

(a) [1, 0, 1/6] (b) [6, 0, 1] (c) [$\overline{3}$,6,1] (d) [1, 1, 1]

107. For a point located at $(-1, 2, 1/3)$, identifying (x, y, z) coordinates, the crystallographic direction is defined as:

(a) [1, 0, 1/6] (b) [6, 0, 1] (c) [$\overline{3}$,6,1] (d) [1, 1, 1]

108. For the family of directions <100>, applicable directions can be:

(a) [100], [101], [001] (b) [110], [001], [010]

(c) [001], [101], [100] (d) [100], [010], [001]

109. Family of directions means the number of crystallographic directions that can be:

(a) parallel but have a similar crystallographic structure

(b) non-parallel but have a dissimilar crystallographic structure

(c) parallel but have a dissimilar crystallographic structure

(d) non-parallel but have a similar crystallographic structure

110. Within the same family of directions, the spacing between the atoms is identical. (T/F)

111. Miller indices are:

(a) reciprocals of the plane axial intercepts

(b) reciprocals of the plane axial intercepts in an integer form

(c) inverse intercepts of the plane along the lattice vectors

(d) reciprocals of the plane axial intercepts

112. Identify Miller indices for the highlighted plane presented in Figure 76.

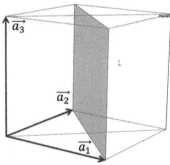

FIGURE 76. Plane in a cubic crystal.

(a) (110) (b) ($\overline{1}$10) (c) (101) (d) (100)

113. Identify Miller indices for the highlighted plane presented in Figure 77.

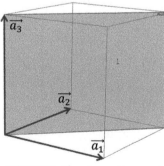

FIGURE 77. Plane in a cubic crystal.

(a) (110) (b) ($\overline{1}$10) (c) (101) (d) (100)

114. Identify Miller indices for the highlighted plane presented in Figure 78.

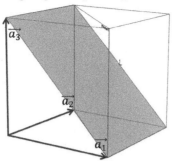

FIGURE 78. Plane in a cubic crystal.

(a) (110)   (b) ($\bar{1}$10)   (c) (101)   (d) (100)

115. Identify Miller indices for the highlighted plane presented in Figure 79.

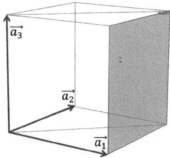

FIGURE 79. Plane in a cubic crystal.

(a) (110)   (b) ($\bar{1}$10)   (c) (101)   (d) (100)

116. Identify the axial intercepts for the plane presented in Figure 80.

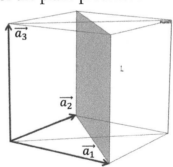

FIGURE 80. Plane in a cubic crystal.

(a) (11∞)   (b) ($\bar{1}$1∞)   (c) (1∞1)   (d) (1$\infty\infty$)

117. Identify the axial intercepts for the plane presented in Figure 81.

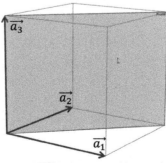

FIGURE 81. Plane in a cubic crystal.

(a) (11∞)   (b) ($\bar{1}$1∞)   (c) (1∞1)   (d) (1$\infty\infty$)

118. Identify the axial intercepts for the plane presented in Figure 82.

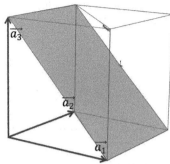

FIGURE 82. Plane in a cubic crystal.

(a) (11∞) (b) $(\bar{1}1\infty)$ (c) $(1\infty1)$ (d) $(1\infty\infty)$

119. Identify the axial intercepts for the plane presented in Figure 83.

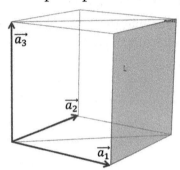

FIGURE 83. Plane in a cubic crystal.

(a) (11∞) (b) $(\bar{1}1\infty)$ (c) $(1\infty1)$ (d) $(1\infty\infty)$

120. Which of the following lists the family planes of $\{100\}$?
(a) (110), (011), (101), $(\bar{1}01)$, $(0\bar{1}1)$, $(01\bar{1})$
(b) (101), (110), (011), $(\bar{1}00)$, $(1\bar{1}0)$, $(10\bar{1})$
(c) (011), (010), (101), $(\bar{1}01)$, $(0\bar{1}0)$, $(11\bar{1})$
(d) (100), (010), (001), $(\bar{1}00)$, $(0\bar{1}0)$, $(00\bar{1})$

121. Reciprocal vectors of the lattice can be determined by:
(a) the diffraction pattern of a crystal lattice
(b) the diffraction of particles
(c) the thermophysical characteristics of the material
(d) the electrical characteristics of the material

122. If a lattice vector and the reciprocal lattice vector are represented by $\vec{r} = n_1\vec{a_1} + n_2\vec{a_2} + n_3\vec{a_3}$ and $\vec{s} = n_1'\vec{b_1} + n_2'\vec{b_2} + n_3'\vec{b_3}$, respectively, $\vec{b_1}, \vec{b_2},$ and $\vec{b_3}$ are calculated from (where V is the lattice volume):

(a) $\vec{b_1} = \dfrac{\vec{a_2} \times \vec{a_3}}{V}, \vec{b_2} = \dfrac{\vec{a_3} \times \vec{a_1}}{V}, \vec{b_3} = \dfrac{\vec{a_1} \times \vec{a_2}}{V}, V = \left| \vec{a_1} \cdot \left(\vec{a_2} \times \vec{a_3} \right) \right|$

(b) $\vec{b_1} = \dfrac{\vec{a_2} \times \vec{a_1}}{V}, \vec{b_2} = \dfrac{\vec{a_3} \times \vec{a_2}}{V}, \vec{b_3} = -\dfrac{\vec{a_1} \times \vec{a_3}}{V}, V = \left| \vec{a_2} \cdot \left(\vec{a_1} \times \vec{a_3} \right) \right|$

(c) $\vec{b_1} = \dfrac{\vec{a_2} \times \vec{a_1}}{V}, \vec{b_2} = \dfrac{\vec{a_3} \times \vec{a_2}}{V}, \vec{b_3} = \dfrac{\vec{a_1} \times \vec{a_3}}{V}, V = \left| \vec{a_3} \cdot \left(\vec{a_2} \times \vec{a_1} \right) \right|$

(d) $\vec{b_1} = \dfrac{\vec{a_2} \times \vec{a_3}}{V}, \vec{b_2} = -\dfrac{\vec{a_3} \times \vec{a_1}}{V}, \vec{b_3} = -\dfrac{\vec{a_1} \times \vec{a_2}}{V}, V = \left| \vec{a_3} \cdot \left(\vec{a_1} \times \vec{a_2} \right) \right|$

123. If the reciprocal lattice vector is represented by $\vec{s} = n_1' \vec{b_1} + n_2' \vec{b_2} + n_3' \vec{b_3}$, the distance between $n_1' n_2' n_3'$ planes is:

   (a) $d_{n_1' n_2' n_3'} = \dfrac{1}{|\vec{s}|}$
   (b) $d_{n_1' n_2' n_3'} = \dfrac{2}{|\vec{s}|}$
   (c) $d_{n_1' n_2' n_3'} = \dfrac{1}{2|\vec{s}|}$
   (d) $d_{n_1' n_2' n_3'} = \dfrac{1}{\sqrt{|\vec{s}|}}$

124. X-rays are used in order to characterize a material since their electromagnetic radiation wavelength is:

   (a) much larger than the distance between atoms
   (b) much smaller than the distance between atoms
   (c) identical to the distance between atoms
   (d) comparable to the distance between atoms

125. Constructive interference means that the emitted wave fields, when overlapped:

   (a) are subtracted and produce weaker peaks
   (b) add up and produce stronger peaks
   (c) interfere with each other to produce more noise
   (d) create newly constructed waves of different wavelengths

126. Bragg's law identifies the:

   (a) path between coherent and incoherent scattering from a crystal lattice
   (b) angle between incoherent scattering from a crystal lattice
   (c) angle between the incoming radiation and the lattice planes which provide the strongest constructive interference for the reflected radiation
   (d) angle between coherent scattering from a crystal lattice

127. The largest constructive interference angle (θ) is calculated from Bragg's law by (where λ is the beam wavelength, n is a positive integer, and d is the interplanar distance):

   (a) $d \sin \theta = n\lambda$
   (b) $d \sin \theta = 2n\lambda$
   (c) $2d \sin \theta = n\lambda$
   (d) $d \sin 2\theta = n\lambda$

128. Using Bragg's law, lattice spacing related to a cubic system (with the side length a) can be obtained from (where $\vec{s} = n_1' \vec{b_1} + n_2' \vec{b_2} + n_3' \vec{b_3}$ is the reciprocal of the lattice vector and $n_1', n_2',$ and n_3' are Miller indices):

   (a) $d = \dfrac{b}{n_1'^2 + n_2'^2 + n_3'^2}$
   (b) $d = \dfrac{c}{n_1'^2 + n_2'^2 + n_3'^2}$
   (c) $d = \dfrac{a}{n_1'^2 + n_2'^2 + n_3'^2}$
   (d) $d = \dfrac{1}{n_1'^2 + n_2'^2 + n_3'^2}$

129. When there is X-ray radiation hitting a crystalline material, it interacts with it and:

   (a) scatters uniformly
   (b) forms a pattern of peaks
   (c) reflects equally in all directions
   (d) is fully absorbed by the material

130. Modes of X-ray-matter interaction when X-ray waves are to be transmitted through a crystalline material are:

   (a) scattering and absorption
   (b) reflection and transmission
   (c) interference and propagation
   (d) (a) and (b)

131. The structure factor for a unit cell describing the intensity of a diffracted beam is represented by (where N is the number of atoms, r_i is the location of a single atom inside the unit cell, $f(r_i)$ is the atom's scattering factor, r_i is the location of each atom within the unit cell, and waves are scattered in the direction of n_1', n_2', and n_3'):

(a) $F\left(n_1' n_2' n_3'\right) = \sum_{i=1}^{N} f(r_i) e^{2\pi i\left(n_1' x_i + n_2' y_i + n_3' z_i\right)}, r_i = \left(a_1 x_i + a_2 y_i + a_3 z_i\right)$

(b) $F\left(n_1' n_2' n_3'\right) = \sum_{i=1}^{N} f(r) e^{2\pi i\left(n_1 \vec{a_1} + n_2 \vec{a_2} + n_3 \vec{a_3}\right)}, r_i = \left(n_1' \vec{a_1} x_i + n_2' \vec{a_2} y_i + n_3' \vec{a_3} z_i\right)$

(c) $F\left(n_1' n_2' n_3'\right) = \sum_{i=1}^{N} f(r) e^{2\pi i\left(n_1' x_i + n_2' y_i + n_3' z_i\right)}, r_i = \left(a_1 x_i + a_2 y_i + a_3 z_i\right),$

(d) $F\left(n_1' n_2' n_3'\right) = \sum_{i=1}^{N} f(r_i) e^{2\pi i\left(n_1 \vec{a_1} + n_2 \vec{a_2} + n_3 \vec{a_3}\right)}, r_i = \left(n_1' \vec{a_1} x_i + n_2' \vec{a_2} y_i + n_3' \vec{a_3} z_i\right)$

132. For a crystalline structure with atomic scattering factors $f(0, 0, 0) = f_1$ and $f(1, 1, 1) = f_2$, the structure scattering factor is equal to (where waves are scattered in the direction of n_1', n_2', and n_3'):

(a) $F\left(n_1', n_2', n_3'\right) = -f_1 + f_2$, if $n_1' + n_2' + n_3'$ is even

(b) $F\left(n_1', n_2', n_3'\right) = f_1 - f_2$, if $n_1' + n_2' + n_3'$ is odd

(c) $F\left(n_1', n_2', n_3'\right) = -f_1 - f_2$, if $n_1' + n_2' + n_3'$ is even

(d) $F\left(n_1', n_2', n_3'\right) = f_1 + f_2$, if $n_1' + n_2' + n_3'$ is even or odd

133. For a crystalline structure with atomic scattering factors $f(0, 0, 0) = f_1$ and $f(1.5, 0.5, 1) = f_2$, the structure scattering factor is equal to (where waves are scattered in the direction of n_1', n_2', and n_3'):

(a) $F\left(n_1', n_2', n_3'\right) = f_1 + f_2$, if $n_1' - n_2'$ is even

(b) $F\left(n_1', n_2', n_3'\right) = -f_1 - f_2$, if $n_1' - n_2'$ is odd

(c) $F\left(n_1', n_2', n_3'\right) = f_1 + f_2,$, if $n_1' - n_2'$ is odd

(d) $F\left(n_1', n_2', n_3'\right) = -f_1 - f_2$, if $n_1' - n_2'$ is even

134. For a crystalline structure with atomic scattering factors $f(0, 0, 0) = f_1, f(1, 1, 1) = f_2$, and $f(1.5, 0.5, 1) = f_3$, the structure scattering factor is equal to (where waves are scattered in the direction of n_1', n_2', and n_3'):

(a) $F\left(n_1', n_2', n_3'\right) = f_1 - f_2 - f_3$, if $n_1' - n_2'$ is even

(b) $F\left(n_1', n_2', n_3'\right) = f_1 + f_2 + f_3$, if $n_1' - n_2'$ is odd

(c) $F\left(n_1', n_2', n_3'\right) = f_1 - f_2 - f_3$, if $n_1' - n_2'$ is even

(d) $F\left(n_1', n_2', n_3'\right) = f_1 - f_2 - f_3$, if $n_1' - n_2'$ is odd

135. When applying a symmetry operation, the material's structure remains the same. (T/F)

136. In case of a translational symmetry, if the lattice is moved along any direction, it will not change the shape of the structure. (T/F)

137. A polar molecule has . . . , . . . a nonpolar molecule.

(a) positive and negative charges / the same as

(b) negative and positive charges / opposite to

(c) positive charges / the same as

(d) negative charges / opposite to

138. . . . requires magnitudes of translational vectors to be the same.

(a) Rotational symmetry
(b) Translational symmetry
(c) Reflectional symmetry
(d) Glide symmetry

139. Reflectional symmetry occurs about a:

(a) line or volume
(b) symmetry line or plane
(c) line or circle
(d) point or line

140. Identify the symmetry type(s) and line in Figure 84.

(a) Mirror and translation, m_1
(b) Translation, t_1
(c) Reflection, m_2
(d) Rotation, m_1

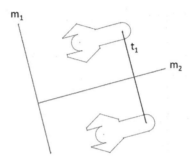

FIGURE 84. Symmetry conditions.

141. Identify the symmetry type(s) in Figure 85.

(a) Glide with periodic pattern

(b) Translation with nonperiodic pattern

(c) Reflection with periodic pattern

(d) Mirror and rotation with nonperiodic pattern

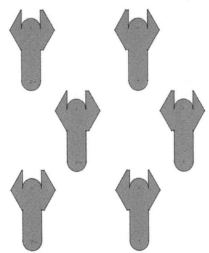

FIGURE 85. Symmetry conditions.

142. The rotational symmetry is identified by (where n is the number of rotational folds and α is the angle):

(a) $n = 2\pi/\alpha$
(b) $n = 4\pi/\alpha$
(c) $n = \pi/\alpha$
(d) $n = 3\pi/\alpha$

143. Identify the rotational symmetry fold for the geometries presented in Figure 86.

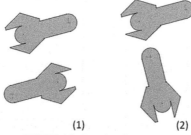

(1)   (2)

FIGURE 86. Rotational symmetry.

(a) (1) 4, (2) 2   (b) (1) 2, (2) 4   (c) (1) 2, (2) 2   (d) (1) 4, (2) 4

144. Identify the rotational symmetry fold for the geometries presented in Figure 87.

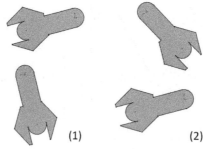

(1)   (2)

FIGURE 87. Rotational symmetry.

(a) (1) 4, (2) 2   (b) (1) 1, (2) 3   (c) (1) 4, (2) 3   (d) (1) 3, (2) 4

145. Identify the rotational symmetry fold for the geometries presented in Figure 88.

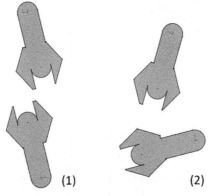

(1)   (2)

FIGURE 88. Rotational symmetry.

(a) (1) 4, (2) 2   (b) (1) 1, (2) 3   (c) (1) 3, (2) 4   (d) (1) 2, (2) 6

146. Demonstrate that a screw symmetry consisting of 8-fold rotational symmetry . . . compatible with crystal translational symmetry (Figure 89).

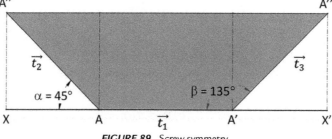

FIGURE 89. Screw symmetry.

(a) is not

(b) occasionally is not

(c) always is

(d) often is

Solution Guide

The original lattice point is located at point A. The vectors AX and A'X' are rotated by α degrees in the clockwise and counterclockwise directions, respectively. Translational symmetry requires the magnitudes of the translational vectors to be the same $\left(\left|\vec{t_1}\right|=\left|\vec{t_2}\right|=\left|\vec{t_3}\right|=\left|\vec{t}\right|\right)$. The new points created after rotating AX and A'X' (A'' and A''') are located on the crystal lattice. The distance between these 2 points should be compatible with the crystal translation symmetry, meaning that it is an integer multiplier of the lattice translation $(A''A'''=n\left|\vec{t}\right|)$, where n is an integer. Employing trigonometric relations between α and β, it can be concluded that $n = 1 + 2\cos\alpha$. Assume the crystal is following an 8-fold rotational symmetry $\left(\alpha = \dfrac{2\pi}{8} = 45°\right)$. This results in $n = 2.41$, which is not an integer. Therefore, 8-fold rotational symmetry is not valid.

147. Demonstrate that a screw symmetry consisting of 3-fold rotational symmetry . . . compatible with crystal translational symmetry (Figure 90).

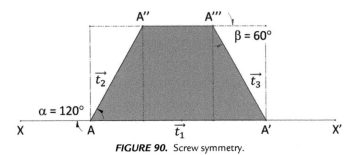

FIGURE 90. Screw symmetry.

(a) occasionally is not

(b) occasionally is

(c) is

(d) is not

Solution Guide

The original lattice point is located at point A. The vectors AX and A'X' are rotated by α degrees in the clockwise and counterclockwise directions, respectively. Translational symmetry requires the magnitudes of the translational vectors to be the same $\left(\left|\vec{t_1}\right|=\left|\vec{t_2}\right|=\left|\vec{t_3}\right|=\left|\vec{t}\right|\right)$. The new points created after rotating AX and A'X' (A'' and A''') are located on the crystal lattice. The distance between these 2 points should be compatible with the crystal translation symmetry, meaning that it is an integer multiplier of the lattice translation $(A''A'''=n\left|t\right|)$, where n is an integer. Employing trigonometric relations between α and β, it can be concluded that $n = 1 + 2\cos\alpha$. Assume that the crystal is following 3-fold rotational symmetry $\left(\alpha = \dfrac{2\pi}{3} = 120°\right)$. This results in $n = 0$, which is an integer. Therefore, 3-fold rotational symmetry is valid.

148. Demonstrate that a screw symmetry consisting of n-fold rotational symmetry . . . more than 3 times the translational length of the original translation (Figure 91).

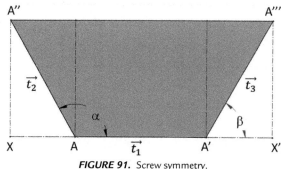

FIGURE 91. Screw symmetry.

(a) cannot have   (b) can have
(c) always has   (d) sometimes has

Solution Guide

The original lattice point is located at point A. The vectors AX and A'X' are rotated by α degrees in the clockwise and counterclockwise directions, respectively. Translational symmetry requires the magnitudes of the translational vectors to be the same $\left(\left|\vec{t_1}\right| = \left|\vec{t_2}\right| = \left|\vec{t_3}\right| = \left|\vec{t}\right|\right)$. The new points created after rotating AX and A'X' (A'' and A''') are located on the crystal lattice. The distance between these 2 points should be compatible with the crystal translation symmetry, meaning that it is an integer multiplier of the lattice translation ($A''A''' = n\left|\vec{t}\right|$), where n is an integer. Employing trigonometric relations between α and β, it can be concluded that $n = 1 - 2 \cos\alpha$. Given that $-1 \le \cos\alpha \le 1$, it can be derived that $-1 \le n \le 3$, meaning that the n maximum value is 3.

149. A point group is a group of symmetrical features:
   (a) with at least a single fixed point   (b) with a single fixed point
   (c) with multiple fixed points   (d) without a fixed point

150. A point group can be formed by:
   (a) nonintersecting reflectional symmetry lines
   (b) intersecting rotational symmetry lines
   (c) intersecting reflectional symmetry lines
   (d) translational symmetry lines

151. A point group can be formed by combining:
   (a) rotational and translational symmetry groups
   (b) rotational and reflectional symmetry groups
   (c) translational and screw symmetry groups
   (d) screw and glide symmetry groups

152. Group theory is applicable to
   (a) a set and operations   (b) a set and numbers
   (c) numbers and operations   (d) letters and numbers

153. A set (G) combines any number of . . . with an . . .
   (a) operations / element   (b) numbers / operation
   (c) elements / operation   (d) operations / number

154. One requirement for compliance with group theory states that there is an identity element whose operation with any set element results in:
(a) the same set element
(b) an inverse element
(c) a positive element
(d) a unit element

155. An example of a group-theory complying set is:
(a) real numbers and operation division
(b) real numbers (excluding zero) and operation multiplication
(c) integer numbers and operation division
(d) integer numbers and operation multiplication

156. Assuming that a vector undergoes a combination of rotation about an axis (α) and translation (t) perpendicular to the axis, the resultant vector can be also recreated from:
(a) translating a point on the bisect of the translated vector and at the distance d from it
(b) rotating a point about the bisect of the translated vector (α) and at the distance d from it
(c) rotating a point on the bisect of the rotated vector and at the distance d from it
(d) translating a point on the bisect of the rotated vector and at the distance d from it

157. A screw symmetry is a combination of (where m and n are integers; n is the number of rotations, and m is the number of translations):
(a) rotational ($\alpha = 2\pi/m$) and translational ($\tau = nT/m$) symmetries, reaching the original lattice orientation, which is located at distance T parallel to the rotational axis
(b) rotational ($\alpha = 2\pi/n$) and translational ($\tau = mT/n$) symmetries, reaching the original lattice orientation, which is located at angle α from the rotational axis
(c) rotational ($\alpha = 2\pi/n$) and translational ($\tau = mT/n$) symmetries, reaching the original lattice orientation, which is located at distance T parallel to the rotational axis
(d) rotational ($\alpha = 2\pi/m$) and translational ($\tau = nT/m$) symmetries, reaching the original lattice orientation, which is located at angle α from the rotational axis

158. Assuming that there are 2 intersecting rotational symmetry axes on a sphere, transporting a point from location A to location B and then to C, it can be concluded that there is:
(a) a third point that is the result of transporting the point by the second rotational symmetry axis
(b) a symmetry plane that is the combination of the 3 rotational symmetry axes
(c) a third rotational symmetry axis that is the combination of the 2 rotational symmetry axes, transiting the point directly from location A to C
(d) a second rotational symmetry axis that is in addition to the first rotational symmetry axes

159. Assuming that there are 2 intersecting rotational symmetry axes on a sphere, transporting a point from location A to B (α), B to C (β), and C to A (γ), the angles between the rotational axes are:
(a)
$$\angle AB = \cos^{-1}\left(\frac{\cos\left(\frac{\gamma}{2}\right)+\cos\left(\frac{\alpha}{2}\right)\cos\left(\frac{\beta}{2}\right)}{\sin\left(\frac{\alpha}{2}\right)\sin\left(\frac{\beta}{2}\right)}\right), \angle BC = \cos^{-1}\left(\frac{\cos\left(\frac{\alpha}{2}\right)+\cos\left(\frac{\beta}{2}\right)\cos\left(\frac{\gamma}{2}\right)}{\sin\left(\frac{\beta}{2}\right)\sin\left(\frac{\gamma}{2}\right)}\right),$$
$$\angle CA = \cos^{-1}\left(\frac{\cos\left(\frac{\beta}{2}\right)+\cos\left(\frac{\alpha}{2}\right)\cos\left(\frac{\gamma}{2}\right)}{\sin\left(\frac{\alpha}{2}\right)\sin\left(\frac{\gamma}{2}\right)}\right)$$

(b)

$$\angle AB = \cos^{-1}\left(\frac{\cos\left(\frac{\gamma}{2}\right)+\sin\left(\frac{\alpha}{2}\right)\sin\left(\frac{\beta}{2}\right)}{\cos\left(\frac{\alpha}{2}\right)\cos\left(\frac{\beta}{2}\right)}\right), \angle BC = \cos^{-1}\left(\frac{\cos\left(\frac{\alpha}{2}\right)+\sin\left(\frac{\gamma}{2}\right)\sin\left(\frac{\beta}{2}\right)}{\cos\left(\frac{\gamma}{2}\right)\cos\left(\frac{\beta}{2}\right)}\right),$$

$$\angle CA = \cos^{-1}\left(\frac{\cos\left(\frac{\beta}{2}\right)+\sin\left(\frac{\gamma}{2}\right)\sin\left(\frac{\alpha}{2}\right)}{\cos\left(\frac{\gamma}{2}\right)\cos\left(\frac{\alpha}{2}\right)}\right)$$

(c)

$$\angle AB = \cos^{-1}\left(\frac{\cos\left(\frac{\gamma}{2}\right)+\sin\left(\frac{\alpha}{2}\right)\cos\left(\frac{\beta}{2}\right)}{\sin\left(\frac{\alpha}{2}\right)\sin\left(\frac{\beta}{2}\right)}\right), \angle BC = \cos^{-1}\left(\frac{\cos\left(\frac{\alpha}{2}\right)+\sin\left(\frac{\gamma}{2}\right)\cos\left(\frac{\beta}{2}\right)}{\sin\left(\frac{\gamma}{2}\right)\sin\left(\frac{\beta}{2}\right)}\right),$$

$$\angle CA = \cos^{-1}\left(\frac{\cos\left(\frac{\beta}{2}\right)+\sin\left(\frac{\gamma}{2}\right)\cos\left(\frac{\alpha}{2}\right)}{\sin\left(\frac{\gamma}{2}\right)\sin\left(\frac{\alpha}{2}\right)}\right)$$

(d)

$$\angle AB = \cos^{-1}\left(\frac{\sin\left(\frac{\gamma}{2}\right)+\sin\left(\frac{\alpha}{2}\right)\sin\left(\frac{\beta}{2}\right)}{\cos\left(\frac{\alpha}{2}\right)\cos\left(\frac{\beta}{2}\right)}\right), \angle BC = \cos^{-1}\left(\frac{\sin\left(\frac{\alpha}{2}\right)+\sin\left(\frac{\gamma}{2}\right)\sin\left(\frac{\beta}{2}\right)}{\cos\left(\frac{\gamma}{2}\right)\cos\left(\frac{\beta}{2}\right)}\right),$$

$$\angle CA = \cos^{-1}\left(\frac{\sin\left(\frac{\beta}{2}\right)+\sin\left(\frac{\gamma}{2}\right)\sin\left(\frac{\alpha}{2}\right)}{\cos\left(\frac{\gamma}{2}\right)\cos\left(\frac{\alpha}{2}\right)}\right)$$

160. If the material's properties remain unchanged along all directions, the material is:
    (a) homogeneous   (b) isotropic
    (c) uniform   (d) consistent

161. Tensors cannot be used to describe a material's properties. (T/F)

162. If a material is exposed to a stimulus, the . . . is related to the . . . by means of the . . .
    (a) response / stimulus / pressure   (b) response / stimulus / temperature
    (c) stimulus / response / property   (d) pressure / response / stimulus

163. Materials respond . . . to the internal or external stimuli.
    (a) the same way   (b) differently
    (c) consistently   (d) randomly

164. Stimulus and response can be scalars or vectors. (T/F)

165. When a material is exposed to an electric field (E), an electric current is created that flows with density J. The relation between the stimulus and response is described by:
(a) Young's modulus
(b) thermal conductivity
(c) elasticity
(d) electrical conductivity

166. When a material is exposed to an electric field (E), an electric current is created that flows with density J. The relation between the two properties is described by:

(a) $\begin{pmatrix} J_3 \\ J_1 \\ J_2 \end{pmatrix} = \begin{pmatrix} \sigma_{11} & \sigma_{12} & \sigma_{13} \\ \sigma_{21} & \sigma_{22} & \sigma_{23} \\ \sigma_{31} & \sigma_{32} & \sigma_{33} \end{pmatrix} \begin{pmatrix} E_1 \\ E_2 \\ E_3 \end{pmatrix}$
(b) $\begin{pmatrix} J_1 \\ J_2 \\ J_3 \end{pmatrix} = \begin{pmatrix} \sigma_{11} & \sigma_{12} & \sigma_{13} \\ \sigma_{21} & \sigma_{22} & \sigma_{23} \\ \sigma_{31} & \sigma_{32} & \sigma_{33} \end{pmatrix} \begin{pmatrix} E_1 \\ E_2 \\ E_3 \end{pmatrix}$

(c) $\begin{pmatrix} J_3 \\ J_2 \\ J_1 \end{pmatrix} = \begin{pmatrix} \sigma_{31} & \sigma_{32} & \sigma_{33} \\ \sigma_{21} & \sigma_{22} & \sigma_{23} \\ \sigma_{11} & \sigma_{12} & \sigma_{13} \end{pmatrix} \begin{pmatrix} E_3 \\ E_2 \\ E_1 \end{pmatrix}$
(d) $\begin{pmatrix} J_3 \\ J_2 \\ J_1 \end{pmatrix} = \begin{pmatrix} \sigma_{21} & \sigma_{22} & \sigma_{23} \\ \sigma_{31} & \sigma_{32} & \sigma_{33} \\ \sigma_{11} & \sigma_{12} & \sigma_{13} \end{pmatrix} \begin{pmatrix} E_3 \\ E_2 \\ E_1 \end{pmatrix}$

167. If a tensor in the original system is shown by (T), the corresponding tensor in the transformed system (T'_{ij}) is related to the original state by the transformation tensor (L) as:
(a) $T'_{ij} = L^T \cdot T \cdot L$
(b) $T'_{ij} = L \cdot T \cdot L^T$
(c) $T'_{ij} = L^T \cdot L \cdot T$
(d) $T'_{ij} = T \cdot L \cdot L^T$

168. Defects can be categorized into point, line, . . . , and volume groups.
(a) circle
(b) square
(c) shape
(d) area

169. Point defects include:
(a) vacancies (missing atoms within a structure), self-interstitials (extra atoms within a structure), and substitutional (different atoms occupy the space)
(b) vacancies (extra atoms within a structure), self-interstitials (missing atoms within a structure), and substitutional (different atoms occupy the space)
(c) vacancies (missing or extra atoms within a structure)
(d) self-interstitials (missing or extra atoms within a structure)

170. Defects can have either . . . natures.
(a) equilibrium or dynamic
(b) kinetic or equilibrium
(c) kinetic or dynamic
(d) dynamic or equilibrium

171. The defects due to the structural equilibrium issues result in:
(a) increased relative free energy
(b) zero relative free energy
(c) reduced relative free energy
(d) zero potential energy

172. The equilibrium fraction of vacant sites (x_v) is represented by (where E_A is activation energy, k is Boltzmann's constant, T is temperature, $n + N$ is the number of sites, and N is the number of atoms with a contagious set of n vacant surface sites):
(a) $x_v = \exp\left(\dfrac{E_A}{kT}\right)$
(b) $x_v = \dfrac{n}{n+N} = \exp\left(\dfrac{-E_A}{kT}\right)$
(c) $x_v = \dfrac{n}{n-N}$
(d) None of the above

173. What is the activation energy for zinc if its equilibrium factor of vacancy at 400°C is 8×10^{-5} atom/m^{-3} ($\rho = 7.13$ g/cm^3, $M = 65.4$ g/mol, $k = 8.62 \times 10^{-5}$ ev/atom.K)?

(a) 0.547 ev/atom (b) 0.189 ev/atom
(c) 0.289 ev/atom (d) 0.489 ev/atom

Solution Guide

$$N = \frac{N}{M} = \frac{6.02 \times 10^{23} \times 7.13}{65.4} = 6.56 \times 10^{22} \text{ atom/cm}^{-3}$$

$$n = \frac{x_v}{1-x_v} N = \left(\frac{8 \times 10^{-5}}{1 - 8 \times 10^{-5}}\right) \times 6.56 \times 10^{22} = 5.25 \times 10^{18} \text{ atom/cm}^{-3}$$

$$q = (-kT) \times Ln\left(\frac{n}{n+N}\right) = -(8.62 \times 10^{-5}) \times (400 + 273.15) \times Ln(8 \times 10^{-5})$$

$$= 0.547 \text{ ev/atom}$$

174. What is the equilibrium factor of vacancy and number of defects for iron whose activation energy is 0.5 ev/atom at 350°C ($\rho = 7.8$ g/cm^3, $M = 55.85$ g/mol, $k = 8.62 \times 10^{-5}$ ev/atom K)?
(a) 4.63×10^{18} atom/cm^{-3} (b) 7.62×10^{12} atom/cm^{-3}
(c) 7.62×10^{18} atom/cm^{-3} (d) 4.63×10^{12} atom/cm^{-3}

Solution Guide

$$x_v = \exp\left(\frac{-q}{kT}\right) = \exp\left(-\frac{0.5}{8.6210^{-5}(350 + 273.15)}\right) = 9.067 \times 10^{-5}$$

$$N = \frac{N}{M} = \frac{6.02 \times 10^{23} \times 7.8}{55.85} = 8.41 \times 10^{22} \text{ atom/cm}^{-3}$$

$$n = \frac{x_v}{1-x_v} N = \left(\frac{9.067 \times 10^{-5}}{1 - 9.067 \times 10^{-5}}\right) \times 8.412 \times 10^{22} = 7.62 \times 10^{18} \text{ atom/cm}^{-3}$$

175. The number of vacancies decreases with increasing temperature. (T/F)

176. The entropy of mixed defects at equilibrium causes the:
(a) presence of point defects (b) absence of point defects
(c) lack of point defects (d) saturation of point defects

177. The enthalpy of point defects formation at equilibrium is responsible for the . . . of the point defects:
(a) variation (b) derivation
(c) saturation (d) concentration

178. The enthalpy of formation in Arrhenius's law can be substituted with:
(a) entropy
(b) Helmholtz free energy
(c) activation energy
(d) Gibbs minus Helmholtz free energy

179. Ionic crystals can be either conductors or insulators, depending on their structure. (T/F)

180. Ionic crystals are insulators when their crystalline structures:
(a) are perfect (b) have defects
(c) are stressed (d) are cooled

181. Ionic crystals are conductors when their crystalline structures:

    (a) are perfect   (b) have defects

    (c) are stressed   (d) are cooled

182. One of the methods to identify point defect lattice positions in crystals is by:

    (a) measuring electric charges, positions, and reactions

    (b) employing the Kröger-Vink notation

    (c) observing the net positive charge due to the cations

    (d) observing the net negative charge due to the anions

183. A Schottky defect is a(an):

    (a) intrinsic point defect, creating vacancies on the lattice sites that are not constrained by the stoichiometric ratio

    (b) line defect, creating vacancies on the cation and anion lattice sites that maintain the stoichiometric ratio

    (c) intrinsic point defect, creating vacancies on the cation and anion lattice sites that maintain the stoichiometric ratio

    (d) surface defect, creating vacancies on the lattice sites that are not constrained by the stoichiometric ratio

184. A Frenkel defect is:

    (a) a line defect where cations or anions move to an interstitial location while creating a vacancy at their former lattice site

    (b) an intrinsic point defect where cations or anions move to an interstitial location while creating a vacancy at their former lattice site

    (c) an intrinsic point defect where cations or anions occupy interstitial locations without creating new lattice vacancies

    (d) a surface defect where cations or anions occupy interstitial locations without creating new lattice vacancies

185. In the Kröger-Vink notation $\left(X_Y^Z\right)$ used to define the ionic crystalline structure, the parameters can be described as:

    (a) X is the lattice site occupied; Y is the occupying species; and Z is the electronic charge relative to the original site

    (b) X is the occupying species; Y is the electronic charge relative to the original site; and Z is the lattice site occupied

    (c) X is the lattice site occupied; Y is the electronic charge relative to the original site; and Z is the occupying species

    (d) X is the occupying species; Y is the lattice site occupied; and Z is the electronic charge relative to the original site

186. Five ionic lattice defect descriptions are given as follows using the Kröger-Vink notation $\left(X_Y^Z\right)$. Also, listed as follows are verbal interpretations of this notation. Select the answer that matches each notation with its correct description.

    (1) Fe_{Fe}^x (2) Cu_{Ni}^x (3) $v_{Cu}^{'}$ (4) $Cl_i^{..}$ (5) $O_i^{''}$

    (A) An oxygen anion on an interstitial lattice site with a double negative charge

    (B) An iron ion sitting on an iron lattice site with a neutral charge

    (C) An oxygen anion on an interstitial lattice site with a double positive charge

(D) A chlorine cation on an interstitial lattice site with a triple positive charge

(E) A nickel ion sitting on a copper lattice site with a single positive charge

(F) A vacancy on a copper lattice site with a single negative charge

(G) A chlorine cation on an interstitial lattice site with a triple negative charge

(H) An iron ion sitting on an iron lattice site with a single positive charge

(I) A copper ion sitting on a vacancy with a single negative charge

(J) A copper ion sitting on a nickel lattice site with a *neutral* charge

(a) (1) H / (2) I / (3) F / 4(D) / (5) C   (b) (1) B / (2) J / (3) F / (4) D / (5) A
(c) (1) H / (2) E / (3) F / (4) G / (5) C   (d) (1) B / (2) E / (3) I / (4) D / (5) C

187. Identify the defect type and formation relation that best match.

(a) Schottky defect, $\varnothing \Leftrightarrow 3v'''_{\text{Fe}} + 3v_{\ddot{\text{O}}}$ (b) Frenkel defect, $\varnothing \Leftrightarrow 2v'''_{\text{Fe}} + 3v_{\ddot{\text{O}}}$

(c) Schottky defect, $\varnothing \Leftrightarrow 2v'''_{\text{Fe}} + 3v_{\ddot{\text{O}}}$ (d) Frenkel defect, $\varnothing \Leftrightarrow 3v'''_{\text{Fe}} + 3v_{\ddot{\text{O}}}$

188. Intrinsic types of defects are based on thermal equilibrium. (T/F)

189. A defect in a crystalline structure facilitates moving atoms within the structure because:

(a) less external energy is required
(b) it is more difficult to exist
(c) the local arrangement interferes with the process
(d) more internal energy is required

190. Mass transport often involves moving atoms from regions of high atomic concentration to regions of low concentration. (T/F)

191. Substitution diffusion rate depends on the:

(a) atoms' concentrations
(b) atomic weight
(c) number of atoms' vacancies and activation energy
(d) thermal diffusivity

192. Diffusion flux (J) and coefficient (D) are defined by (where C is concentration, x is distance, E_A is activation energy, T is temperature, and R is gas constant):

(a) $J = -D \dfrac{dC_0}{dx}, D = D_0 \exp\left(\dfrac{E_A}{RT}\right)$ (b) $J = D \dfrac{dC}{dC_0}, D = C_0 \exp\left(-\dfrac{E_A}{RT}\right)$

(c) $J = -D \dfrac{dC}{dx}, D = D_0 \exp\left(-\dfrac{E_A}{RT}\right)$ (d) $J = D \dfrac{dC_0}{dC}, D = C_0 \exp\left(\dfrac{E_A}{RT}\right)$

193. The diffusion involving diffusing corrosive gases through oxidative layers toward a metal surface is defined by (where D is the diffusion coefficient, C is concentration, C_0 is concentration at time 0 s, t is time, and x is distance from the surface):

(a) $dC = C - C_0 = erfc\left(\dfrac{x}{2\sqrt{\pi Dt}}\right)$ (b) $\dfrac{dC_x}{dC_s} = \dfrac{C_x - C_0}{C_s - C_0} = erfc\left(\dfrac{x}{2\sqrt{Dt}}\right)$

(c) $dC_0 = C_s - C_0 = erfc\left(\dfrac{x}{2\pi\sqrt{Dt}}\right)$ (d) $\dfrac{dC_s}{dC} = \dfrac{C_s - C_0}{C_x - C_0} = erfc\left(\dfrac{x}{2\sqrt{Dt}}\right)$

194. An electronic equipment enclosure made of steel is exposed to sulfur gases at 97 °C. If the sulfur concentration at the steel surface is 100 ppm (0.01 wt%), what is the concentration (C_x) and diffusion length (x) at 0.5 mm away from its surface, on the steel interior surface after 30 min? Note that the diffusion coefficient for sulfur gases at 97 °C is 1.862×10^{-9} m²/s.

(a) $C_x = 0.01$ wt%, $x = 0.0037$ m   (b) $C_x = 0.001$ wt%, $x = 0.0013$ m

(c) $C_x = 0.009$ wt%, $x = 0.0052$ m   (d) $C_x = 0.008$ wt%, $x = 0.0021$ m

Solution Guide

$$\frac{dC_x}{dC_s} = \frac{C_x - C_0}{C_s - C_0} = erfc\left(\frac{x}{2\sqrt{Dt}}\right) = 1 - erf\left(\frac{x}{2\sqrt{Dt}}\right)$$

$$\therefore \frac{C_x - C_0}{C_s - C_0} = 1 - erf\left(\frac{x}{2\sqrt{Dt}}\right) = 1 - erf\left(\frac{0.5}{2\sqrt{1.862\times10^{-9}\times30\times60}}\right) = 0.999$$

$$\frac{C_x - C_0}{C_s - C_0} = \frac{C_x - C_0}{0.01 - C_0}\bigg|_{C_0=0} = \frac{C_x}{0.01} = 0.999 \therefore C_x = 0.01\, wt\%$$

diffusion length $= 2\sqrt{Dt} = 2\sqrt{1.862\times10^{-9}\times30\times60} = 0.0037\,\text{m}$

195. An electronic equipment enclosure made of copper is exposed to sulfur gases at 52 °C. If the sulfur concentration in the copper interior 1 mm away from its surface, after 1 hour, is 0.01 wt%, what is the sulfur concentration and diffusion length at the copper surface? Note that the diffusion coefficient for sulfur gases at 52 °C is 1.891×10^{-9} m²/s.

(a) $C_x = 0.001$ wt%, $x = 0.0013$ m   (b) $C_x = 0.009$ wt%, $x = 0.0052$ m

(c) $C_x = 0.008$ wt%, $x = 0.0021$ m   (d) $C_s = 0.01$ wt%, $x = 0.0052$ m

Solution Guide

$$\frac{dC_x}{dC_s} = \frac{C_x - C_0}{C_s - C_0} = erfc\left(\frac{x}{2\sqrt{Dt}}\right) = 1 - erf\left(\frac{x}{2\sqrt{Dt}}\right)$$

$$\therefore \frac{C_x - C_0}{C_s - C_0} = 1 - erf\left(\frac{x}{2\sqrt{Dt}}\right) = 1 - erf\left(\frac{1}{2\sqrt{1.891\times10^{-9}\times1\times3600}}\right) = 0.999$$

$$\frac{C_x - C_0}{C_s - C_0} = \frac{0.01 - C_0}{C_s - C_0}\bigg|_{C_0=0} = \frac{0.01}{C_s} = 0.999$$

$\therefore C_x = 0.01$ wt%

diffusion length $= 2\sqrt{Dt} = 2\sqrt{1.891\times10^{-9}\times1\times3600} = 0.0052\,\text{m}$

196. Interstitial diffusion is . . . vacancy diffusion because . . .

(a) faster than / bonding of interstitials to the surrounding atoms is generally weaker, and there are more interstitial sites than vacancy sites to jump to

(b) slower than / smaller atoms are more mobile, with a larger probability for a vacant interstitial site

(c) faster than / larger atoms are less mobile, with a smaller probability for a vacant interstitial site

(d) slower than / larger atoms are less mobile, with a larger probability for a vacant interstitial site

197. Line defects are classified as:

    (a) edge flaws   (b) vacancy and interstitial

    (c) substitutional and vacancy   (d) dislocations

198. From the following list, which are classified as *surface* defects?

    (1) Grain boundaries; (2) Stacking faults; (3) Substitutes; and (4) Vacancies

    (a) (1) and (3)   (b) (2) and (3)   (c) (3) and (4)   (d) (1) and (2)

199. From the following list, which are classified as *volume* defects?

    (1) Grain boundaries; (2) Voids; (3) Vacancies; and (4) Porosities

    (a) (1)   (b) (2) and (4)   (c) (2) and (3)   (d) (1) and (3)

200. Deformation is the change of a material's shape under load. (T/F)

201. Elastic deformation is . . . , while plastic deformation is . . .

    (a) permanent / reversible   (b) reversible / permanent

    (c) stretchable / soft   (d) rubbery / viscous

202. Elastic deformation is:

    (a) irreversible and linearly related to the applied forces

    (b) irreversible and independent of the applied forces

    (c) reversible and linearly related to the applied forces

    (d) reversible and independent of the applied forces

203. Plastic deformation depends linearly on the forces applied to an object. (T/F)

204. When identifying dislocations, the slip planes can be parallel or perpendicular to the applied forces. (T/F)

205. By strengthening a material, the structure:

    (a) is more tolerant to stresses, increasing critical stress values

    (b) is more tolerant to forces, increasing the maximum applicable forces

    (c) can be plastically deformed, given specific loads

    (d) all of the above

206. Introducing defects can strengthen the material. (T/F)

207. Plastically deforming the material can make it stronger. (T/F)

208. If the Burgers relocation vectors have the same direction and magnitude, the edge dislocation is:

    (a) developed by propagating through the material

    (b) prevented from propagating further through the material

    (c) not prevented from propagating further through the material

    (d) quickly developed by propagating through the material and stopping in the middle

209. The dangling bonds of a broken surface (2D defects) as well as energy to create new bonds depend on the:

    (a) cutting volume and surface per atom's density

    (b) cutting plane and surface per atom's density

    (c) cutting plane and surface density per atom's surface area

    (d) cutting volume and surface density per atom's surface area

210. If there are no chemical entities that can bond to free-hanging bonds, surface atoms bond with one another and reconstruct surfaces. (T/F)

211. Point defects are thermodynamically unstable. (T/F)

212. Dislocations are precise random marginal dislocations. (T/F)

213. The descriptor(s) used to identify crystal disclination is(are):
(a) the Frank-Nabarro circuit
(b) the Burgers vector
(c) the diffusion length
(d) both (a) and (b)

214. Total deformation consists of . . . deformation.
(a) plastic plus elastic
(b) plastic
(c) elastic
(d) permanent

215. When material is subjected to tensile stress . . . ; when it is subjected to shear stress . . .
(a) deforming force is applied perpendicular to the surface / deforming force is applied perpendicular to the surface
(b) deforming force is applied parallel to the surface / deforming force is applied parallel to the surface
(c) deforming force is applied parallel to the surface / deforming force is applied perpendicular to the surface
(d) deforming force is applied perpendicular to the surface / deforming force is applied parallel to the surface

216. The ratio of the tensile stress to deformation per unit length of the material in the stress direction is the:
(a) shear stress
(b) modulus of elasticity
(c) shear strain
(d) strain

217. By increasing the material's modulus of elasticity, its impact fatigue decreases. (T/F)

218. By increasing the material's temperature, its modulus of elasticity increases. (T/F)

219. A plastic band that is originally rigid softens with increasing temperature. The normal displacement field is described by U. Determine normal strain (ε_x), assuming that $U = U_x = U_0 \sin\left(\dfrac{2\pi}{C}x\right)$, where C is a constant and x is the position along the x-coordinate.

(a) $\varepsilon_x = U_0 \dfrac{2\pi}{C}\cos\left(\dfrac{2\pi}{C}x\right)$

(b) $\varepsilon_x = U_0 \dfrac{2\pi}{C}\sin\left(\dfrac{2\pi}{C}x\right)$

(c) $\varepsilon_x = -U_0 \dfrac{2\pi}{C}\cos\left(\dfrac{2\pi}{C}x\right)$

(d) $\varepsilon_x = -U_0 \dfrac{2\pi}{C}\sin\left(\dfrac{2\pi}{C}x\right)$

220. A plastic band that is originally rigid softens with increasing temperature. The shear displacement field is described by U. Determine shear strain (γ_x), assuming that $U = U_x = \dfrac{2A\pi}{C}U_0 \sin\left(\dfrac{2\pi}{C}x\right)\cos\left(\dfrac{2\pi}{C}y\right)$, where A and C are constants and x and y are the positions along the respective coordinates.

(a) $\gamma_x = AU_0 \dfrac{8\pi^2}{C^2}\sin\left(\dfrac{2\pi}{C}(x+y)\right)$

(b) $\gamma_x = AU_0 \dfrac{4\pi^2}{C^2}\cos\left(\dfrac{4\pi}{C}(x+y)\right)$

(c) $\gamma_x = AU_0 \dfrac{8\pi^2}{C^2}\cos\left(\dfrac{2\pi}{C}(x+y)\right)$

(d) $\gamma_x = AU_0 \dfrac{4\pi^2}{C^2}\sin\left(\dfrac{4\pi}{C}(x+y)\right)$

221. A plastic band that is originally rigid softens with increasing temperature. The shear displacement field is described by U. Determine shear stress (τ_{xy}), assuming that $U = U_x = U_y = AU_0 \cos\left(\dfrac{2\pi}{C}xy\right)$, where A and C are constants, G is shear modulus, and x and y are the positions along the respective coordinates.

(a) $\tau_{xy} = -GAU_0\dfrac{2\pi}{C}\sin\left(\dfrac{2\pi}{C}(x+y)\right)$ (b) $\tau_{xy} = GAU_0\dfrac{2\pi}{C}\sin\left(\dfrac{2\pi}{C}xy\right)$

(c) $\tau_{xy} = -GAU_0\dfrac{2\pi}{C}\cos\left(\dfrac{2\pi}{C}(x+y)\right)$ (d) $\tau_{xy} = GAU_0\dfrac{2\pi}{C}\cos\left(\dfrac{2\pi}{C}xy\right)$

222. A plastic band that is originally rigid softens with increasing temperature. The shear displacement field is described by U. Determine shear stress (τ_{xy}), assuming that $U = U_x = U_y = \cos x \cos y$, where G is shear modulus, and x and y are the positions along the respective coordinates.

(a) $\tau_{xy} = -G \sin(x-y)$ (b) $\tau_{xy} = G \sin(x+y)$
(c) $\tau_{xy} = -G \sin(x+y)$ (d) $\tau_{xy} = G \sin(x-y)$

223. A plastic band that is originally rigid softens with increasing temperature. The displacement field is described by U. Determine normal strain (ε_x), assuming that $U = U_x = U_y = \cos x \cos y$, where G is shear modulus and x and y are the positions along the respective coordinates.

(a) $\varepsilon_{xy} = -\sin(x+y)$ (b) $\varepsilon_{xy} = \sin(x+y)$

(c) $\varepsilon_{xy} = \dfrac{1}{2}\sin(x+y)$ (d) $\varepsilon_{xy} = -\dfrac{1}{2}\sin(x+y)$

224. Two types of dislocations are:
    (a) edge (slip is perpendicular to the edge) and screw (slip is parallel to the edge)
    (b) edge (slip is parallel to the edge) and screw (slip is perpendicular to the edge)
    (c) reflection (slip is perpendicular to the edge) and glide (slip is parallel to the edge)
    (d) reflection (slip is parallel to the edge) and glide (slip is perpendicular to the edge)

225. Dislocations consist of tangent (aligned with the direction of the defect) and Burgers (aligned with the dislocation line) vectors. (T/F)

226. If the defect line movement is parallel to the direction of the stress, the dislocation is the edge type; (T/F) If the Burgers vector and tangent dislocation components are parallel, the dislocation is the edge type. (T/F)
    (a) T / F   (b) T / T   (c) F / T   (d) F / F

227. If the defect line movement is perpendicular to the direction of the stress, the dislocation is the screw type; (T/F) If the Burgers vector and tangent dislocation components are perpendicular, the dislocation is the screw type. (T/F)
    (a) T / F   (b) T / T   (c) F / T   (d) F / F

228. In a dislocation loop, the dislocation makes a(an) . . . , the Burgers vector . . . , and the tangent vector . . .
    (a) closed curve / is a constant / remains the same
    (b) open curve / is a constant / varies
    (c) closed curve / is a constant / varies
    (d) open curve / varies / varies

229. Critical resolved shear stress (*CRSS*) is the:

(a) magnitude of shear stress perpendicular to slip, required to initiate slip in a grain
(b) component of shear stress in the direction of slip, required to initiate slip in a grain
(c) magnitude of shear stress in the direction of slip, required to terminate slip in a grain
(d) component of shear stress perpendicular to slip, required to initiate slip in a grain

230. The ratio of the critical resolved shear stress (*CRSS*) to applied stress is the:

(a) Peclet number
(b) Schmidt factor
(c) Eckert number
(d) Knudsen number

231. The ratio of the critical resolved shear stress (*CRSS*) to applied stress is (where ϕ is the angle between the normal of the slip plane and the direction of the applied force and λ is the angle between the slip direction and the direction of the applied force):

(a) $\cos\phi \sin\lambda$
(b) $\sin\phi \cos\lambda$
(c) $\sin\phi \sin\lambda$
(d) $\cos\phi \cos\lambda$

232. In order to strengthen the material's structure, the following methods may be employed:

(a) warm work, grain size enlargement, forming liquid solutions, and precipitation
(b) cold work, grain size reduction, forming solid solutions, and precipitation
(c) cold work, grain size enlargement, forming solid solutions, and solidification
(d) warm work, grain size reduction, forming liquid solutions, and solidification

233. Grain boundaries . . . slip and therefore . . . strength.

(a) prevent / increase
(b) enhance / decrease
(c) increase / increase
(d) decrease / decrease

234. Dislocations interact by means of repulsion and attraction. (T/F)

235. Dislocation density is the:

(a) number of dislocations in a unit length
(b) number of dislocations in a unit area
(c) number of dislocations in a unit volume
(d) total number of dislocations in an object

236. Dislocation density can be obtained from a number of dislocation lines crossing a unit area. (T/F)

237. If dislocations annihilate, they form flawed crystals. (T/F)

238. In order to minimize the surface free energy, the exposed:

(a) surface can be increased and volume construction may take place
(b) surface can be increased or surface construction may take place
(c) surface can be decreased or surface reconstruction may take place
(d) volume can be decreased and volume reconstruction may take place

239. Wulff construction is:

(a) obtained by maximizing the total surface free energy of the crystal-medium interface
(b) when surface free energy is maximized by assuming a shape of low surface energy
(c) when surface free energy is maximized by assuming a shape of high surface energy
(d) used to determine the equilibrium shape of a crystal with fixed volume inside a medium

240. Surface defects can act as . . . . They . . . the reaction rate without . . . in it.

    (a) catalysts / increase / participating
    (b) decelerators / increase / contributing
    (c) catalysts / decrease / taking part
    (d) decelerators / decrease / sharing

241. For the same-phase grain boundaries presented in Figure 92, calculate the length L required for a crystal equilibrium state, taking into account the balance of forces ($a = b$).

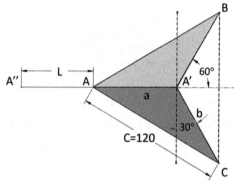

FIGURE 92. Same-phase grain boundaries (dimensions are in micrometers).

    (a) 34.6 μm   (b) 46.2 μm   (c) 115.5 μm   (d) 69.3 μm

Solution Guide

$$c = \sqrt{a^2 + b^2 - 2ab\cos\theta}$$

$$\therefore 120 = \sqrt{2a^2\left(1 - \cos 120°\right)}$$

$$\therefore \quad a = b = 69.3\,m$$

$$L = 69.3 \times 2 \times \cos 60° = 69.3 \text{ m}$$

242. For the dissimilar-phase grain boundaries presented in Figure 93, calculate the length L required for a crystal equilibrium state, taking into account the balance of forces.

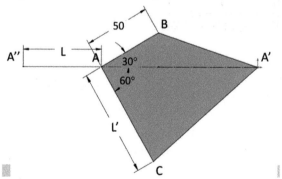

FIGURE 93. Dissimilar-phase grain boundaries (dimensions are in micrometers).

    (a) 60.51 μm   (b) 57.74 μm   (c) 12.35 μm   (d) 63.14 μm

Solution Guide

$$50 \times \sin 30° - L' \times \sin 60° = 0$$

$$\therefore L' = \frac{50 \times \sin 30°}{\sin 60°} = 28.87\,m$$

$$L = 50 \times \cos 30° + 28.87 \times \cos 60° = 57.74\,m$$

243. Tensile stress ($\sigma = 100$ MPa) is applied along the direction [010] (Figure 94) to a body-centered cubic (BCC) crystal. Assuming that the slip occurs along the Plane (110), calculate the Schmid factor (m) and the resolved shear stress (τ_{RSS}).

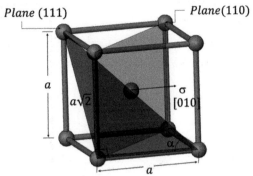

Plane (111)   Plane(110)

a

$a\sqrt{2}$

σ
[010]

α

a

FIGURE 94. A body-centered cubic (BCC) crystal under a slip.

(a) $m = 0.41$, $\tau_{RSS} = 40.8$ MPa   (b) $m = 0.22$, $\tau_{RSS} = 20.4$ MPa
(c) $m = 0.41$, $\tau_{RSS} = 10.2$ MPa   (d) $m = 0.22$, $\tau_{RSS} = 81.2$ MPa

Solution Guide

To calculate τ_{RSS}, one needs to obtain the angles between the tensile stress $[\vec{a}]$ and the slip plane $(\vec{b})$, which is α, as well as the angle of the tensile stress $[\vec{a}]$ and the normal to the slip plane $[\vec{c}]$, which is β. Employing the cosine law, the angle between the two vectors is the inverse cosine of their internal dot products divided by the product of their magnitudes. Note that since $[\vec{c}]$ is normal to $(\vec{b})$, their internal dot product is zero, and therefore one of the possible combinations complying with this condition may be chosen.

$$\vec{a}[0 \quad 1 \quad 0] = a_1\vec{i} + a_2\vec{j} + a_3\vec{k} = j$$

$$\vec{b}(1 \quad 1 \quad 0) = \vec{i} + \vec{j}$$

$$\vec{c}[c_1 \quad c_2 \quad c_3] = c_1\vec{i} + c_2\vec{j} + c_3\vec{k}$$

$$\cos\alpha = \cos(\vec{a},\vec{b}) = \frac{\vec{a}.\vec{b}}{|\vec{a}|\cdot|\vec{b}|} = \frac{1}{\sqrt{1}\times\sqrt{2}} = \frac{1}{\sqrt{2}}$$

$$\therefore \alpha = 45°$$

$$\cos \quad = \cos(b,c) = \frac{b\,c}{|b|\cdot|c|} = \frac{c_1 + c_2}{\sqrt{1+1}\times\sqrt{c_1^2 + c_2^2 + c_3^2}} = \frac{c_1 + c_2}{\sqrt{2}\times\sqrt{c_1^2 + c_2^2 + c_3^2}} = 0$$

$$\therefore c_1 = -c_2$$

$$\therefore \vec{c}[c_1 \quad c_2 \quad c_3] = \vec{c}[-1 \quad 1 \quad 1]$$

Other possible combinations for the slip systems can be $[-1\ 1\ 0]$, $[1\ -1\ 0]$, $[1\ -1\ 1]$.

$$\cos\beta = \cos(\vec{a},\vec{c}) = \frac{\vec{a}.\vec{c}}{|\vec{a}|\cdot|\vec{c}|} = \frac{1}{\sqrt{1}\times\sqrt{1+1+1}} = \frac{1}{\sqrt{1}\times\sqrt{3}} = \frac{1}{\sqrt{3}}$$

$$\therefore \beta = 54.7°$$

$$m = \cos\alpha\times\cos\beta = \cos 45°\times\cos 54.7° = 0.41$$

$$\tau_{RSS} = \cos\alpha\times\cos\beta\times\sigma = \cos 45°\times\cos 54.7°\times 100 = 40.8\,\text{MPa}$$

244. Tensile stress (σ) is applied along the direction [111] to a body-centered cubic (BCC) crystal (Figure 95). Assuming that the slip occurs along the Plane (101), and resolved shear stress is known (τ_{RSS} = 100 MPa), calculate the Schmid factor (m) and the tensile stress (σ).

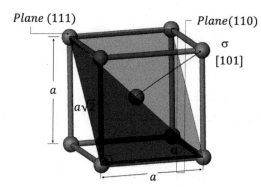

FIGURE 95. A body-centered cubic (BCC) crystal under a slip.

(a) $m = 0.27$, $\sigma = 184.3$ MPa   (b) $m = 0.54$, $\sigma = 82.5$ MPa
(c) $m = 0.27$, $\sigma = 367.4$ MPa   (d) $m = 0.54$, $\sigma = 253.2$ MPa

Solution Guide

$$\vec{a}[1\quad 1\quad 1] = a_1\vec{i} + a_2\vec{j} + a_3\vec{k} = \vec{i} + \vec{j} + \vec{k}$$

$$\vec{b}(1\quad 0\quad 1) = b_1\vec{i} + b_2\vec{j} + b_3\vec{k} = \vec{i} + \vec{k}$$

$$\vec{c}[c_1\quad c_2\quad c_3] = c_1\vec{i} + c_2\vec{j} + c_3\vec{k}$$

$$\cos\alpha = \cos(\vec{a},\vec{b}) = \frac{\vec{a}.\vec{b}}{|\vec{a}|\cdot|\vec{b}|} = \frac{1+1}{\sqrt{1+1+1}\times\sqrt{1+1}} = \frac{2}{\sqrt{3}\times\sqrt{2}} = \frac{2}{\sqrt{6}} = 0.82$$

$$\alpha = 35.3°$$

$$\cos\gamma = \cos(\vec{b},\vec{c}) = \frac{\vec{b}.\vec{c}}{|\vec{b}|\cdot|\vec{c}|} = \frac{c_1+c_3}{\sqrt{1+1}\times\sqrt{c_1^2+c_2^2+c_3^2}} = \frac{c_1+c_3}{\sqrt{2}\times\sqrt{c_1^2+c_2^2+c_3^2}} = 0$$

$$\therefore c_1 = -c_3$$

$$\therefore \vec{c}\begin{bmatrix}c_1 & c_2 & c_3\end{bmatrix} = \vec{c}\begin{bmatrix}-1 & 1 & 1\end{bmatrix}$$

Other possible combinations for the slip systems can be $[1\ 1\ -1]$, $[-1\ 0\ 1]$, $[1\ 0\ -1]$.

$$\cos\beta = \cos(\vec{a},\vec{c}) = \frac{\vec{a}.\vec{c}}{|\vec{a}|\cdot|\vec{c}|} = \frac{-1+1+1}{\sqrt{1+1+1}\times\sqrt{1+1+1}} = \frac{1}{\sqrt{3}\times\sqrt{3}} = \frac{1}{3} = 0.33$$

$$\therefore \beta = 70.5°$$

$$m = \cos\alpha\times\cos\beta = \cos 35.3°\times\cos 70.5° = 0.27$$

$$\tau_{RSS} = \cos\alpha\times\cos\beta\times\sigma \therefore \sigma = \frac{100}{\cos 35.3°\times\cos 70.5°} = 367.4\,\text{MPa}$$

245. An aluminum cylinder undergoes a 1-mm elongation in the direction of the force that is applied parallel to its axis (Figure 96). Assuming that the cylinder's original length was 100 mm, calculate the force causing this elongation. Assume the modulus of elasticity of aluminum (E) is 69 GPa, the cylinder diameter is 35 mm, and the deformation is elastic.

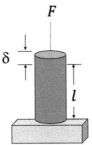

FIGURE 96. An elongated aluminum cylinder.

(a) 332.3 kN   (b) 663.8 kN   (c) 223.5 kN   (d) 456.3 kN

Solution Guide

$$\varepsilon = \frac{\delta}{l} = \frac{1}{100} = 0.01$$

$$\sigma = E\varepsilon = 69 \times 0.01 = 0.69 \, \text{GPa}$$

$$F = \sigma A = 0.69 \times 10^9 \times 3.14 \times \left(\frac{17.5}{1000}\right)^2 = 663.8 \, \text{kN}$$

246. A copper grain undergoes plastic deformation (Figure 97). If the yield stress is defined by $\sigma_y = \sigma_0 + \frac{k}{\sqrt{d}}$, where d is the grain size and k is a constant, what is the minimum force (F) required to create the plastic deformation, if the original length is 30 μm and the elongated length before breakage is 80 μm? Assume the reference stress (σ_0) is 25 MPa and the constant k is 0.11 m$^{0.5}$ MPa. The modulus of elasticity of copper (E) is 117 GPa and its yield tensile strength is 33.3 MPa. Assuming that the grain can have a rectangular cross section, calculate the rectangle height (a) if its height (a) is much smaller than its length (b).

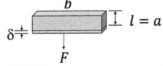

FIGURE 97. An elongated copper grain.

(a) $F = 1.14$ kN, $x = 0.022$ mm   (b) $F = 1.14$ kN, $x = 0.166$ mm
(c) $F = 2.27$ kN, $x = 0.088$ mm   (d) $F = 2.27$ kN, $x = 0.044$ mm

Solution Guide

$$\sigma_y = \sigma_0 + \frac{k}{\sqrt{d}}$$

$$\sigma_y = \sigma_0 + \frac{k}{\sqrt{d}}$$

$$\therefore d = \left(\frac{k}{\sigma_y - \sigma_0}\right)^2 = 0.176 \, \text{mm}$$

$$\varepsilon = \frac{\delta}{l} = \frac{50-30}{30} = 0.67$$

$$\sigma = E\varepsilon = 117 \times 0.67 = 78\,\text{GPa}$$

$$F = \sigma A = 78 \times 10^9 \times 3.14 \times \left(\frac{0.176}{2000}\right)^2 = 1.89\ \text{KN}$$

To calculate the rectangle length, it is assumed that the circular cross section is the hydraulic diameter of the rectangle. Since the rectangle length is much larger than its height, the height can be ignored in an addition operation.

$$d_h = \frac{4A}{P} = 0.176\,\text{mm}$$

$$\therefore d_h = \frac{a \times b}{2 \times (a+b)} = \frac{a \times b}{2 \times b} = \frac{a}{2} = 0.176\,\text{mm}$$

$$\therefore a = 0.35\ \text{mm}$$

Answer Key									
1. (d)	**2.** (a)	**3.** (c)	**4.** (a)	**5.** (b)	**6.** (d)	**7.** (c)	**8.** T	**9.** T	**10.** (a)
11. (b)	**12.** (a)	**13.** (a)	**14.** (b)	**15.** (c)	**16.** (a)	**17.** (c)	**18.** (b)	**19.** T	**20.** F
21. F	**22.** (d)	**23.** (a)	**24.** (b)	**25.** (d)	**26** (c)	**27.** (c)	**28.** (a)	**29.** (b)	**30.** (b)
31. T	**32.** (d)	**33.** (a)	**34.** (b)	**35.** (d)	**36.** (a)	**37.** (c)	**38.** (d)	**39.** (b)	**40.** (a)
41. (a)	**42.** (d)	**43.** (a)	**44.** (c)	**45.** (a)	**46.** (c)	**47.** T	**48.** F	**49.** F	**50.** F
51. (a)	**52.** (a)	**53.** (c)	**54.** (a)	**55.** (b)	**56.** (a)	**57.** (d)	**58.** (a)	**59.** (b)	**60.** (a)
61. (b)	**62.** (c)	**63.** (d)	**64.** (d)	**65.** (c)	**66.** (a)	**67.** (b)	**68.** T	**69.** (a)	**70.** (d)
71. (a)	**72.** (c)	**73.** (d)	**74.** F	**75.** T	**76.** (d)	**77.** (a)	**78.** (b)	**79.** (b)	**80.** (a)
81. (c)	**82.** (b)	**83.** (a)	**84.** (d)	**85.** (c)	**86.** F	**87.** (a)	**88.** T	**89.** (a)	**90.** F
91. T	**92.** (a)	**93.** (b)	**94.** (d)	**95.** (b)	**96.** (a)	**97.** T	**98.** (c)	**99.** (c)	**100.** (a)
101. (a)	**102.** (b)	**103.** (d)	**104.** (a)	**105.** (a)	**106.** (b)	**107.** (c)	**108.** (d)	**109.** (d)	**110.** T
111. (c)	**112.** (a)	**113.** (b)	**114.** (c)	**115.** (d)	**116.** (a)	**117.** (b)	**118.** (c)	**119.** (d)	**120.** (d)
121. (a)	**122.** (a)	**123.** (a)	**124.** (d)	**125.** (b)	**126.** (c)	**127.** (c)	**128.** (c)	**129.** (b)	**130.** (d)
131. (a)	**132.** (d)	**133.** (a)	**134.** (c)	**135.** T	**136.** T	**137.** (b)	**138.** (b)	**139.** (b)	**140.** (c)
141. (a)	**142.** (a)	**143.** (b)	**144.** (c)	**145.** (d)	**146.** (a)	**147.** (c)	**148.** (a)	**149.** (b)	**150.** (c)
151. (b)	**152.** (a)	**153.** (c)	**154.** (a)	**155.** (b)	**156.** (b)	**157.** (c)	**158.** (c)	**159.** (a)	**160.** (b)
161. F	**162.** (c)	**163.** (b)	**164.** T	**165.** (d)	**166.** (b)	**167.** (b)	**168.** (d)	**169.** (a)	**170.** (b)
171. (c)	**172.** (b)	**173.** (a)	**174.** (c)	**175.** F	**176.** (a)	**177.** (d)	**178.** (c)	**179.** T	**180.** (a)
181. (b)	**182.** (a)	**183.** (c)	**184.** (b)	**185.** (d)	**186.** (b)	**187.** (c)	**188.** T	**189.** (a)	**190.** F
191. (c)	**192.** (c)	**193.** (b)	**194.** (a)	**195.** (d)	**196.** (a)	**197.** (d)	**198.** (d)	**199.** (b)	**200.** T
201. (b)	**202.** (c)	**203.** F	**204.** T	**205.** (d)	**206.** T	**207.** T	**208.** (b)	**209.** (b)	**210.** T
211. F	**212.** F	**213.** (a)	**214.** (a)	**215.** (d)	**216.** (b)	**217.** T	**218.** F	**219.** (a)	**220.** (b)
221. (a)	**222.** (c)	**223** (d)	**224.** (a)	**225.** T	**226.** (a)	**227.** (a)	**228.** (c)	**229.** (b)	**230.** (b)
231. (d)	**232.** (b)	**233.** (a)	**234.** T	**235.** (c)	**236.** T	**237.** F	**238.** (c)	**239.** (d)	**240.** (a)
241. (d)	**242.** (b)	**243.** (a)	**244.** (c)	**245.** (b)	**246.** (c)				

CHAPTER 5

DYNAMICS AND CONTROL

1. A system is:
   - (a) any human-made machine
   - (b) any natural machine
   - (c) either (a) or (b)
   - (d) none of the above

2. Among the following, which one is not considered an example of a system?
   - (a) Powered desk
   - (b) Screw
   - (c) Bird
   - (d) Space shuttle

3. A system's information is carried by:
   - (a) signals
   - (b) waves
   - (c) wavelengths
   - (d) frequencies

4. Among the following, signals are generated by:
   - (a) radio frequency remote controls
   - (b) thermal cameras
   - (c) video cameras
   - (d) all of the above

5. Among the following, which one describes most accurately how a thermal imager observes temperature?
   - (a) Pressure responses
   - (b) Thermal responses in the form of object signals
   - (c) Statistical data
   - (d) Object color

6. Signals carry:
   - (a) information
   - (b) temperature
   - (c) data
   - (d) wavelength

7. Signal versus system:
   - (a) A signal describes how one parameter changes with the other one; a system is any process that produces an output signal in response to an input signal.
   - (b) A signal is any process that produces an output signal in response to an input signal; a system describes how one parameter changes with the other one.
   - (c) A signal is any machine that produces an input signal in response to an output signal; a system describes how one parameter changes with the other one.
   - (d) They are the same.

8. System feedback provides information on a system's:
   (a) performance   (b) parameters
   (c) deviation from the reference point (d) all of the above

9. A control system measures the operational data and compares them with . . .; it also issues control signals to . . .
   (a) internal requirements / desired outputs
   (b) external requirements / correct the output
   (c) ideal requirements / desired outputs
   (d) desired outputs / correct the output

10. A system's output is defined as the:
   (a) output plus the evolution of the inputs given the external disturbances and internal variations
   (b) inputs given the external disturbances and internal variations
   (c) processed and controlled inputs given the external disturbances and internal variations
   (d) variation of the external disturbances and inputs

11. Some advantages of a control system are:
   (a) security, performance, and precision
   (b) performance, security, and applicability
   (c) applicability, security, and accuracy
   (d) performance, security, accuracy, and precision

12. Control systems do the following for the processes with no exception:
   (a) observe, record, and report   (b) observe and record
   (c) observe   (d) record

13. The difference between dynamic versus static in a control system is that a dynamic system is:
   (a) time-oriented while a static system is not
   (b) not time-oriented while a static system is
   (c) dependent on the previous responses while a static system is not
   (d) dependent on the previous responses and so is a static system

14. A system that moves is always a dynamic control system. (T/F)

15. Among the following, which one is best described as a controlled dynamic system?
   (a) Boiling hot water to make green tea in a teapot
   (b) Accidentally moving a teapot on a table
   (c) Eating cheese puffs and yogurt
   (d) Stirring sugar in tea

16. A control system's objective is to achieve the desired:
   (a) outputs   (b) inputs
   (c) disturbances   (d) all of the above

17. Disturbances are always fully controllable. (T/F)

18. Control systems can be found both in nature and in human-made machines. (T/F)

19. Among the following, which one does not include a control system:
   (a) Giraffe   (b) An open book sitting on a table
   (c) Car   (d) Flying birds

20. Among the following, which one does not involve a control system?

    (a) Picking up a book from a shelf

    (b) Heating a room to a desired temperature

    (c) Scratching your nose

    (d) Scarecrow staring at a bird

21. A system can be made up of isolated subsystems. (T/F)

22. When walking to your doorway, reaction feedback is required to ensure you step in the right direction and reach the destination. (T/F)

23. When reaching for a cup of hot tea, reaction feedback:

    (a) ensures you pick up the cup without spilling any tea

    (b) allows you to see the cup

    (c) provides power for hand movement

    (d) is not required

24. A feedback system synchronizes:

    (a) intentions and actions   (b) intentions, actions, and results

    (c) actions and results   (d) none of the above

25. A system described by equation $y(t) = \dfrac{x(t)}{a}$ is dynamic. (T/F)

26. Raising water in a tank to a specified level is a dynamic system. (T/F)

27. From the following, select the correct statement:

    (a) Qualitative variables can be measured.

    (b) Quantitative variables can be measured.

    (c) Qualitative variables may be measurable.

    (d) Quantitative variables may not be measurable.

28. Identify the set of characteristics with correct qualitative versus quantitative descriptors:

    (a) Care (quantitative), odor (quantitative), temperature (quantitative), and loudness (quantitative)

    (b) Excitement (qualitative), odor (quantitative), temperature (qualitative), and loudness (quantitative)

    (c) Sadness (quantitative), odor (qualitative), temperature (qualitative), and loudness (qualitative)

    (d) Happiness (qualitative), odor (qualitative), temperature (quantitative), and loudness (quantitative)

29. Which one best describes the characteristics of continuous, discrete, and binary feedback systems in the order presented?

    (a) Connected, separate, on/off   (b) Separate, on/off, connected

    (c) On/Off, connected, separate   (d) All of the above

30. Among the following, which one best represents artificial systems?

    (a) Human-made and natural systems such as aircraft and birds

    (b) Natural systems such as insects

    (c) Human-made systems such as cellphones

    (d) Active systems such as robots and non-active entities such as books

31. The continuous-time signals are characterized by . . . and describe a continuous set of values.
    (a) dependent parameters that are discrete
    (b) independent parameters that are discrete
    (c) dependent parameters that are continuous
    (d) independent parameters that are continuous

32. All control systems include both signals and feedback. (T/F)

33. Any system that sends a signal:
    (a) is controllable   (b) is not controllable
    (c) can be changed   (d) cannot be changed

34. Which of the following is a signal (and not a system)?
    (a) Image of milky way picked up by a telescope
    (b) Aircraft
    (c) Reaching for a pen
    (d) Driving a car

35. The main purpose of a control system is to:
    (a) achieve the best performance
    (b) revise the system's behavior
    (c) connect the dots
    (d) shut down the system

36. Among the following items, identify the signals, respectively:
    (a) Temporal temperature, musical sounds, and spatial images
    (b) Transient temperature, musical sounds, and disturbances
    (c) Voltage from the input port, light intensity level, and appearance
    (d) Glucose metabolism, heat generation, and facial expressions

37. A signal is:
    (a) a vector   (b) a scalar quantity
    (c) occasionally a vector or a scalar   (d) always in integer quantity

38. A signal:
    (a) is never accompanied by noise
    (b) has an associated physical quantity that shows temporal or spatial variations
    (c) is a vector that changes with time and position to a reference point
    (d) is a function that conveys limited information

39. A signal can be shown by graphs, tables, other data, or combination of them. (T/F)

40. Signals can be described by any of the following terms: Binary, digital, continuous, and stochastic. (T/F)

41. A control signal depends on the system's:
    (a) outputs   (b) disturbances
    (c) inputs   (d) all of the above

42. In a control system, the following devices usually exist: Transducers, receivers, signal generators, processors, and sensors. (T/F)

43. Signals are measured by:
    (a) generators   (b) sensors   (c) processors   (d) sensors or transducers

44. What can be transferred in a system?

(a) Motion
(b) Heat and work
(c) Emotion and motion
(d) Signal, material, and energy

45. A control system can be simulated by a mechanical system consisting of a:

(a) spring, damper, and mass
(b) gear, screw, and nut
(c) pump, valve, and pipe
(d) ramp, weight, and string

46. An electric circuit in which resistance does not depend on the time, can be described as:

(a) dynamic
(b) static
(c) predictable
(d) consistent

47. An electric circuit in which resistance does not depend on the time, but the inductance depends on the current integral (accumulated charge over time), can be described as:

(a) dynamic
(b) static
(c) predictable
(d) consistent

48. Among the following control models, which one is the correct representation of a PID control system? Assume K_i, K_d, and K_p are, respectively, integral, derivative, and proportional control operators.

(a) $K_{.}e + K \int_0^t edt + K \dfrac{de}{dt}$

(b) $K_i e + K_p \int_0^t edt + K_d \dfrac{de}{dt}$

(c) $K_d e + K_p \int_0^t edt + K_i \dfrac{de}{dt}$

(d) $K_i e + K_d \int_0^t edt + K_p \dfrac{de}{dt}$

49. Among the following equations, which one does not apply to a dynamic system?

(a) $V(t) = \dfrac{Q(t)}{C} = \dfrac{1}{C} \int_{t_0}^t I(\tau)dt + V(t_0)$

(b) $\varepsilon(t) = -L\dfrac{dI(t)}{dt}$

(c) $V(t) = RI(t)$

(d) $V(t) = \int_{t_0}^t a(\tau)dt + V(t_0)$

50. Among the following, which one is the most accurate regarding the system's constants and variables?

(a) constants do not change over time, but the variables can change
(b) both constants and variables change over time
(c) car acceleration is a constant while its engine horsepower is a variable
(d) aircraft speed is a constant while its wingspan is a variable

51. Analyze the forces for the spring-mass system presented in Figure 98 (assume acceleration is a constant).

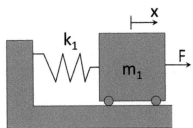

FIGURE 98. Spring-mass system.

(a) $F = -k_1 x_1 + \dfrac{d^2}{dt^2} m_1 x_1$

(b) $F = k_1 x_1 + \dfrac{d^2}{dt^2} m_1 x_1$

(c) $F = k_1 x_1 - \dfrac{d^2}{dt^2} m_1 x_1$

(d) $F = -k_1 x_1 - \dfrac{d^2}{dt^2} m_1 x_1$

52. Analyze the forces for the spring-mass system presented in Figure 99 (assume acceleration is a constant).

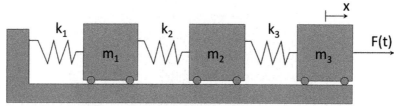

FIGURE 99. Spring-mass system.

(a) $F(t)\begin{bmatrix} 0 \\ 0 \\ 1 \end{bmatrix} - \begin{bmatrix} k_1 & k_2 & 0 \\ 0 & k_2 & k_3 \\ 0 & 0 & -k_3 \end{bmatrix}\begin{bmatrix} x_1 \\ x_2 \\ x_3 \end{bmatrix} - \dfrac{d^2}{dt^2}\begin{bmatrix} m_1 & 0 & 0 \\ 0 & m_2 & 0 \\ 0 & 0 & m_3 \end{bmatrix}\begin{bmatrix} x_1 \\ x_2 \\ x_3 \end{bmatrix} = 0$

(b) $F(t)\begin{bmatrix} 0 \\ 0 \\ 1 \end{bmatrix} + \begin{bmatrix} k_1 & k_2 & 0 \\ 0 & -k_2 & k_3 \\ 0 & 0 & k_3 \end{bmatrix}\begin{bmatrix} x_1 \\ x_2 \\ x_3 \end{bmatrix} + \dfrac{d^2}{dt^2}\begin{bmatrix} m_1 & 0 & 0 \\ 0 & m_2 & 0 \\ 0 & 0 & m_3 \end{bmatrix}\begin{bmatrix} x_1 \\ x_2 \\ x_3 \end{bmatrix} = 0$

(c) $F(t)\begin{bmatrix} 0 \\ 0 \\ 1 \end{bmatrix} - \begin{bmatrix} -k_1 & k_2 & 0 \\ 0 & k_2 & k_3 \\ 0 & 0 & k_3 \end{bmatrix}\begin{bmatrix} x_1 \\ x_2 \\ x_3 \end{bmatrix} - \dfrac{d^2}{dt^2}\begin{bmatrix} m_1 & 0 & 0 \\ 0 & m_2 & 0 \\ 0 & 0 & m_3 \end{bmatrix}\begin{bmatrix} x_1 \\ x_2 \\ x_3 \end{bmatrix} = 0$

(d) $F(t)\begin{bmatrix} 0 \\ 0 \\ 1 \end{bmatrix} + \begin{bmatrix} -k_1 & k_2 & 0 \\ 0 & -k_2 & k_3 \\ 0 & 0 & -k_3 \end{bmatrix}\begin{bmatrix} x_1 \\ x_2 \\ x_3 \end{bmatrix} - \dfrac{d^2}{dt^2}\begin{bmatrix} m_1 & 0 & 0 \\ 0 & m_2 & 0 \\ 0 & 0 & m_3 \end{bmatrix}\begin{bmatrix} x_1 \\ x_2 \\ x_3 \end{bmatrix} = 0$

Solution Guide

Note that the total force has to resist the spring forces and move masses.

$F(t) - k_3 x_3 = m_3 a_3 = m_3 \ddot{x}_3$

$k_3 x_3 - k_2 x_2 = m_2 a_2 = m_2 \ddot{x}_2$

$k_2 x_2 - k_1 x_1 = m_1 a_1 = m_1 \ddot{x}_1$

$\therefore F(t) - k_1 x_1 = \sum_{i=1}^{3} m_i a_i = \dfrac{d^2}{dt^2}\sum_{i=1}^{3} m_i a_i = \sum_{i=1}^{3} m_i \ddot{x}_i$

$\therefore F(t) - k_1 x_1 - \sum_{i=1}^{3} m_i \ddot{x}_i = 0$

53. Figure 100 shows a tank being filled with water, with water exiting it through an opening near the bottom with velocity V_1. We are interested in maintaining a particular water level

h in the tank; ρ is density, A is area, and V is flow velocity (1 and 2 subscripts stand for the status of the systems at times t_1 and t_2). State variable is:

(a) water temperature (b) water level

(c) water density (d) water flow rate

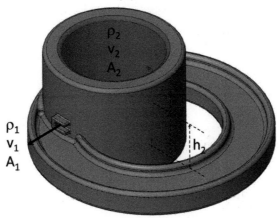

FIGURE 100. Water pond.

54. In Figure 100, the system is . . . because the water level depends on the . . .

(a) dynamic / past water levels (b) static / temperature

(c) static / past water levels (d) dynamic / temperature

55. The designer is interested in maintaining the water level for the water pond presented in Figure 100 as a function of time so it doubles as time increases every 1 s (Figure 101). Design a flow control model to represent the system. Assume the water level at the initial time (t_0) is h_0. Assume flow is incompressible (density is a constant). Note that h is the water level, ρ is density, A is area, and V is flowvelocity; and 1 and 2 subscripts stand for the status of the systems at times t_1 and t_2.

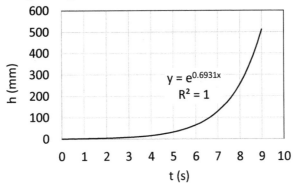

FIGURE 101. Water level control diagram.

(a) $\dot{m}_1 = \rho_1 A_1 V_1 = 0.69 \rho_1 A_2 e^{0.69t}$ (b) $\dot{m}_1 = \rho_2 A_2 V_2 = 0.69 \rho_1 A_1 e^{1.69t}$

(c) $\dot{m}_1 = \rho_1 A_1 V_1 = 1.69 \rho_2 A_2 e^{0.69t}$ (d) $\dot{m}_1 = \rho_2 A_2 V_1 = 1.69 \rho_1 A_1 e^{0.69t}$

Solution Guide

$$h(t_0) = h_0$$

$$h(t_0 + 1) = 2h_0$$

$$h(t_0 + 2) = 2(2h_0) = 4h_0 = 2^2 h_0$$

$$h(t_0 + 3) = 2(4h_0) = 8h_0 = 2^3 h_0$$

$$h(t_0 + n) = h(t) = 2^n h_0$$

$$\therefore \text{for } n = 9, 2^9 = 512$$

$$h(t) = e^{0.69t}$$

$$\dot{m}_2 = \rho_2 A_2 V_2 = \rho_2 A_2 \frac{dh(t)}{dt} = \rho_2 A_2 (0.69 e^{0.69t}) = \dot{m}_1 = \frac{dm_1}{dt} = \rho_1 A_1 V_1$$

$$V_1 = (0.69 e^{0.69t}) \frac{A_2}{A_1}$$

$$\therefore Q_1 = A_1 V_1 = (0.69 e^{0.69t}) A_2$$

$$\therefore \dot{m}_1 = \rho_1 A_1 V_1 = 0.69 \rho_1 A_2 e^{0.69t}$$

56. A signal contains both useful information and noise. (T/F)

57. In Figure 102, the control signal is . . . and parameters affecting it are . . .
    (a) temperature / internal and external variables
    (b) heat flow / internal variables
    (c) heat flux / external variables
    (d) temperature gradient / internal and external variables

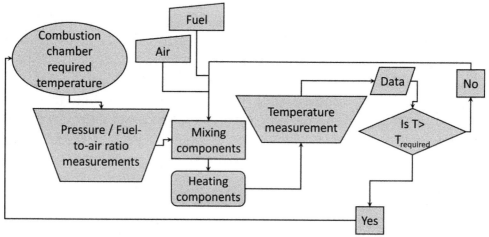

FIGURE 102. Control system for a combustion chamber.

58. Control system output is not affected by the variables external to the system. (T/F)

59. A hybrid system is a:
    (a) process that can be discrete locally, but event-based or continuous at the higher level
    (b) process that can be discrete locally, but event-based or discrete at the higher level
    (c) dynamical system that can show both continuous and discrete dynamic behaviors
    (d) dynamical system that can behave discretely

60. The control system presented in Figure 103 consists of:
    (a) one loop, two series, and one parallel structure
    (b) two loops, one series, and one parallel structure
    (c) two loops, two series, and two parallel structures
    (d) one loop, one series, and one parallel structure

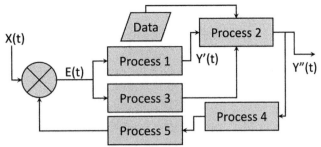

FIGURE 103. Closed-loop control system.

61. In an autonomous control system, the signal cannot be changed. (T/F)

62. In a forced control system, signal variation is:
(a) avoidable
(b) unavoidable
(c) not necessarily the case
(d) not applicable

63. If a control system's time-domain response can be described by an exponential function $(y(t) = y_0 e^{at})$, its frequency-domain response using the Laplace transform is:

(a) $y(s) = -\dfrac{y_0}{s-a}$ 　(b) $y(s) = \dfrac{y_0}{s^2 - a^2}$ 　(c) $y(s) = \dfrac{y_0}{s-a}$ 　(d) $y(s) = -\dfrac{y_0}{s^2 - a^2}$

64. If a control system's time-domain response can be described by a sinusoidal function $y(t) = y_0 \sin(\omega t + \phi)$, its frequency-domain response using the Laplace transform is:

(a) $y(s) = y_0 \dfrac{\omega s}{s^2 - \omega^2}$

(b) $y(s) = -y_0 \dfrac{\omega s}{s^2 - \omega^2}$

(c) $y(s) = -y_0 \dfrac{\omega s}{s^2 + \omega^2}$

(d) $y(s) = y_0 \dfrac{\omega s}{s^2 + \omega^2}$

65. Among the control systems presented in Figure 104, which one is the representative of the control model $y(a) = x(a)(G_1(a) + G_2(a))G_3(a)$?

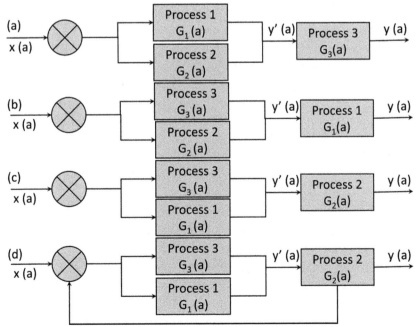

FIGURE 104. Block diagrams presenting control systems and their operators.

66. For the control system presented in Figure 105, identify the model that best describes it.

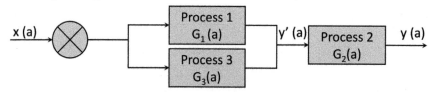

FIGURE 105. A block diagram presenting a control system.

(a) $y(a) = x(a)(G.(a) + G (a))G$

(b) $y(a) = \dfrac{x(a)(G_1(a) + G_3(a))G_2(a)}{1 + x(a)(G_1(a) + G_3(a))G_2(a)}$

(c) $y(a) = x(a)(G_1(a) + G_2(a))G_3$

(d) $y(a) = \dfrac{x(a)(G_1(a) + G_3(a))G_2(a)}{1 - x(a)(G_1(a) + G_3(a))G_2(a)}$

67. For the control system presented in Figure 106, identify the model that best describes it.

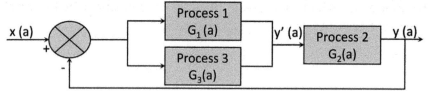

FIGURE 106. A block diagram presenting a control system.

(a) $y(a) = x(a)(G_1(a) + G_3(a))G_2$

(b) $y(a) = \dfrac{x(a)(G_1(a) + G_3(a))G_2(a)}{1 + x(a)(G_1(a) + G_3(a))G_2(a)}$

(c) $y(a) = x(a)(G_1(a) - G_2(a))G_3$

(d) $y(a) = \dfrac{x(a)(G_1(a) + G_3(a))G_2(a)}{1 - x(a)\left(G_1(a) + G_3(a)\right)G_2(a)}$

68. Considering the signals generated by the system, a continuous system to a logic one is like a(an):

(a) discrete system to a binary one
(b) deterministic system to a stochastic one
(c) approximated system to a stochastic one
(d) single-variable system to a multivariable one

69. Considering the signals generated by the system, a static system to a dynamic one is like a:

(a) discrete system to a binary one
(b) linear system to a static one
(c) stationary system to a transient one
(d) concentrated system to a stochastic one

70. Among the following models, which list includes all models that are applicable to control systems?

(a) Input-output relations, internal-external relations, Buckingham π, and experimental correlations
(b) experimental correlations, and Likert scale
(c) Input-output relations, Likert scale, and correlations
(d) Behavioral correlations and input-output

71. If $y(a) = x(a)G(a)$, $G(a)$ is:

(a) a transformer
(b) an input-output relator
(c) an operator
(d) none of the above

72. If $y(a) = x(a)G(a)$, $y(a)$ is:
    (a) transformed   (b) transferred
    (c) transmitted   (d) transverse

73. If the response of a system is defined by $y(t) = y_0 \sin(\omega t + \phi)$, y_0 is:
    (a) magnitude   (b) transverse frequency
    (c) transmitted phase   (d) transferred magnitude

74. If the response of a system is defined by $y(t) = y_0 \sin(\omega t + \phi)$, ϕ is:
    (a) magnitude   (b) phase
    (c) transmitted phase   (d) transferred magnitude

75. If the response of a system is defined by $y(t) = y_0 \sin(\omega t + \phi)$, ω is:
    (a) magnitude   (b) phase
    (c) frequency   (d) transverse frequency

76. Among the following, which one gives the Laplace transform that correctly matches the time-domain function?

    (a) $y(t-\tau) \underset{\text{Laplace}}{\Leftrightarrow} e^{ts}y(s)$

    (b) $ay_1(t) + by_2(t) \underset{\text{Laplace}}{\Leftrightarrow} ay_1(s) + by_2(s)$

    (c) $a\dfrac{dy_1(t)}{dt} + b\dfrac{dy_2(t)}{dt} \underset{\text{Laplace}}{\Leftrightarrow} a(sy_1(s) - y_1(0)) + b(sy_2(s) - y_2(0))$

    (d) $\displaystyle\int_0^\tau y(t)dt \underset{\text{Laplace}}{\Leftrightarrow} \tau\dfrac{y(s)}{s}$

77. The Laplace transform for $y(t) = \cos(at)$ is:

    (a) $y(s) = \dfrac{1}{s^2 + a^2}$ (b) $y(s) = \dfrac{1}{s+a}$ (c) $y(s) = \dfrac{1}{s^2 - a^2}$ (d) $y(s) = \dfrac{s}{a}$

78. The Laplace transform for $y(t) = y_0(e^{-at} + e^{bt})$ is:

    (a) $y(s) = \dfrac{1}{s+a+b}$ (b) $y(s) = \dfrac{1}{sab}$

    (c) $y(s) = \dfrac{s}{a+b}$ (d) $y(s) = \dfrac{2s+a-b}{(s+a)(s-b)}$

79. The inverse Laplace transform for $y(s) = 5s + (s-1)^{-1} - 2(s+2)^{-1} + 3(s-3)^{-1}$ is:

    (a) $y(t) = 5 + e^t - 2e^{-2t} + 3e^{3t}$ (b) $y(t) = -5t + e^t + 2e^{-2t} - 3e^{3t}$
    (c) $y(t) = 5t - e^{-t} - 2e^{2t} + 3e^{-3t}$ (d) $y(t) = 5 - e^{-t} + 2e^{2t} - 3e^{-3t}$

80. For the control system shown in Figure 107, select the best model.

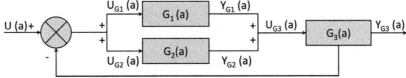

FIGURE 107. A block diagram presenting a thermostat system.

(a)
$$\begin{bmatrix} U_{G1}(a)G_1(a) = Y_{G1}(a) \\ U_{G2}(a)G_2(a) = Y_{G2}(a) \\ Y_{G1}(a) + Y_{G2}(a) = U_{G3}(a) \\ U_{G3}(a)G_3(a) = Y_{G3}(a) \\ U(a) - Y_{G3}(a) = U_{G1}(a) + U_{G2}(a) \end{bmatrix}$$

(b)
$$\begin{bmatrix} U(a) - U_{G2}(a) = U_{G1}(a) \\ U_{G3}(a)G_3(a) = Y_{G3}(a) \end{bmatrix}$$

(c)
$$\begin{bmatrix} U_{G1}(a)G_1(a) = Y_{G1}(a) \\ U_{G2}(a)G_2(a) = Y_{G2}(a) \\ Y_{G1}(a) + Y_{G2}(a) = U_{G3}(a) \end{bmatrix}$$

(d)
$$\begin{bmatrix} U_{G1}(a)G_1(a) = Y_{G1}(a) \\ U_{G2}(a)G_2(a) = Y_{G2}(a) \\ Y_{G1}(a) + Y_{G2}(a) = U_{G3}(a) \\ U_{G3}(a)G_3(a) = Y_{G3}(a) \end{bmatrix}$$

81. A designer would like to vary the air humidity level according to an exponential function shown in Figure 108 ($\phi(\tau) = 0.001e^{2\tau}$). If the humidity level is equal to ϕ_0 at time τ_0, which Laplace transform corresponds to this function?

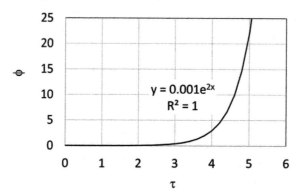

FIGURE 108. Dimensionless humidity level control diagram.

(a) $\phi(s) = 0.001 \dfrac{1}{s-1}$

(b) $\phi(s) = 0.001 \dfrac{1}{s-2}$

(c) $\phi(s) = 0.001 \dfrac{2}{s-2}$

(d) $\phi(s) = 0.001 \dfrac{2}{s-1}$

82. Servosystems correct system state disturbances by using the feedback to maintain the desired setpoint. (T/F)

83. Regulation modifies processes in order to control them and result in certain outputs. (T/F)

84. A stable controlled system will have a limited output for a given limited input. (T/F)

85. When an unstable system which is initially at equilibrium is disturbed, it always returns to the original state. (T/F)

86. Among the process block diagrams presented in Figure 109, the one that most accurately represents a weather system is:

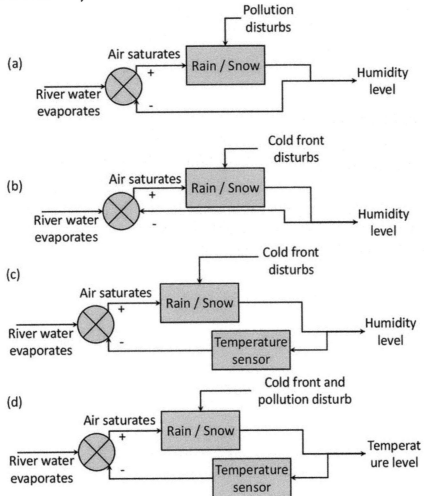

FIGURE 109. Control models presenting a weather system.

87. Among the control responses representing the block diagram shown in Figure 110, the one that most accurately represents a model for the given identified (G_i) processes is (subscript i represents any of the indicated processes):

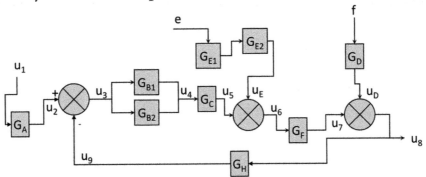

FIGURE 110. A control model presenting a rocket launch system.

(a) $u_8 = \dfrac{(G_{B1}+G_{B2})G_C G_F G_A}{1+(G_{B1}+G_{B2})G_C G_F G_H} u_1 + \dfrac{G_{E1}G_{E2}G_F}{1+(G_{B1}+G_{B2})G_C G_F G_H} e + \dfrac{G_D}{1+(G_{B1}+G_{B2})G_C G_F G_H} f$

(b) $u_8 = G_A \dfrac{(G_{B1}+G_{B2})G_C G_F}{1+(G_{B1}+G_{B2})G_C G_F G_H} u_1 + (G_{E1}G_{E2}) \dfrac{G_F}{1+(G_{B1}+G_{B2})G_C G_F G_H} e$

(c) $u_8 = G_A \dfrac{(G_{B1}+G_{B2})G_C G_F}{1+(G_{B1}+G_{B2})G_C G_F G_H} u_1 + G_D \dfrac{(G_{B1}+G_{B2})G_C G_F}{1+(G_{B1}+G_{B2})G_C G_F G_H} f$

(d) $u_8 = G_A \dfrac{(G_{B1}+G_{B2})G_C G_F}{1+(G_{B1}+G_{B2})G_C G_F G_H} u_2 + \dfrac{G_F}{1+(G_{B1}+G_{B2})G_C G_F G_H} u_E$

$\qquad + \dfrac{(G_{B1}+G_{B2})G_C G_F G_H}{1+(G_{B1}+G_{B2})G_C G_F G_H} u_D$

88. Among the following, the one that most accurately describes the steps a control mission takes to land a satellite on the Mars is:

    (a) detect and modify the position, and avoid obstacles
    (b) avoid obstacles, modify the position, and land
    (c) avoid obstacles and modify the position
    (d) detect and modify the position, avoid obstacles, repeat the cycle, and land

89. With regard to the block diagram shown in Figure 110, if both the process control operators B (G_{Bi}) (subscript i represents any of the indicated B-related processes) and C (G_C) have very large magnitudes, select the simplified control model that can best represent this diagram:

    (a) $u_8 = G_A \dfrac{G_F}{1+G_F G_H} u_1 + (G_{E1}G_{E2}) \dfrac{G_F}{1+G_F G_H} e$

    (b) $u_8 = \dfrac{G_A}{G_H} u_2 + \dfrac{1}{1+G_F G_H} u_E + G_D \dfrac{G_F G_H}{1+G_F G_H} f$

    (c) $u_8 = \dfrac{G_A}{G_H} u_1$

    (d) $u_8 = \dfrac{G_A}{G_H} u_2 + (G_{E1}G_{E2}) \dfrac{1}{1+G_F G_H} u_E$

90. In normal feedback control with integral, steady errors may be reduced, while with proportional and derivative controls, steady errors cause output to have a constant error relative to the set input. (T/F)

91. Select the diagram which would be a simplified representation of the diagram shown in Figure 110 if either of the process control operators B (G_{Bi}) (subscript i represents any of the indicated B-related processes) or C (G_C) have very large magnitudes.

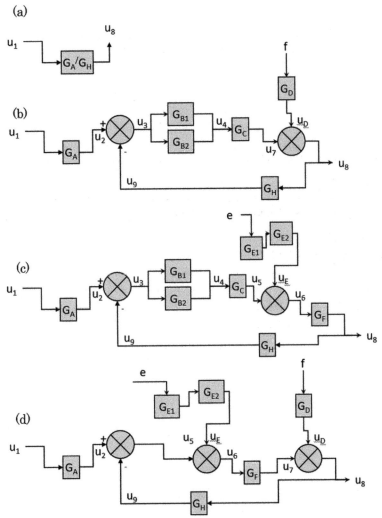

FIGURE 111. Control models presenting a simplified rocket launch system.

92. Performance of a position-control system in a steady state scenario cannot be evaluated by:

(a) best performance at its final position
(b) peak performance at a specific time
(c) time to reach its peak performance
(d) time to reach its optimized performance

93. An open-loop control system includes a regulatory system. (T/F)

94. Characteristics of proportional control are:

(a) faster response, more oscillations, decrease in steady-state error
(b) larger oscillations, elimination of steady-state error
(c) higher noise, transient error
(d) all of the above

95. Characteristics of integral control are:

(a) faster response, more oscillations, decrease in steady-state error
(b) larger oscillations, elimination of steady-state error
(c) higher noise, transient trend
(d) all of the above

96. Characteristics of derivative control are:
   (a) faster response, more oscillations, decrease in steady-state error
   (b) larger oscillations, elimination of steady-state error
   (c) higher noise, transient trend
   (d) all of the above

97. The error signal is used by the control system to generate its control input. (T/F) You can have zero error signal but nonzero control input. (T/F)
   (a) T / T   (b) T/ F   (c) F / T   (d) F/ F

98. Control signal is proportional to:
   (a) cumulative control error   (b) proportional control error
   (c) integrated control error   (d) time-derivative control error

99. Among the following statements, which one describes everything that an observer measures in a control system?
   (a) Raw variables, filtered variables, and disturbances
   (b) Disturbances only
   (c) Raw variables only
   (d) Non-filtered variables only

100. With regard to the system stability, which statement best applies:
   (a) The more stability, the slower response time
   (b) The less stability, the more trajectory control
   (c) The more stability, the less maneuverability control
   (d) All of the above

101. To operate an inherently unstable system, one may . . . to use control systems.
   (a) agree   (b) prefer not
   (c) wish   (d) decide

102. What are some of challenges of control systems?
   (a) Temporal and spatial limitations
   (b) Resource and communication limitations
   (c) Uncertainty, insecurity, and tolerance
   (d) All of the above

103. The accurate representation of the Laplace transformation for a control system is (where y' is the process variable deviation from its original value, y is the process variable original value, t' is the time deviation from a reference time, and t is the reference time point):

   (a) $Y(s) = \hat{y}(s) = \int_{t=0}^{t=\infty} e^{-s't'} y'(t)dt$ (b) $Y(s) = \hat{y}(s') = \int_{t'=0}^{t'=\infty} e^{-st'} y'(t')dt'$

   (c) $Y(s) = \hat{y}(s) = \int_{t'=0}^{t'=\infty} e^{-st'} y'(t')dt'$ (d) $Y(s) = \hat{y}(s) = \int_{t'=0}^{t'=\infty} e^{-s't'} y(t)dt'$

104. The exponential decay term in the Laplace transformation for a control system is (where t' is the time deviation from a reference time, and t is the reference time point):
   (a) e^{st}, from one to infinity   (b) $e^{st'}$, from one to infinity
   (c) e^{-st}, from zero to one   (d) $e^{-st'}$, from zero to one

105. The end of an arm (1-m radius) travels 1 m is 1 s. Angular velocity of the arm is:

(a) 60 rpm   (b) 30 rpm   (c) 0.11 rpm   (d) 9.55 rpm

106. Present the displacement vector for the spring-mass system presented in Figure 112 (assume masses are moving at constant speeds).

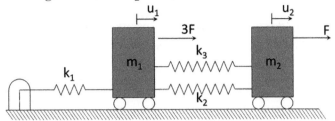

FIGURE 112. Spring-mass system.

(a) $\begin{bmatrix} -k_1-k_2-k_3 & k_2+k_3 \\ k_2+k_3 & -k_2-k_3 \end{bmatrix}\begin{bmatrix} u_1 \\ u_2 \end{bmatrix} = F\begin{bmatrix} 0 \\ -1 \end{bmatrix}$ (b) $\begin{bmatrix} k_1+k_2+k_3 & k_2+k_3 \\ -k_2-k_3 & -k_2-k_3 \end{bmatrix}\begin{bmatrix} u_1 \\ u_2 \end{bmatrix} = -F\begin{bmatrix} 0 \\ -1 \end{bmatrix}$

(c) $\begin{bmatrix} -k_1+k_2-k_3 & -k_2-k_3 \\ -k_2+k_3 & k_2+k_3 \end{bmatrix}\begin{bmatrix} u_1 \\ u_2 \end{bmatrix} = F\begin{bmatrix} 0 \\ 1 \end{bmatrix}$ (d) $\begin{bmatrix} k_1+k_2-k_3 & k_2+k_3 \\ k_2-k_3 & -k_2-k_3 \end{bmatrix}\begin{bmatrix} u_1 \\ u_2 \end{bmatrix} = F\begin{bmatrix} 0 \\ -1 \end{bmatrix}$

Solution Guide

Note that the total force has to resist and be equal to the spring forces and move masses at a constant velocity, resulting in zero accelerations.

$$\sum_{i=1}^{n} F_i = \sum_{i=1}^{n} m_i \ddot{x}_i, \ddot{x}_i = 0$$

$$\therefore \begin{cases} -F_1 + F_2 + F_3 = 0 \\ -F_2 - F_3 + F = 0 \end{cases}$$

$$\therefore \begin{cases} -k_1 u_1 + k_2(u_2 - u_1) + k_3(u_2 - u_1) = 0 \\ -k_2(u_2 - u_1) - k_3(u_2 - u_1) + F = 0 \end{cases}$$

$$\therefore \begin{cases} -(k_1 + k_2 + k_3)u_1 + (k_2 + k_3)u_2 = 0 \\ (k_2 + k_3)u_1 - (k_2 + k_3)u_2 = 0 \end{cases}$$

$$\therefore \begin{bmatrix} -k_1-k_2-k_3 & k_2+k_3 \\ k_2+k_3 & -(k_2+k_3) \end{bmatrix}\begin{bmatrix} u_1 \\ u_2 \end{bmatrix} = F\begin{bmatrix} 0 \\ -1 \end{bmatrix}$$

107. Present the rate of energy (u) for the pressure-flow system presented in Figure 113. Pressure (P) as a function of flow (Q) is presented as $P = -0.04Q^2 + 40Q$ (assume $0 < Q < 500$ m³/s).

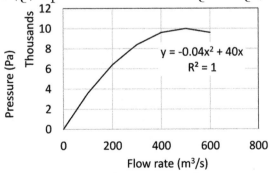

FIGURE 113. Pressure versus flow.

(a) 3.33 MW   (b) 1.67 MW   (c) 6.76 MW   (d) 5.1 MW

Solution Guide

Note that the area under the *P-Q* diagram should be equal to the rate of energy.

$$u = \int_a^b PdQ = \int_0^{500}(-0.04Q^2+40Q)dQ = \left[\frac{-0.04}{3}Q^3 + \frac{40}{2}Q^2\right]_0^{500}$$

108. Present the rate of co-energy (*u**) for the pressure-flow system presented in Figure 113. Pressure (*P*) as a function of flow (*Q*) is presented as $P = -0.04Q^2 + 40Q$ (assume $0 < Q <$ 500 m³/s).

(a) 3.33 MW   (b) 1.67 MW   (c) 6.76 MW   (d) 5.1 MW

Solution Guide

Note that the area under the Q-P diagram should be equal to the rate of co-energy.

$$u* = \int_a^b QdP = \int_0^{500}Q(-0.08Q+40)dQ = \left[\frac{-0.08}{3}Q^3 + \frac{40}{2}Q^2\right]_0^{500}$$

109. For a control system causing a displacement (δ) in the opposite direction to the force (*F*), energy (*u*) and co-energy (*u**) are:

(a) $u = F\delta, u* = F\delta$ (b) $u = 0, u* = F\delta$
(c) $u = -F\delta, u* = 0$ (c) $u = -F\delta, u* = -F\delta$

110. For a spring-mass system, causing a displacement (δ) in the direction of the force (*F*), energy (*u*), and co-energy (*u**) are (spring constant is *k*):

(a) $u = \frac{F^2}{2k}, u* = \frac{k\delta^2}{2}$ (b) $u = \frac{k\delta^2}{2}, u* = \frac{k\delta^2}{2}$

(c) $u = \frac{F^2}{2k}, u* = \frac{F^2}{2k}$ (d) $u = \frac{k\delta^2}{2}, u* = \frac{F^2}{2k}$

111. For the spring-mass system presented in Figure 114, using the co-energy approach, present a relationship between k_1, k_2, k_3, and forces applied by springs 1 (F_1), 2 (F_2), and 3 (F_3). Assume the system is at equilibrium.

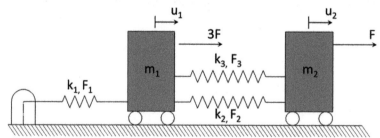

FIGURE 114. Spring-mass system.

(a) $F_1 = 4F, F_2 = \frac{k_2}{k_2+k_3}F, F_3 = \frac{k_3}{k_2+k_3}F$ (b) $F_1 = 4F, F_2 = \frac{k_2}{k_2-k_3}F, F_3 = \frac{k_3}{k_2-k_3}F$

(c) $F_1 = 2F, F_2 = -\frac{k_1}{k_2+k_3}F, F_3 = -\frac{k_2}{k_2+k_3}F$ (d) $F_1 = 2F, F_2 = -\frac{k_1}{k_2-k_3}F, F_3 = -\frac{k_2}{k_2+k_3}F$

Solution Guide

Spring forces need to balance the applied forces. For the spring-mass system with displacement (δ) in the direction of the force source (F) and spring constant k, the co-energy is $u_i* = \dfrac{F_i^2}{2k_i}$. The total co-energy ($u*$) is equal to the summation of the co-energies of the springs.

$$\sum_{i=1}^{n} u^*_i = \sum_{i=1}^{n} \frac{F_i^2}{2k_i}$$

$$\therefore \begin{cases} -F_1 + F_2 + F_3 + 3F = 0 \\ -F_2 - F_3 + F = 0 \end{cases}$$

$$\therefore \begin{cases} F_1 = F_2 + F_3 + 3F \\ F_2 + F_3 = F \end{cases}$$

$$\therefore \begin{cases} F_1 = 4F \\ F_3 = F - F_2 \end{cases}$$

$$\therefore u* = \frac{F_2^2}{2k_2} + \frac{F_3^2}{2k_3} + \frac{(F_2 + F_3)^2}{2k_1}$$

$$\therefore u* = \frac{F_2^2}{2k_2} + \frac{(F - F_2)^2}{2k_3} + \frac{(4F)^2}{2k_1}$$

$$\therefore \frac{\partial u*}{\partial F_2} = \frac{F_2}{k_2} - \frac{(F - F_2)}{k_3} = 0$$

$$\therefore F_2(k_3 + k_2) - Fk_2 = 0$$

$$\therefore F_2 = \frac{k_2}{k_2 + k_3} F$$

112. For the spring-mass system presented in Figure 115, calculate eigenvalues λ_1, λ_2, and λ_3 (assume the system's frequency and amplitude are ω and x, $u = xe^{iwt}$).

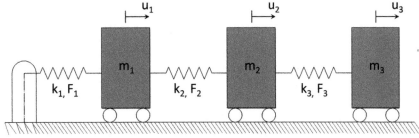

FIGURE 115. Spring-mass system.

(a) $\lambda_1 = 0.7, \lambda_2 = 0.9, \lambda_3 = 4.1$ (b) $\lambda_1 = 0.2, \lambda_2 = 1.6, \lambda_3 = 3.3$

(c) $\lambda_1 = 0.4, \lambda_2 = 2.1, \lambda_3 = 6.5$ (d) $\lambda_1 = 0.9, \lambda_2 = 0.8, \lambda_3 = 1.6$

Solution Guide

The balance of forces for each mass determines its acceleration. The next step is to define new displacement variables to replace the current ones ($u_1 = x_1 e^{iwt}, u_2 = x_2 e^{iwt}, u_3 = x_3 e^{iwt}$, where $e^{iwt} \neq 0$).

$$\sum_{i=1}^{n} F_i = \sum_{i=1}^{n} m_i \ddot{u}_i$$

$$\therefore \begin{cases} -k_1 u_1 + k_2(u_2 - u_1) = m_1 \ddot{u}_1 \\ -k_2(u_2 - u_1) + k_3(u_3 - u_2) = m_2 \ddot{u}_2 \\ -k_3(u_3 - u_2) = m_3 \ddot{u}_3 \end{cases}$$

$$\therefore \begin{cases} -(k_1 + k_2)u_1 + k_2 u_2 = m_1 \ddot{u}_1 \\ k_2 u_1 - (k_2 + k_3)u_2 + k_3 u_3 = m_2 \ddot{u}_2 \\ k_3 u_2 - k_3 u_3 = m_3 \ddot{u}_3 \end{cases}$$

$$\therefore \begin{bmatrix} -(k_1 + k_2) & k_2 & 0 \\ k_2 & -(k_2 + k_3) & k_3 \\ 0 & k_3 & -k_3 \end{bmatrix} \begin{bmatrix} u_1 \\ u_2 \\ u_3 \end{bmatrix} = \frac{d^2}{dt^2} \begin{bmatrix} m_1 & 0 & 0 \\ 0 & m_2 & 0 \\ 0 & 0 & m_3 \end{bmatrix} \begin{bmatrix} u_1 \\ u_2 \\ u_3 \end{bmatrix}$$

Assume: $u_1 = x_1 e^{iwt}, u_2 = x_2 e^{iwt}, u_3 = x_3 e^{iwt}, e^{iwt} \neq 0$

$$\therefore \begin{bmatrix} -m_1 w^2 + (k_1 + k_2) & -k_2 & 0 \\ -k_2 & -m_2 w^2 + (k_2 + k_3) & -k_3 \\ 0 & -k_3 & -m_3 w^2 + k_3 \end{bmatrix} \begin{bmatrix} x_1 \\ x_2 \\ x_3 \end{bmatrix} e^{iwt} = \begin{bmatrix} 0 \\ 0 \\ 0 \end{bmatrix}$$

$$\therefore e^{iwt} \neq 0$$

$$\therefore \begin{bmatrix} -m_1 w^2 + (k_1 + k_2) & -k_2 & 0 \\ -k_2 & -m_2 w^2 + (k_2 + k_3) & -k_3 \\ 0 & -k_3 & -m_3 w^2 + k_3 \end{bmatrix} \begin{bmatrix} x_1 \\ x_2 \\ x_3 \end{bmatrix} = \begin{bmatrix} 0 \\ 0 \\ 0 \end{bmatrix}$$

$$\therefore \begin{bmatrix} \dfrac{-m_1 w^2 + (k_1 + k_2)}{k_1} & -k_2/k_1 & 0 \\ -1 & \dfrac{-m_2 w^2 + (k_2 + k_3)}{k_2} & -k_3/k_2 \\ 0 & -1 & \dfrac{-m_2 w^2 + k_3}{k_3} \end{bmatrix} \begin{bmatrix} x_1 \\ x_2 \\ x_3 \end{bmatrix} = \begin{bmatrix} 0 \\ 0 \\ 0 \end{bmatrix}$$

$$\therefore \begin{cases} \dfrac{-m_1 w^2 + (k_1 + k_2)}{k_1} x_1 - \dfrac{k_2}{k_1} x_2 = 0 \\ -x_1 + \dfrac{-m_2 w^2 + (k_2 + k_3)}{k_2} x_2 - \dfrac{k_3}{k_2} x_3 = 0 \\ -x_2 + \dfrac{-m_3 w^2 + k_3}{k_3} x_3 = 0 \end{cases}$$

$$\therefore \lambda_i = \frac{m_i w^2}{k_i}$$

$$\therefore \begin{cases} \left[-\lambda_1 + \left(1 + \dfrac{k_2}{k_1}\right)\right]x_1 - \dfrac{k_2}{k_1}x_2 = 0 \\[4mm] -x_1 + \left[-\lambda_2 + \left(1 + \dfrac{k_3}{k_2}\right)\right]x_2 - \dfrac{k_3}{k_2}x_3 = 0 \\[4mm] -x_2 + (-\lambda_3 + 1)x_3 = 0 \end{cases}$$

$$\therefore k_1 = k_2 = k_3, m_1 = m_2 = m_3$$

$$\therefore \begin{bmatrix} -\lambda+2 & -1 & 0 \\ -1 & -\lambda+2 & -1 \\ 0 & -1 & -\lambda+1 \end{bmatrix} \begin{bmatrix} x_1 \\ x_2 \\ x_3 \end{bmatrix} = \begin{bmatrix} 0 \\ 0 \\ 0 \end{bmatrix}$$

Trivial solutions:

$$\therefore x_1 = \frac{\begin{vmatrix} 0 & -1 & 0 \\ 0 & -\lambda+2 & -1 \\ 0 & -1 & -\lambda+1 \end{vmatrix}}{\begin{vmatrix} -\lambda+2 & -1 & 0 \\ -1 & -\lambda+2 & -1 \\ 0 & -1 & -\lambda+1 \end{vmatrix}} = 0$$

$$\therefore x_2 = \frac{\begin{vmatrix} -\lambda+2 & 0 & 0 \\ -1 & 0 & -1 \\ 0 & 0 & -\lambda+1 \end{vmatrix}}{\begin{vmatrix} -\lambda+2 & -1 & 0 \\ -1 & -\lambda+2 & -1 \\ 0 & -1 & -\lambda+1 \end{vmatrix}} = 0$$

$$\therefore x_3 = \frac{\begin{vmatrix} -\lambda+2 & -1 & 0 \\ -1 & -\lambda+2 & 0 \\ 0 & -1 & 0 \end{vmatrix}}{\begin{vmatrix} -\lambda+2 & -1 & 0 \\ -1 & -\lambda+2 & -1 \\ 0 & -1 & -\lambda+1 \end{vmatrix}} = 0$$

To calculate the non-trivial solution, set the determinant to zero to calculate the eigenvalues.

$$\therefore \begin{vmatrix} -\lambda+2 & -1 & 0 \\ -1 & -\lambda+2 & -1 \\ 0 & -1 & -\lambda+1 \end{vmatrix} = 0$$

$$\therefore (-\lambda+2)(-\lambda+2)(-\lambda+1)-(-\lambda+1)-(-\lambda+2)=0$$
$$\therefore -\lambda^3+5\lambda^2-6\lambda+1=0$$
$$\therefore \begin{cases} \lambda_1=0.198 \\ \lambda_2=1.56 \\ \lambda_3=3.25 \end{cases}$$

The ratio of the non-trivial solutions, given each eigenvalue, can be calculated.

$$\frac{x_1}{x_2}=\frac{1}{-\lambda+2}, \frac{x_3}{x_2}=\frac{1}{-\lambda+1}$$

Answer Key									
1. (c)	**2.** (b)	**3.** (a)	**4.** (d)	**5.** (b)	**6.** (a)	**7.** (a)	**8.** (d)	**9.** (a)	**10.** (c)
11. (d)	**12.** (c)	**13.** (c)	**14.** F	**15.** (a)	**16.** (a)	**17.** F	**18.** T	**19.** (b)	**20.** (d)
21. F	**22.** T	**23.** (a)	**24.** (b)	**25.** F	**26.** T	**27.** (b)	**28.** (d)	**29.** (a)	**30.** (c)
31. (d)	**32.** F	**33.** (b)	**34.** (a)	**35.** (b)	**36.** (b)	**37.** (b)	**38.** (b)	**39.** T	**40.** T
41. (d)	**42.** T	**43.** (d)	**44.** (d)	**45.** (a)	**46.** (b)	**47.** (a)	**48.** (a)	**49.** (c)	**50.** (a)
51. (b)	**52.** (d)	**53.** (b)	**54.** (a)	**55.** (a)	**56.** T	**57.** (a)	**58.** F	**59.** (c)	**60.** (a)
61. F	**62.** (b)	**63.** (c)	**64.** (d)	**65.** (a)	**66.** (a)	**67.** (b)	**68.** (a)	**69.** (c)	**70.** (a)
71. (c)	**72.** (a)	**73.** (a)	**74.** (b)	**75.** (c)	**76.** (c)	**77.** (a)	**78.** (d)	**79.** (a)	**80.** (a)
81. (b)	**82.** T	**83.** T	**84.** T	**85.** F	**86.** (b)	**87.** (a)	**88.** (d)	**89.** (c)	**90.** T
91. (a)	**92.** (a)	**93.** F	**94.** (a)	**95.** (b)	**96.** (c)	**97.** (a)	**98.** (a)	**99.** (a)	**100.** (a)
101. (b)	**102.** (d)	**103.** (c)	**104.** (d)	**105.** (d)	**106.** (a)	**107.** (a)	**108.** (b)	**109.** (c)	**110.** (d)
111. (a)	**112.** (b)								

CHAPTER 6

THERMODYNAMICS

1. Work is defined as a product of the changes in:
   (a) pressure and temperature (b) temperature and density
   (c) pressure and volume (d) volume and density

2. If the work done by an ideal gas is the same for an isobaric and isochoric process where the system state changes from 1 to 2, the following relation is valid (P is pressure and V is volume):
   (a) $P_1V_1 = P_2V_2$ (b) $P_1V_2 = P_2V_1$ (c) $P_1V_1^\gamma = P_2V_1^\gamma$ (d) $P_1V_2^\gamma = P_2V_1^\gamma$

3. An ideal gas system state changes from 1 to 2 while performing zero work. If P is pressure and V is volume, the following relation is valid:
   (a) $P_1V_1 = P_2V_2$ (b) $P_1V_2 = P_2V_1$ (c) $P_1V_1^\gamma = P_2V_1^\gamma$ (d) $P_1V_2^\gamma = P_2V_1^\gamma$

4. You decide to store rice in a vacuum container. To achieve a partial vacuum, you use a handpump. If the container's air pressure is initially at 1 atm and 25 °C, the final temperature is expected to be:
   (a) 25 °C (b) below 25 °C (c) above 25 °C (d) 0 °C

5. You are given a helium balloon for your birthday and decide to carry out some experiments. You attach a camera to the balloon and let it go inside a building with large open space with perfectly controlled constant temperature throughout. As it rises toward the ceiling, you will observe that the balloon:
   (a) becomes smaller
   (b) becomes larger
   (c) remains the same
   (d) becomes larger or smaller, depending on the temperature change rate with altitude

6. You are given a helium balloon for your birthday and decide to carry out some experiments. You attach a camera to the balloon and let it go in the open air this time. From your studies, you know that temperature decreases as you go up in the atmosphere. Thus, as it goes up, you expect that the balloon (assume pressure is directly related to the temperature):
   (a) becomes smaller
   (b) becomes larger

(c) remains the same size

(d) becomes larger or smaller, depending on the temperature change rate with altitude

7. You take an aluminum can of soft drink at standard atmospheric conditions and submerge it into liquid nitrogen. You observe that the can:

(a) explodes   (b) collapses

(c) remains the same   (d) expands

8. If volume and pressure of an ideal gas in an enclosure are doubled in a rapid process, its temperature is:

(a) quadrupled   (b) the same   (c) halved   (d) quartered

9. If the pressure and temperature of an ideal gas are doubled in a rapid process, its volume is:

(a) quadrupled   (b) the same   (c) halved   (d) doubled

10. Two ideal gases at thermal equilibrium are kept in two different containers, with one container double the size of the other. Assuming that they have the same pressure, the number of moles in the larger container is . . . that of the smaller one.

(a) double   (b) half   (c) quarter   (d) the same as

11. For a gas occupying a large container, lowering the temperature causes the molecular velocity to . . . ; furthermore, . . . collisions are made between the molecules and the pressure . . .

(a) decrease / more / decreases   (b) increase / fewer / increases

(c) increase / more / increases   (d) decrease / fewer / decreases

12. Ten moles of an ideal gas at 100 °C fill a spherical ball with the 1-m radius. Gas pressure is:

(a) 7.4 kPa   (b) 7.4 Pa   (c) 2.0 kPa   (d) 2.0 Pa

13. Twenty moles of an ideal gas at 100 K fill a cylindrical container with the 0.5-m diameter and 1-m height. The total magnitude of forces applied by the gas on the two circular ends of the cylinder is:

(a) 166.3 N   (b) 33.26 kN   (c) 84.7 kN   (d) 84.7 N

14. One hundred moles of an ideal gas at 200 K fill a cubic container with the 2-m side length. If the work done by gas on the system is 100 kJ, pressure (P), volume change (ΔV), and temperature increase (ΔT) are (assume pressure is kept constant):

(a) $P = 0.21$ Pa, $\Delta V = 0.05$ m^3, $\Delta T = 1.2$ K

(b) $P = 2.1$ kPa, $\Delta V = 0.5$ m^3, $\Delta T = 12.0$ K

(c) $P = 20.8$ kPa, $\Delta V = 4.8$ m^3, $\Delta T = 120.3$ K

(d) $P = 208$ kPa, $\Delta V = 48.1$ m^3, $\Delta T = 1202.8$K

15. There are 5×10^9 atoms in a container with 2 square faces (1 mm × 1 mm × 0.5 mm). If pressure is kept at 0.3 Pa, temperature is (Boltzmann constant is 1.38×10^{-23} J/K):

(a) 21,739.0 K   (b) 217.4 K   (c) 21.7 K   (d) 2,173.9 K

16. The number of atoms in 200 moles of an ideal gas is:

(a) 1.205×10^{26} (b) 1.205×10^{23} (c) 6.025×10^{26} (d) 6.025×10^{23}

17. Energy associated with each quadratic degree of freedom for a single degree of freedom of gas at 1,000 K is:

(a) 2.76×10^{-20} J   (b) 1.38×10^{-20} J   (c) 2.07×10^{-20} J   (d) 6.9×10^{-21} J

18. For an isothermal system, the total change in internal energy, where heat capacity remains constant, is:

(a) zero

(b) twice the work done

(c) half of the total energy exchanged

(d) reduced by the amount of the work done on the system

19. A thermodynamic process with no change in pressure is:

(a) isobaric   (b) isothermal   (c) isochoric   (d) isentropic

20. In an isobaric process, work is defined by (P is pressure and V is volume):

(a) PdV (b) VdP (c) $d(PV)$ (d) PV

21. A thermodynamic process with no change in volume is:

(a) isobaric   (b) isothermal   (c) isochoric   (d) isentropic

22. A thermodynamic process with no change in temperature is:

(a) isobaric   (b) isothermal   (c) isochoric   (d) isentropic

23. A quasi-static, adiabatic, and reversible process is:

(a) isobaric   (b) isothermal   (c) isochoric   (d) isentropic

24. In an adiabatic process, there is:

(a) no heat transfer   (b) some work done by the system

(c) energy transfer   (d) heat transfer

25. If objects A and B are at thermal equilibrium with object C, objects A and B are . . . equilibrium with one another.

(a) not at thermal   (b) at thermal

(c) not at mechanical   (d) at mechanical

26. The zeroth law of thermodynamics states that if objects A and B are:

(a) at thermal equilibrium with object C, objects A and B are at thermal equilibrium with one another

(b) at thermal equilibrium with object C, objects A and B are not at thermal equilibrium with one another

(c) not at thermal equilibrium with object C, objects A and B are at thermal equilibrium with one another

(d) not at thermal equilibrium with object C, objects A and B are not at thermal equilibrium with one another

27. Entropy is a measure of:

(a) temperature   (b) energy   (c) pressure   (d) disorder

28. The second law of thermodynamics states that thermal energy flows from a region with:

(a) lower temperature to that of higher temperature

(b) higher temperature to that of lower temperature

(c) lowest temperature to that of lower temperature

(d) higher temperature to that of highest temperature

29. Based on the second law of thermodynamics, all mechanical systems are highly efficient (over 90%). (T/F)

30. An isolated system has . . . while an adiabatic system has . . . crossing their boundaries.

(a) neither energy nor mass / mass but no energy

(b) both energy and mass / neither mass nor energy

(c) energy but no mass / no mass but energy

(d) no energy but mass / both mass and energy

31. A 114-g mass of nitrogen is contained in a cylinder with a moving piston. Gas is initially at 1 atm and 300 K within a 0.1 m³ volume. The piston moves in an isobaric process to expand the volume to 0.2 m³. For this process to take place, 35.4 kJ of heat is added to the system. The added heat raises the molecular temperature from the initial temperature. If specific heat capacity at constant pressure is 1.04 kJ/kgK, final temperature (T) and change of internal energy (dU) are:

(a) $dT = 298$ K, $dU = 23.5$ kJ   (b) $dT = 305$ K, $dU = 27.5$ kJ
(c) $dT = 303$ K, $dU = 21.3$ kJ   (d) $dT = 300$ K, $dU = 25.3$ kJ

32. The work done by gas in the process from A to B shown in Figure 116 is:

(a) 1 kJ   (b) 1.75 kJ   (c) 0.75 kJ   (d) 0 kJ

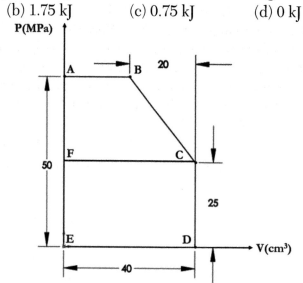

FIGURE 116. Pressure versus volume.

33. The work done by gas in the process from A to C shown in Figure 116 is:

(a) 1 kJ   (b) 1.75 kJ   (c) 0.25 kJ   (d) 0.40 kJ

34. The work done by gas in the process from A to F to E shown in Figure 116 is:

(a) 1 kJ   (b) 1.75 kJ   (c) 0.25 kJ   (d) 0 kJ

35. The work done by gas in the process from A to B to C to F to A shown in Figure 116 is:

(a) 1 kJ   (b) 1.75 kJ   (c) 0.75 kJ   (d) 25 kJ

36. Tires lose on the average 1.5 PSI pressure for every 10 °F temperature increase. Assume you have pressurized your 2020 Ferrari 488 Pista tires based on the recommended values suggested for Alaska (30 PSI and 29 PSI for front and rear tires). Front and rear tire sizes are 245/35R20 (26.8 × 9.6 × R20) and 305/30R20 (27.2 × 12 × R20). Tire dimensions given in brackets are in inches: (d_o, d, d_i), where d_o and d_i are outside and inside diameters and d is the tire width. If the initial adjustment was made in winter, when the temperature was −60 °F, the pressure in summer, when the temperature is 60 °F, will adhere to the recommended value if each front tire is . . . by . . . moles and each rear tire is . . . by . . . moles (1 in³ = 0.00001639 m³, 1 PSI = 6,894.76 Pa) [6,7].

(a) deflated / 3 / deflated / 4   (b) inflated / 3 / deflated / 40
(c) deflated / 30 / inflated / 40   (d) inflated / 30 / inflated / 4

37. Tires lose on the average 1.5 PSI pressure for every 10 °F temperature change. Assume you have pressurized your 2015 Honda Civic (P195/65R15, 25 × 7.9 × R15) tires based on the recommended values suggested for Los Angeles (30 PSI for front and rear tires). Tire dimensions given in brackets are in inches: (d_o, d, d_i), where d_o and d_i are outside and inside diameters and d is the tire width. If the initial adjustment was made in summer, when the temperature was 75 °F, the pressure in winter, when the temperature is 57 °F, will adhere to the recommended value if each tire is ... by ... moles (1 in^3 = 0.00001639 m^3, 1 PSI = 6894.76 Pa).

(a) deflated / 0.44   (b) inflated / 0.44
(c) deflated / 4.4   (d) inflated / 4.4

38. Tires lose on the average 1.5 PSI pressure for every 10 °F temperature change. Assume you have pressurized your 2005 Volvo XC70 (215/65R16, 27 × 8.7 × R16), with tires based on the recommended values suggested for New York (38 PSI for front and rear tires). Tire dimensions given in brackets are in inches: (d_o, d, d_i), where d_o and d_i are outside and inside diameters and d is the tire width. The initial adjustment was made in June, when the temperature was 77 °F. When the season changed, to maintain the recommended tire pressure, the owner inflated each tire by 1.6 moles. The temperature in which this adjustment is made is (1 in^3 = 0.00001639 m^3, 1 PSI = 6894.76 Pa):

(a) 55 °C   (b) 25 °C   (c) 0 °C   (d) –25 °C

39. Tires lose on the average 1.5 PSI pressure for every 10 °F temperature change. Assume you have pressurized your 1997 Aston Martin DB7 Volante tires based on the recommended values suggested for the state of Washington (34 PSI for front and rear tires). Front and rear tire sizes are 245/40ZR18 (25.7 × 9.8 × R18) and 265/35ZR18 (25.3 × 10.7 × R18). Tire dimensions given in brackets are in inches: (d_o, d, d_i), where d_o and d_i are outside and inside diameters and d is the tire width. If this initial adjustment was made in December, when the temperature was 38 °F, to maintain the tire pressures in mid-July (80 °F), the owner should . . . each tire by 1.1 mole. The front tire recommended pressure is . . . and the back tire recommended pressure is . . . (1 in^3 = 0.00001639 m^3, 1 PSI = 6894.76 Pa).

(a) inflate / 27.7 PSI / 31.5 PSI   (b) deflate / 31.5 PSI / 27.7 PSI
(c) inflate / 40.3 PSI / 40.3 PSI   (d) deflate / 27.7 PSI / 27.7 PSI

40. An Otto cycle consists of:
(a) two isochoric and two isentropic processes
(b) two isobaric and two isentropic processes
(c) two isobaric and two isochoric processes
(d) two isobaric and two isothermal processes

41. An Otto cycle consists of, in order:
(a) isentropic expansion, isochoric increase in pressure, isentropic compression, and isochoric decrease in pressure
(b) isentropic compression, isochoric decrease in pressure, isentropic compression, and isochoric increase in pressure
(c) isentropic expansion, isochoric decrease in pressure, isentropic expansion, and isochoric increase in pressure
(d) isentropic compression, isochoric increase in pressure, isentropic expansion, and isochoric decrease in pressure

42. Internal combustion engines follow the:
(a) Otto cycle   (b) Brayton cycle   (c) Rankine cycle   (d) Carnot cycle

43. Identify each cycle in Figure 117.

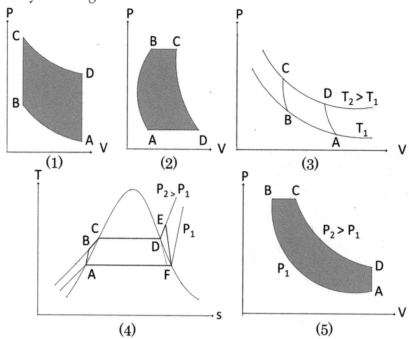

FIGURE 117. Thermodynamic cycles.

(a) (1) Otto, (2) Brayton, (3) Carnot, (4) Rankine, (5) Diesel
(b) (2) Otto, (1) Brayton, (3) Carnot, (4) Rankine, (5) Diesel
(c) (3) Otto, (4) Brayton, (5) Carnot, (1) Rankine, (2) Diesel
(d) (4) Otto, (5) Brayton, (1) Carnot, (2) Rankine, (3) Diesel

44. An Otto cycle is presented in Figure 118. Processes A to B and C to D, are adiabatic and reversible. Using the data presented in Table 46, find the temperature, volume, and pressure values missing from the table. There are 0.054 moles of air participating in the process ($K = 1.4$).

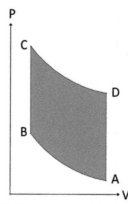

FIGURE 118. Pressure versus volume in an Otto cycle.

TABLE 46. Pressure, temperature, and volume data for the Otto cycle presented in Figure 118.

Process	T (K)	V (m³)	P (kPa)
A		0.000975	101
B			1,615
C			
D	700		

Process	T (K)	V (m³)	P (kPa)
A	12	0.000975	101
B	26	0.000135	1,615
C	1,545	0.000135	95,445
D	700	0.000975	5,969

(a)

Process	T (K)	V (m³)	P (kPa)
A	219	0.000975	101
B	484	0.000135	1,615
C	1,545	0.000135	5,154
D	700	0.000975	322

(b)

Process	T (K)	V (m³)	P (kPa)
A	24	0.000975	101
B	52	0.000135	1,615
C	1,545	0.000135	47,723
D	700	0.000975	2,985

(c)

Process	T (K)	V (m³)	P (kPa)
A	16	0.000975	101
B	35	0.000135	1,615
C	1,545	0.000135	71,584
D	700	0.000975	4,477

(d)

45. Identify heat and work for the four processes of the Otto cycle presented in Figure 118, using the data presented in Table 47. Processes A to B and C to D, are adiabatic and reversible. There is 1 mole of air participating in the process ($K = 1.4$).

TABLE 47. Pressure, temperature, and volume data for the Otto cycle presented in Figure 118.

Process	T (K)	V (m³)	P (kPa)
A		0.00195	202
B			3,230
C			
D	1,050		

Process	dW (kJ)	dU (kJ)	dQ (kJ)
A-B	-1.189	1.189	0.000
B-C	0.000	33.963	33.963
C-D	19.769	-19.769	0.000
D-A	0.000	-15.383	-15.383

(a)

Process	dW (kJ)	dU (kJ)	dQ (kJ)
A-B	-1.189	1.189	0.000
B-C	0.000	21.918	21.918
C-D	13.180	-13.180	0.000
D-A	0.000	-9.927	-9.927

(b)

Process	dW (kJ)	dU (kJ)	dQ (kJ)
A-B	-1.189	1.189	0.000
B-C	0.000	46.009	46.009
C-D	26.359	-26.359	0.000
D-A	0.000	-20.840	-20.840

(c)

Process	dW (kJ)	dU (kJ)	dQ (kJ)
A-B	-1.189	1.189	0.000
B-C	0.000	9.872	9.872
C-D	6.590	-6.590	0.000
D-A	0.000	-4.471	-4.471

(d)

46. Identify the total heat and work for the four processes of the Otto cycle presented in Figure 118, using the data presented in Table 48. Processes A to B and C to D, are adiabatic and reversible. There are 0.5 moles of air participating in the process ($K = 1.4$).

TABLE 48. Pressure, temperature, and volume data for the Otto cycle presented in Figure 118.

Process	T (K)	V (m³)	P (kPa)
A		0.0013	300
B			1,200
C			
D	1,000		

(a) 9.63 J, 9.63 J   (b) 4.58 kJ, 4.58 kJ
(c) 4.58 kJ, 9.63 kJ   (d) 9.63 kJ, 4.58 kJ

47. The efficiency of the Otto cycle presented in Table 49 (Figure 118) is . . . . There are 0.4 moles of air participating in the process ($K = 1.4$).

TABLE 49. Pressure, temperature, and volume data for the Otto cycle presented in Figure 118.

Process	T (K)	V (m³)	P (kPa)
A	117	0.0013	299
B	212	0.000294	2,398
C	1,200	0.000294	13,574
D	1,000	0.001300	2,558

(a) 10.6%   (b) 91.7%   (c) 81.1%   (d) 21.9%

48. The compression ratio of the Otto cycle presented in Table 50 (Figure 118) is . . . . There are 0.25 moles of air participating in the process ($K = 1.4$).

TABLE 50. Pressure, temperature, and volume data for the Otto cycle presented in Figure 118.

Process	T (K)	V (m³)	P (kPa)
A	47	0.000975	160
B	105	0.000135	2,587
C	1,100	0.000135	27,097
D	700	0.000975	2,388

(a) 7.2   (b) 1   (c) 0.14   (d) 0

49. Heat is an indication of flow of the . . . energy within a medium; temperature is related to the . . . energy of the molecules.
(a) thermal / kinetic   (b) kinetic / thermal
(c) thermal / thermal   (d) kinetic / kinetic

50. Select the option which lists temperatures in order from the coldest to the warmest:
(a) 100 K < 150 °R < −100 °C < 0 °F   (b) 150 °R < 100 K < −100 °C < 0 °F
(c) 150 °R < 100 K < 0 °F < −100 °C   (d) −100 °C < 0 °F < 100 K < 150 °R

51. Construct a PV diagram for the data presented in Table 51.

TABLE 51. Pressure, temperature, and volume data.

Process	P (kPa)	V (m³)
A	110	0.002
B	220	0.004
C	110	0.008
D	55	0.016

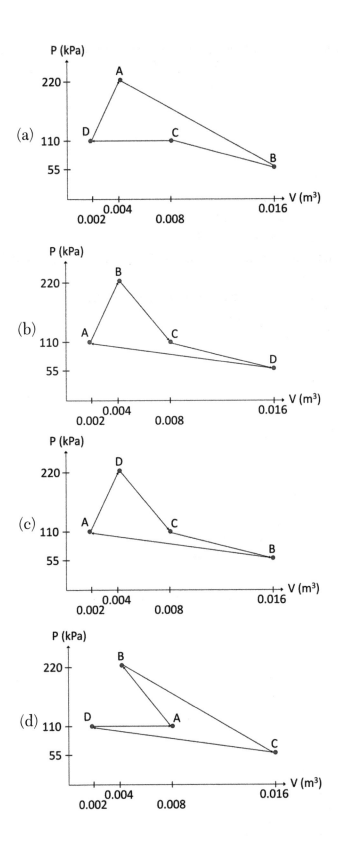

52. For the PV diagram presented in Figure 119, calculate the work done by gas on the piston.

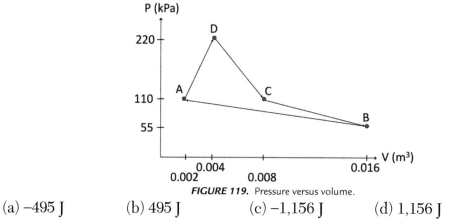

FIGURE 119. Pressure versus volume.

(a) −495 J   (b) 495 J   (c) −1,156 J   (d) 1,156 J

53. For the PV diagram presented in Figure 120, calculate the work done by gas on the piston.

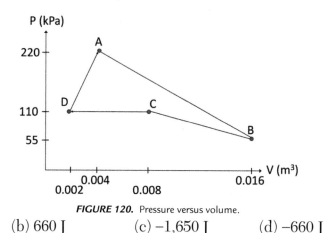

FIGURE 120. Pressure versus volume.

(a) 1,650 J   (b) 660 J   (c) −1,650 J   (d) −660 J

54. For the PV diagram presented in Figure 121, calculate the work done by gas on the piston.

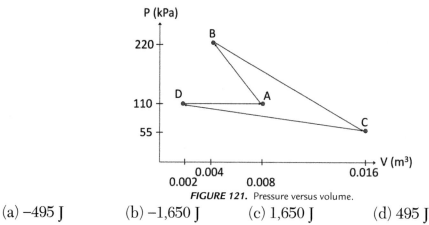

FIGURE 121. Pressure versus volume.

(a) −495 J   (b) −1,650 J   (c) 1,650 J   (d) 495 J

55. A Brayton cycle consists of:
(a) two isochoric and two isentropic processes
(b) two isobaric and two isentropic processes
(c) two isobaric and two isochoric processes
(d) two isobaric and two isothermal processes

56. A Brayton cycle consists of, in order:

(a) adiabatic compression, isobaric expansion, isentropic expansion, and isobaric compression

(b) isentropic expansion, isochoric expansion, isentropic compression, and isochoric compression

(c) adiabatic expansion, isochoric compression, isentropic compression, and isochoric expansion

(d) isentropic compression, isochoric expansion, isentropic expansion, and isochoric expansion

57. A Brayton cycle is presented in Figure 122. Processes A to B and C to D, are adiabatic and reversible. Using the data presented in Table 52, find temperature, volume, and pressure values missing from the table. There are 0.054 moles of methane participating in the process ($K = 1.3$).

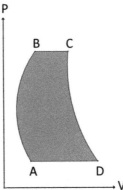

FIGURE 122. Pressure versus volume in a Brayton cycle.

TABLE 52. Pressure, temperature, and volume data for the Brayton cycle presented in Figure 122.

Process	T (K)	V (m³)	P (kPa)
A		0.000975	101
B			1,615
C			
D	700		

(a)

Process	T (K)	V (m³)	P (kPa)
A	219	0.000975	101
B	416	0.000116	1,615
C	1,327	0.000369	1,615
D	700	0.003112	101

(b)

Process	T (K)	V (m³)	P (kPa)
A	24	0.000975	101
B	52	0.000135	1,615
C	1,545	0.003978	1,615
D	700	0.028811	101

(c)

Process	T (K)	V (m³)	P (kPa)
A	16	0.000975	101
B	35	0.000135	1,615
C	1,545	0.005967	1,615
D	700	0.043216	101

(d)

Process	T (K)	V (m³)	P (kPa)
A	12	0.000975	101
B	26	0.000135	1,615
C	1,545	0.007956	1,615
D	700	0.057622	101

58. Identify heat and work for the four processes of the Brayton cycle presented in Figure 122, using the data presented in Table 53. Processes A to B and C to D, are adiabatic and reversible. There is 1 mole of methane ($K = 1.3$).

TABLE 53. Pressure, temperature, and volume data for the Brayton cycle presented in Figure 122.

Process	T (K)	V (m³)	P (kPa)
A		0.00195	202
B			3,230
C			
D	1,050		

(a)

Process	dU (kJ)	dW (kJ)	dQ (kJ)
A-B	1.2	-1.2	0.0
B-C	11.3	3.4	14.7
C-D	-6.5	6.5	0.0
D-A	-6.0	-1.8	-7.8

(b)

Process	dU (kJ)	dW (kJ)	dQ (kJ)
A-B	1.2	-1.2	0.0
B-C	25.1	7.5	32.6
C-D	-13.0	13.0	0.0
D-A	-13.2	-4.0	-17.2

(c)

Process	dU (kJ)	dW (kJ)	dQ (kJ)
A-B	1.2	-1.2	0.0
B-C	38.9	11.7	50.6
C-D	-19.6	19.6	0.0
D-A	-20.5	-6.2	-26.7

(d)

Process	dU (kJ)	dW (kJ)	dQ (kJ)
A-B	1.2	-1.2	0.0
B-C	52.7	15.8	68.5
C-D	-26.1	26.1	0.0
D-A	-27.8	-8.3	-36.1

59. Identify the total heat and work for the four processes of the Brayton cycle presented in Figure 122, using the data presented in Table 54. Processes A to B and C to D, are adiabatic and reversible. There are 0.5 moles of methane participating in the process ($K = 1.3$).

TABLE 54. Pressure, temperature, and volume data for the Brayton cycle presented in Figure 122.

Process	T (K)	V (m³)	P (kPa)
A		0.0013	300
B			2,400
C			
D	2,000		

(a) 21 kJ, 21 kJ   (b) 43 kJ, 43 kJ   (c) 21 kJ, 43 kJ   (d) 43 kJ, 21 kJ

60. Modern gas turbine engines follow the:

(a) Otto cycle   (b) Brayton cycle
(c) Rankine cycle   (d) Carnot cycle

61. Airbreathing and ram compression jet engines follow the:

(a) Otto cycle   (b) Brayton cycle
(c) Rankine cycle   (d) Carnot cycle

62. In jet engines, the . . . is highest as the . . . approaches the . . . for creation of the . . . residual kinetic energy.

(a) propulsive efficiency / exhaust jet velocity / vehicle speed / smallest

(b) fuel efficiency / intake jet velocity / gas speed / largest

(c) propulsive efficiency / exhaust jet velocity / vehicle speed / smallest

(d) fuel efficiency / intake jet velocity / gas speed / largest

63. The Brayton cycle is also known as the:

(a) Otto cycle   (b) Joule cycle

(c) Rankine cycle   (d) Carnot cycle

64. The efficiency of the Brayton cycle presented in Table 55 (Figure 122) is . . . . There are 0.75 moles of methane participating in the process ($K = 1.3$).

TABLE 55. Pressure, temperature, and volume data for the Brayton cycle presented in Figure 122.

Process	T (K)	V (m³)	P (kPa)
A	16	0.00098	102
B	30	0.00012	1,613
C	1,327	0.00513	1,613
D	700	0.04266	102

(a) 46.7%   (b) 97.7%   (c) 89.6%   (d) 43.3%

65. The pressure ratio of the Brayton cycle presented in Table 56 (Figure 122) is . . . . There are 0.6 moles of methane participating in the process ($K = 1.3$).

TABLE 56. Pressure, temperature, and volume data for the Brayton cycle presented in Figure 122.

Process	T (K)	V (m³)	P (kPa)
A	20	0.00098	102
B	37	0.00012	1,591
C	1,327	0.00416	1,591
D	700	0.03413	102

(a) 0.063   (b) 1   (c) 15.5   (d) 0

66. A Carnot cycle consists of:

(a) two isobaric and two isentropic processes

(b) two isobaric and two isochoric processes

(c) two isobaric and two isothermal processes

(d) two isothermal and two isentropic processes

67. A Carnot cycle consists of, in order:

(a) isothermal expansion, isentropic expansion, isothermal compression, and isentropic compression

(b) isobaric expansion, isochoric expansion, isentropic compression, and isochoric compression

(c) isothermal expansion, isentropic expansion, isothermal expansion, and isentropic expansion

(d) isentropic compression, isochoric compression, isentropic expansion, and isochoric compression

68. A Carnot cycle is presented in Figure 123 ($T_2 > T_1$). Processes A to B and C to D, are isothermal. Processes B to C and D to A are adiabatic and reversible. Using the data presented in Table 57, find temperature, volume, and pressure values missing from the table. There are 0.25 moles of hydrogen participating in the process ($K = 1.6$).

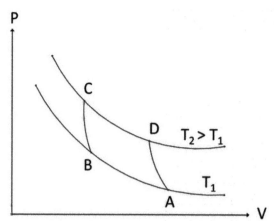

FIGURE 123. Pressure versus volume in a Carnot cycle.

TABLE 57. Pressure, temperature, and volume data for the Carnot cycle presented in Figure 123.

Process	T (K)	V (m³)	P (kPa)
A		0.000975	101
B			1,615
C			
D	700		

(a)

Process	T (K)	V (m³)	P (kPa)
A	15.79	0.000975	101
B	15.79	0.000061	1,615
C	700.00	0.0000001	39,739,410
D	700.00	0.000002	2,485,251

(b)

Process	T (K)	V (m³)	P (kPa)
A	23.69	0.000975	101
B	23.69	0.000061	1,615
C	700.00	0.0000002	13,478,598
D	700.00	0.000003	842,934

(c)

Process	T (K)	V (m³)	P (kPa)
A	47.38	0.000975	101
B	47.38	0.000061	1,615
C	700.00	0.000001	2,122,746
D	700.00	0.000011	132,754

(d)

Process	T (K)	V (m³)	P (kPa)
A	11.84	0.000975	101
B	11.84	0.000061	1,615
C	700.00	0.0000001	85,583,764
D	700.00	0.000001	5,352,297

69. Identify heat and work for the four processes of the Carnot cycle presented in Figure 123, using the data presented in Table 58. Processes A to B and C to D, are adiabatic and reversible. There is 1 mole of hydrogen ($K = 1.6$).

TABLE 58. Pressure, temperature, and volume data for the Carnot cycle presented in Figure 123.

Process	T (K)	V (m³)	P (kPa)
A		0.00195	202
B			3,230
C			
D	1,050		

(a)

Process	dU (kJ)	dW (kJ)	dQ (kJ)
A-B	0.000	-1.092	-1.092
B-C	2.981	-2.981	0.000
C-D	0.000	6.050	6.050
D-A	-2.981	2.981	0.000

(b)

Process	dU (kJ)	dW (kJ)	dQ (kJ)
A-B	0.000	-1.092	-1.092
B-C	6.618	-6.618	0.000
C-D	0.000	12.099	12.099
D-A	-6.618	6.618	0.000

(c)

Process	dU (kJ)	dW (kJ)	dQ (kJ)
A-B	0.000	-1.092	-1.092
B-C	10.256	-10.256	0.000
C-D	0.000	18.149	18.149
D-A	-10.256	10.256	0.000

(d)

Process	dU (kJ)	dW (kJ)	dQ (kJ)
A-B	0.000	-1.092	-1.092
B-C	13.893	-13.893	0.000
C-D	0.000	24.198	24.198
D-A	-13.893	13.893	0.000

70. Identify the total heat and work for the four processes of the Carnot cycle presented in Figure 123, using the data presented in Table 59. Processes A to B and C to D, are adiabatic and reversible. There are 0.5 moles of hydrogen participating in the process ($K = 1.6$).

TABLE 59. Pressure, temperature, and volume data for the Carnot cycle presented in Figure 123.

Process	T (K)	V (m^3)	P (kPa)
A		0.0013	300
B			2,400
C			
D	2,000		

(a) 25.12 kJ, 25.12 kJ   (b) 16.48 kJ, 16.48 kJ
(c) 16.48 kJ, 25.22 kJ   (d) 25.22 kJ, 16.48 kJ

71. Theoretical ideal thermodynamic cycles follow the:

(a) Otto cycle   (b) Brayton cycle
(c) Rankine cycle   (d) Carnot cycle

72. The efficiency of the Carnot cycle presented in Table 60 (Figure 123) is . . . . There are 0.75 moles of hydrogen participating in the process ($K = 1.6$).

TABLE 60. Pressure, temperature, and volume data for the Carnot cycle presented in Figure 123.

Process	T (K)	V (m^3)	P (kPa)
A	16	0.009750	10
B	16	0.006100	16
C	700	0.000011	396,482
D	700	0.000018	249,917

(a) 97.7%   (b) 22.6%   (c) 89.6%   (d) 43.3%

73. A Rankine cycle consists of, in order:

(a) two isobaric and two potentially work-generating (consuming) isentropic processes
(b) two isochoric and two isobaric processes

(c) two isobaric and two work-consuming isentropic processes

(d) two isochoric and two work-generating processes

74. An ideal Rankine cycle consists of, in order:

(a) isentropic heating from a saturated liquid state to generate supercooled liquid, isobaric heating to form superheated gas, isentropic cooling to generate dry saturated vapor, and isobaric cooling to form saturated liquid

(b) isentropic compression, isochoric expansion, adiabatic compression, and isochoric compression

(c) isothermal compression, adiabatic expansion, isothermal expansion, and adiabatic expansion

(d) dry vapor compression, isentropic compression to generate dry saturated vapor, isentropic expansion to wet fluid, and isobaric condensation to form saturated liquid

75. Thermal power generation plants follow the:

(a) Otto cycle   (b) Brayton cycle

(c) Carnot cycle   (d) Rankine cycle

76. Using the enclosed steam tables and diagrams (Table 61 and Table 62), identify if the pressure, temperature, and volume for states A to F related to the Rankine cycle presented in Figure 124 ($P_2 > P_1$) are correctly represented by the data presented in Table 63. Note that processes A to B and E to F are not isentropic. The operating fluid is water. (T/F)

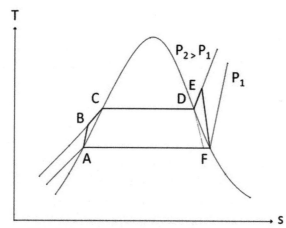

FIGURE 124. Pressure versus volume in a Rankine cycle.

TABLE 61. Pressure, temperature, and volume data for the Rankine cycle presented in Figure 124.

State	Temperature °C	Pressure MPa
A		0.10142
B	120	5
C		
D		
E	300	5
F		

TABLE 62. Thermodynamic properties of moist air.

State	Temperature	Pressure	Volume (m³/kg)			Energy (kJ/kg)		
	°C	MPa	vf	vfg	vg	uf	ufg	ug
A	100	0.1	0.00104	1.67076	1.67180	419.06	2,086.94	2,506.00
B	120	5.0		0.00106			501.90	
C	263	5.0	0.00129	0.03987	0.04115	1,144.25	1,451.55	2,595.80
D	263	5.0	0.00129	0.03987	0.04115	1,144.25	1,451.55	2,595.80
E	300	5.0		0.04540			2,699.00	
F	100	0.1	0.00104	1.67076	1.67180	419.06	2,086.94	2,506.00

State	Temperature	Pressure	Enthalpy (kJ/kg)			Entropy (kJ/kgK)			Quality
	°C	MPa	hf	hfg	hg	sf	sfg	sg	x
A	100	0.1	419.20	2,256.40	2,675.60	1.31	6.05	7.35	0.00
B	120	5.0		507.20			1.52		
C	263	5.0	1,150.70	1,642.00	2,792.70	2.91	3.07	5.98	0.00
D	263	5.0	1,150.70	1,642.00	2,792.70	2.91	3.07	5.98	1.00
E	300	5.0		2,925.70			6.21		
F	100	0.1	419.20	2,256.40	2,675.60	1.31	6.05	7.35	1.00

TABLE 63. Thermodynamic properties of moist air for the data presented in Table 62.

State	Temperature	Volume	Energy	Enthalpy	Entropy
	°C	v (m³/kg)	u (kJ/kg)	h (kJ/kg)	s (kJ/kgK)
A	100	0.0010	419.1	419.2	1.31
B	120	0.0011	501.9	507.2	1.52
C	263	0.0013	1,144.3	1,150.7	2.91
D	263	0.0412	2,595.8	2,792.7	5.98
E	300	0.0454	2,699.0	2,925.7	6.21
F	100	1.6718	2,506.0	2,675.6	7.35

77. Using the data presented in Table 64, and the enclosed steam tables and diagrams (Figure 125 to Figure 128), identify if pressure, temperature, and volume for the Rankine cycle presented in Figure 124 ($P_2 > P_1$) are correctly represented by the data in Table 65 and Table 66. For an ideal cycle, processes A to B and E to F are adiabatic and reversible. The operating fluid is water. Efficiency of the turbine (process E to F) and pump (process A to B) as well as thermal efficiency of the cycle are provided. (T/F)

TABLE 64. Pressure, temperature, and volume data for the Rankine cycle presented in Figure 124.

State	Temperature	Pressure
	°C	MPa
A		0.1
B	120	5
C		
D		
E	300	5
F		

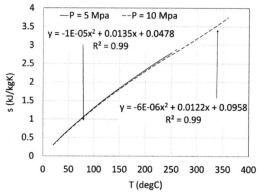

FIGURE 125. Entropy versus temperature for compressed liquid (water).

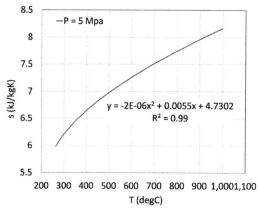

FIGURE 126. Entropy versus temperature for superheated gas (water).

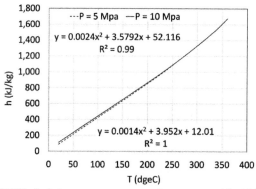

FIGURE 127. Enthalpy versus temperature for compressed liquid (water).

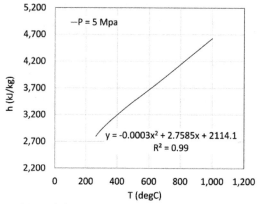

FIGURE 128. Enthalpy versus temperature for superheated gas (water).

TABLE 65. Thermodynamic properties of moist air for the data presented in Table 64.

State	Real: Temperature	Pressure	Real: Entropy	Real: Enthalpy	Ideal: Temperature	Ideal: Entropy	Ideal: Enthalpy	Quality
	T (°C)	P (Mpa)	s (kJ/kgK)	h (kJ/kg)	T (°C)	s (kJ/kgK)	h (kJ/kg)	x
A	100	0.1	1.31	419.20	100	1.31	419.20	0
B	120	5.0	1.52	507.20	101	1.31	424.29	-
C	263	5.0	2.91	1,150.70	263	2.91	1,150.70	0
D	263	5.0	5.98	2,792.70	263	5.98	2,792.70	1
E	300	5.0	6.21	2,925.70	634	7.35	3,741.34	-
F	100	0.1	7.35	2,675.60	100	7.35	2,675.60	1

TABLE 66. Thermodynamic properties of moist air for the data presented in Table 64.

Process	Real: Enthalpy change	Ideal: Enthalpy change	Real: Entropy change	Ideal: Entropy change	Real: dQ or dW	Ideal: dQ or dW	Real: Etta thermal	Ideal: Etta thermal	Etta
	dh (kJ/kg)	dh (kJ/kg)	ds (kJ/kgK)	ds (kJ/kgK)	(kJ/kg)	(kJ/kg)			
A-B	88.0	5.1	0.22	0.00	88.0	5.1			Pump
B-C	643.5	726.4	1.39	1.61					5.8%
C-D	1,642.0	1,642.0	3.07	3.07	2,418.5	3,317.0	6.7%	32.0%	
D-E	133.0	948.6	0.23	1.38					Turbine
E-F	-250.1	-1,065.7	1.14	0.00	-250.1	-1,065.7			23.5%
F-A	-2,256.4	-2,256.4	-6.05	-6.05	-2,256.4	-2,256.4			

78. A Diesel cycle is presented in Figure 129. Processes A to B and C to D, are reversible and adiabatic. Using the data presented in Table 67, find the pressure, temperature, and volume values missing from the table. There are 0.35 moles of natural gas participating in the process ($K = 1.32$).

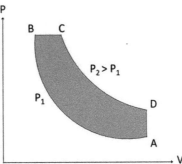

FIGURE 129. Pressure versus volume in a Diesel cycle.

TABLE 67. Pressure, temperature, and volume data for the Diesel cycle presented in Figure 129.

Process	T (K)	V (m³)	P (kPa)
A		0.000975	101
B			1,615
C			
D	700		

(a)

Process	T (K)	V (m³)	P (kPa)
A	338.41	0.000975	101
B	662.66	0.000119	1,615
C	1,149.26	0.000207	1,615
D	700.00	0.000975	209

(b)

Process	T (K)	V (m³)	P (kPa)
A	219.34	0.000975	101
B	429.50	0.000119	1,615
C	1,034.58	0.000288	1,615
D	700.00	0.000975	322

(c)

Process	T (K)	V (m³)	P (kPa)
A	23.69	0.000975	101
B	46.39	0.000119	1,615
C	603.17	0.001553	1,615
D	700.00	0.000975	2,985

(d)

Process	T (K)	V (m³)	P (kPa)
A	33.84	0.000975	101
B	66.27	0.000119	1,615
C	657.65	0.001185	1,615
D	700.00	0.000975	2,089

79. Identify heat and work for states A to F related to the Diesel cycle presented in Figure 129, using the data presented in Table 68. Note that processes A to B and C to D, are reversible and adiabatic. There are 1.5 moles of natural gas ($K = 1.32$).

TABLE 68. Pressure, temperature, and volume data for the Diesel cycle presented in Figure 129.

Process	T (K)	V (m³)	P (kPa)
A		0.00195	202
B			3,230
C			
D	1,050		

(a)

Process	dU (kJ)	dW (kJ)	dQ (kJ)
A-B	0.807	-0.807	0.000
B-C	15.096	4.831	19.927
C-D	3.326	-3.326	0.000
D-A	-19.229	0.000	-19.229

(b)

Process	dU (kJ)	dW (kJ)	dQ (kJ)
A-B	0.807	-0.807	0.000
B-C	0.297	0.095	0.392
C-D	-0.861	0.861	0.000
D-A	-0.242	0.000	-0.242

(c)

Process	dU (kJ)	dW (kJ)	dQ (kJ)
A-B	1.179	-1.179	0.000
B-C	31.857	10.194	42.052
C-D	6.653	-6.653	0.000
D-A	-39.690	0.000	-39.690

(d)

Process	dU (kJ)	dW (kJ)	dQ (kJ)
A-B	0.807	-0.807	0.000
B-C	10.565	3.381	13.946
C-D	1.038	-1.038	0.000
D-A	-12.409	0.000	-12.409

80. Identify the total heat and work for states A to F related to the Diesel cycle presented in Figure 129. Note that processes A to B and C to D, are reversible and adiabatic. Use the data presented in Table 69. There are 0.45 moles of natural gas participating in the process ($K = 1.32$).

TABLE 69. Pressure, temperature, and volume data for the Diesel cycle presented in Figure 129.

Process	T (K)	V (m³)	P (kPa)
A		0.0013	300
B			1,615
C			
D	2,000		

(a) 11.63 kJ, 11.63 kJ   (b) 11.63 kJ, 1.90 kJ
(c) 1.90 kJ, 11.63 kJ   (d) −1.90 kJ, −1.90 kJ

81. Reciprocating internal combustion engines follow the:
(a) Jet engine cycle   (b) Brayton cycle
(c) Rankine cycle   (d) Diesel cycle

82. The efficiency of the Diesel cycle presented in Table 70 (Figure 129) is . . . . There are 0.75 moles of natural gas participating in the process ($K = 1.32$).

TABLE 70. Pressure, temperature, and volume data for the Diesel cycle presented in Figure 129.

Process	T (K)	V (m³)	P (kPa)
A	15.79	0.000975	101
B	30.92	0.000119	1,620
C	546.71	0.0021041	1,620
D	700.00	0.000975	4,477

(a) 97.1%   (b) 97.7%   (c) 89.6%   (d) 43.3%

83. The compression and cut-off ratios for the Diesel cycle presented in Table 71 (Figure 129) are . . . . There are 0.65 moles of natural gas participating in the process ($K = 1.32$).

TABLE 71. Pressure, temperature, and volume data for the Diesel cycle presented in Figure 129.

Process	T (K)	V (m³)	P (kPa)
A	18.22	0.000975	101
B	35.68	0.000119	1,620
C	566.01	0.0018878	1,620
D	700.00	0.000975	3,880

(a) 15.9, 0.1   (b) 8.2, 15.9   (c) 8.2, 0.1   (d) 15.9, 8.2

84. Identify whose primitive fields the following properties are: Pressure, volume, and mass?
(a) Mechanics, mechanics, and geometry
(b) Geometry, mechanics, and geometry
(c) Geometry, mechanics, and mechanics
(d) Mechanics, geometry, and mechanics

85. Identify whose primitive fields the following properties are: Velocity, current, area, and energy?
(a) Mechanics, electricity, geometry, and electricity
(b) Mechanics, electricity, geometry, and mechanics
(c) Mechanics, electricity, mechanics, and mechanics
(d) Mechanics, electricity, geometry, and geometry

86. Thermodynamic systems are defined by their:

(a) boundaries   (b) space confinement

(c) limits   (d) freedoms

87. Thermodynamic systems are regions of space with . . . boundaries.

(a) rigid   (b) flexible   (c) no   (d) defined

88. The interaction between thermodynamic systems is . . . and . . . .

(a) a one-way process / happens at the borders

(b) a two-way process / crosses the boundaries

(c) a three-way process / crosses the borders

(d) not a process / happens at the boundaries

89. The system to the secondary system is like the:

(a) primary system to the surroundings

(b) secondary system to the surroundings

(c) primary system to the system

(d) secondary system to the system

90. If a system lets mass cross its boundaries, it is an open system. (T/F)

91. Energy transfer is possible for both a closed and an open system. (T/F)

92. In an isolated system:

(a) both energy and mass are transferred

(b) energy is transferred but not mass

(c) energy is not transferred but mass is

(d) neither energy nor mass is transferred

93. Shape of a boundary is not relevant when defining a system. (T/F)

94. Shape of boundaries is rigid when defining a system. (T/F)

95. For a piston-cylinder system, if all valves are fully closed, the system is a(an) . . . system. If only one of the valves is open, the system is a(an) . . . system. If all valves are open, the system is a(an) . . . system.

(a) closed / open / open   (b) open / open / open

(c) closed / closed / closed   (d) open / closed / closed

96. State variables are a set of:

(a) measurable properties sufficient to determine all other properties

(b) non-measurable properties needed to determine all other properties

(c) measurable properties not required to calculate all other properties

(d) non-measurable properties that are not necessarily used to predict all other properties

97. In most thermodynamic cases, state variables . . . as systems interact.

(a) remain the same   (b) slightly change

(c) do not change   (d) change

98. In thermodynamics, a state function:

(a) is a function that relates several state variables or state quantities and depends on the current equilibrium state of the system

(b) does not depend on the path which the system undergoes before arriving at its current state

(c) describes the equilibrium state of a system

(d) all of the above

99. Internal energy, enthalpy, and entropy are:

(a) state quantities

(b) variable quantities

(c) state variables

(d) state functions

100. Mechanical work and heat are:

(a) state quantities

(b) variable quantities

(c) state or path variables

(d) state or path functions

101. Main thermodynamic properties used to identify the system state of equilibrium are:

(a) state quantities and volume

(b) temperature, pressure, and volume

(c) state quantities and pressure

(d) path quantities and temperature

102. In thermodynamics, the state variables are identified using either . . . or . . . variables.

(a) microscopic (particles collision-based) / macroscopic (average-based)

(b) microscopic (average-based) / macroscopic (particles collision-based)

(c) microscopic (particles moving-based) / macroscopic (particles stationary-based)

(d) microscopic (particles stationary-based) / macroscopic (particles moving-based)

103. In thermodynamics, thermal energy and entropy are:

(a) primitive properties

(b) basic properties

(c) not properties

(d) microscopic properties

104. In thermodynamics, derived properties are:

(a) a combination of thermodynamic properties

(b) a combination of primitive thermodynamic properties

(c) uncombined thermodynamic properties

(d) uncombined thermodynamic macroscopic properties

105. Extensive to intensive properties are like:

(a) additive to subtractive properties

(b) additive to constant properties

(c) constant to additive properties

(d) constant to subtractive properties

106. In thermodynamics, the specific property is an:

(a) intensive property divided by an extensive property

(b) extensive property divided by an extensive property

(c) intensive property divided by an intensive property

(d) extensive property divided by an intensive property

107. The minimum set of properties required to identify the state of a system are:

(a) three (pressure, volume, and temperature)

(b) two (pressure and mass)

(c) three (pressure, mass, and density)

(d) two (pressure and temperature)

108. In thermodynamics, if a state can be identified by a point in state space, the state is at equilibrium. (T/F)

109. Thermodynamic equilibrium consists of:

(a) thermal, mechanical, chemical, and radiative equilibrium states

(b) mechanical, chemical, and radiative equilibrium states

    (c) thermal and chemical equilibrium states

    (d) thermal and mechanical equilibrium states

110. A thermodynamic process happens when:

    (a) state functions change

    (b) system state changes

    (c) system state remains unchanged

    (d) state variables remain constant

111. For any thermodynamic process to happen:

    (a) all state variables must change

    (b) at least two state variables should change

    (c) at least one state variable should change

    (d) none of state variables need to change

112. If all the intermediate states of a thermodynamic process are at thermodynamic equilibrium, the process is:

    (a) a quasi-static process   (b) a non-quasi-static process

    (c) mechanically at equilibrium   (d) chemically at equilibrium

113. If all the intermediate states of a thermodynamic process are at thermodynamic equilibrium, a connecting path in state space can be defined. (T/F)

114. If a path cannot be defined for a thermodynamic process, the process is:

    (a) non-quasi-static   (b) dynamic

    (c) quasi-static   (d) reversible

115. Intensive thermodynamic properties can be only defined for:

    (a) dynamic systems   (b) quasi-static processes

    (c) stationary systems   (d) reversible systems

116. A thermodynamic cycle consists of a sequence of:

    (a) processes that start from the initial state and end at the final state

    (b) heat-generating steps that start and end at the final state

    (c) work-generating steps that start and end at the initial state

    (d) processes that start from and end at the initial state

117. For a thermodynamic cycle to be possible, the processes should be:

    (a) reversible   (b) at thermal equilibrium

    (c) quasi-static   (d) any of the above

118. When defining the change in property, exact differentials only depend on the:

    (a) initial and final states

    (b) partial change in the initial and final states

    (c) initial state

    (d) final state

119. To derive an exact differential variable:

    (a) variable integral over the path is calculated

    (b) temperature over the entropy change is considered

    (c) pressure over volume change is predicted

    (d) enthalpy over temperature change is determined

120. When you cook spaghetti sauce in a pot covered by a spill-stopper lid, the system can be considered as a closed system. (T/F)

121. If an isolated system consists of multiple sub-systems, each of the sub-systems is also isolated. (T/F)

122. Work and heat are primitive thermodynamic properties. (T/F)

123. In a piston-cylinder system, three types of work are independently carried out by (1) an electric motor, (2) a stirring spoon, and (3) a mechanical source driving the cylinder. The total work is calculated by (v, I, t, τ, θ, P, and V are voltage, current, time, torque, rotational angle, pressure, and volume):
(a) $vIDt - \tau d\theta + PdV$ (b) $Idt + t\theta - PV$
(c) $vdt - \tau d\theta + dPdV$ (d) $vI + d\tau d\theta - VdP$

124. In most cases, when calculating work using $dW = XdY$, X is an . . . property and Y is an . . . property.
(a) intensive / intensive   (b) extensive / extensive
(c) extensive / intensive   (d) intensive / extensive

125. Work conducted by a system or on a system can be:
(a) one-way   (b) two-way
(c) either one-way or two-way (d) reversible

126. An example of two-way work is rechargeable cells. (T/F)

127. The method of summation can be used to calculate the thermodynamic work for a quasi-static process. (T/F)

128. Work . . . an exact differential because it . . . on the path:
(a) is / depends   (b) is not / depends
(c) is / does not depend   (d) is not / does not depend

129. Expansion of a gas can be an example of a closed system. (T/F)

130. If the internal energy of a thermodynamic system does not change:
(a) cyclic heat and work are identical
(b) instanteneous heat and work are identical
(c) cyclic heat and work are not identical
(d) instanteneous heat and work are not identical

131. The number of independent intensive properties needed to define the state of a system is equal to the:
(a) number of two-way work plus two (b) number of one-way work plus one
(c) number of two-way work plus one (d) number of one-way work plus two

132. Which single property is applicable to all thermodynamic systems with only one independent intensive property?
(a) pressure   (b) temperature
(c) volume   (d) internal energy

133. Freezing and evaporation states are when . . . and . . . are at equilibrium at . . .
(a) ice-water / water-vapor / 1 atm   (b) ice-water / water-ice / 1 kPa
(c) vapor-ice / water-ice / 1 atm   (d) water-vapor / water-ice / 1 kPa

134. The relation between the length of the column of water (L_G) in a thermometer, and its associated temperature (T_G) with the length L_0 at the reference temperature T_0 and the length L_{100} at temperature T_{100}, is (assume T_0 and T_{100} are 0 and 100 °C):

(a) $T_G = \left(\dfrac{L_G}{100}\right)(T_{100} - T_0) + T_0$

(b) $L_G = \left(\dfrac{T_G}{100}\right)(L_{100} + L_0) - L_0$

(c) $L_G = \left(\dfrac{T_G}{100}\right)(L_{100} - L_0) + L_0$

(d) $T_G = \left(\dfrac{L_G}{100}\right)(T_{100} + T_0) - T_0$

135. If the system mass does not vary at specific time intervals, the system can be closed, open, or isolated. (T/F)

136. Gas thermometry is more accurate than liquid thermometry. (T/F)

137. Isotherms on a PV diagram form ... curves, facing ..., and follow ...
(a) convex / outward, increasing outward / Boyle's law
(b) concave / inward, decreasing outward / Boyle's law
(c) convex / outward, increasing inward / Gay-Lussac's law
(d) concave / inward, decreasing inward / Gay-Lussac's law

138. In an ideal gas, isochores follow ... and isobars follow ...
(a) Charles's law / Gay-Lussac's law
(b) Gay-Lussac's law / Charles's law
(c) Boyle's law / Gay-Lussac's law
(d) Charles's law / Boyle's law

139. At its triple point of ... , water has ... that are ... with one another.
(a) 273.16 K / three phases of solid, liquid, and gas / at equilibrium
(b) 273.15 K / two phases of solid and liquid / at equilibrium
(c) 273.16 K / three phases of solid, liquid, and gas / not at equilibrium
(d) 273.15 K / two phases of solid and liquid / not at equilibrium

140. The Kelvin temperature scale is based on ... ; the Celsius temperature scale is based on ... temperature.
(a) the ideal gas law / Kelvin
(b) Kelvin / the ideal gas law
(c) the ideal gas law / the ideal gas law
(d) Kelvin / Kelvin

141. The state of a thermodynamic system is defined using the:
(a) state variables
(b) intensive properties
(c) extensive properties
(d) primitive variables

142. Enthalpy is a function of ... based on the ... relation (h is enthalpy, u is internal energy, R is gas constant, T is temperature, p is pressure, and v is volume).
(a) temperature / $h = u + RT$
(b) temperature / $h = u + pv$
(c) pressure / $h = u + pv$
(d) volume / $h = u + pv$

143. Internal energy and enthalpy are functions of:
(a) pressure
(b) volume
(c) temperature
(d) all of the above

144. Heat capacity at constant pressure is ... than that at a constant volume by a factor of ... (γ is heat capacity ratio).
(a) smaller / γ
(b) larger / γ
(c) smaller / $1/\gamma$
(d) larger / $1/\gamma$

145. Rank the compressibility of three states of matter:
(a) gas > liquid > solid
(b) liquid > gas > solid
(c) solid > liquid > gas
(d) gas > solid > liquid

146. Simplifying the internal energy relations presented as follows, results in equation . . . , which shows that the internal energy is a function of (P is pressure, V is volume, T is temperature, u is internal energy, and c_p and c_v are specific heat capacities at constant pressure and volume) . . .

(1) $u = f(T, P)$

$$\therefore du = \left(\frac{\partial u}{\partial T}\right)_P dT + \left(\frac{\partial u}{\partial P}\right)_T dP$$

(2) $u = f(T, V)$

$$\therefore du = \left(\frac{\partial u}{\partial T}\right)_V dT + \left(\frac{\partial u}{\partial V}\right)_T dV$$

(a) $du = c_v dT$ / temperature
(b) $du = c_p dT$ / temperature
(c) $du = (c_p + c_v)dT$ / pressure and temperature
(d) $du = (c_p/c_v)dT$ / pressure and temperature

147. Simplifying the enthalpy relations presented as follows, results in equation . . . , which shows that the enthalpy is a function of (P is pressure, V is volume, T is temperature, h is enthalpy, and c_p and c_v are specific heat capacities at constant pressure and volume) . . .

(1) $h = f(T, P)$

$$\therefore dh = \left(\frac{\partial h}{\partial T}\right)_P dT + \left(\frac{\partial h}{\partial P}\right)_T dP$$

(2) $h = f(T, V)$

$$\therefore dh = \left(\frac{\partial h}{\partial T}\right)_V dT + \left(\frac{\partial h}{\partial V}\right)_T dV$$

(a) $dh = c_v dT$ / temperature
(b) $dh = c_p dT$ / temperature
(c) $dh = (c_p + c_v)dT$ / pressure and temperature
(d) $dh = (c_p / c_v)dT$ / pressure and temperature

148. In the Van der Waals equation, the term modifying pressure is due to the . . . and the term modifying volume is due to the . . .

(a) gas molecules interaction / molecules themselves
(b) molecules themselves / gas molecules interaction
(c) gas molecules interaction / gas molecules interaction
(d) molecules themselves / molecules themselves

149. In pure materials, the chemical composition:

(a) does not remain the same for all phases
(b) is not independent of the state of matter
(c) varies based on the state of matter
(d) remains the same for all phases

150. The Mollier diagram presents:

(a) entropy versus enthalpy
(b) entropy versus temperature
(c) pressure versus temperature
(d) pressure versus enthalpy

151. The Mollier diagram is a graphic representation of:

(a) temperature, moisture content, pressure, and volume
(b) temperature, moisture content, pressure, volume, enthalpy, and entropy
(c) temperature, pressure, quality, and entropy
(d) moisture content, pressure, volume, and enthalpy

152. The pressure-volume-temperature diagram of water consists of:

(a) six regions   (b) four regions   (c) five regions   (d) three regions

153. The . . . is located at the end of the solid-vapor line in a water pressure-temperature diagram.

(a) critical point   (b) triple point   (c) melting point   (d) freezing point

154. To identify water state using thermodynamic tables, at least . . . properties are required.

(a) 3   (b) 2   (c) 1   (d) 0

155. If steam tables are used to draw the pressure-temperature diagram, the point at which the three phases (ice, water, and vapor) meet is:

(a) critical point   (b) melting point
(c) triple point   (d) sublimation point

156. The . . . is located at the end of the liquid-vapor line in a water pressure-temperature diagram.

(a) triple point   (b) melting point
(c) freezing point   (d) dcritical point

157. The liquid-vapor line in a water pressure-temperature diagram starts from the . . . and extends to the . . .

(a) critical point / triple point   (b) triple point / critical point
(c) critical point / critical point   (d) triple point / triple point

158. Beyond the critical point, the state of the matter is . . . and the latent heat . . .

(a) gas / can be calculated   (b) gas / is not applicable
(c) liquid-solid / can be calculated (d) solid-gas / is not applicable

159. States of matter transitions for melting-freezing are . . . , for vaporization-condensation are . . . , and for sublimation-deposition are . . .

(a) liquid-to-solid and solid-to-liquid / vapor-to-liquid and liquid-to-vapor / vapor-to-solid and solid-to-vapor
(b) vapor-to-liquid and liquid-to-vapor / vapor-to-solid and solid-to-vapor / liquid-to-solid and solid-to-liquid
(c) solid-to-liquid and liquid-to-solid / liquid-to-vapor and vapor-to-liquid / solid-to-vapor and vapor-to-solid
(d) vapor-to-liquid and solid-to-vapor / vapor-to-solid and solid-to-liquid / liquid-to-solid and liquid-to-vapor

160. You observe that the concrete in your exterior stairs keeps crumbling more and more after every winter. You conclude that the cracks in concrete grow due to the:

(a) water freezing and expanding inside the cracks
(b) water freezing and contracting inside the cracks
(c) rainwater accumulating in the cracks
(d) melting snow accumulating in the cracks

161. When freezing, water . . . and carbon dioxide . . .
(a) expands / contracts
(b) contracts / expands
(c) expands / expands
(d) contracts / contracts

162. At the saturation point, either pressure or temperature is sufficient to identify the state of the matter. (T/F)

163. If the pressure of matter in a given state is less than that of the saturation one, the matter is:
(a) superheated gas
(b) compressed liquid
(c) saturated liquid
(d) saturated vapor

164. If temperature of matter in a given state is less than that of the saturation one, the matter is:
(a) superheated gas
(b) compressed liquid
(c) saturated liquid
(d) saturated vapor

165. With increasing temperature, pressure . . . , partial volume of fluid . . . , and partial volume of gas . . .
(a) decreases / increases / increases
(b) increases / decreases / decreases
(c) decreases / decreases / increase
(d) increases / increases / decreases

166. With increasing temperature, internal energy . . . , enthalpy . . . , entropy of fluid . . . , and entropy of gas . . .
(a) decreases / decreases / increases / increases
(b) increases / increases / decreases / increases
(c) decreases / decreases / increases / decreases
(d) increases / increases / increases / decreases

167. Heating water during an isobaric process starts from the . . . state to the . . . state.
(a) superheated / supercooled
(b) supercooled / superheated
(c) supercooled / supercooled
(d) superheated / superheated

168. If temperature and pressure of matter are both above those of the critical point, the matter is in the:
(a) superheated zone
(b) compressed liquid zone
(c) supercritical zone
(d) subcooled zone

169. On the pressure-temperature diagram for water, the matter on the left side of the liquid line is . . . , the matter on the right side of the vapor line is . . . , and the matter in between the two lines is . . .
(a) steam / liquid water / two-phase water-vapor
(b) liquid water / steam / two-phase water-vapor
(c) two-phase water-vapor / steam / liquid water
(d) liquid water / two-phase water-vapor / steam

170. If the steam table data is missing the exact information for the desired pressure or temperature, the data is calculated by:
(a) data cannot be obtained
(b) best guess
(c) looking for a more complete table
(d) interpolation or extrapolation

171. At very low pressures, where temperature is lower than the saturation temperature at the given pressure, entropy is equal to that of the:
(a) superheated gas
(b) saturated liquid
(c) saturated vapor
(d) wet vapor

172. At very low pressures, where temperature is lower than the saturation temperature at the given pressure, enthalpy is equal to the internal energy:

(a) of the saturated vapor
(b) of the saturated liquid
(c) plus pressure by volume of the saturated liquid
(d) minus temperature by entropy of the saturated vapor

173. Vapor quality represents the percentage of the:

(a) vapor in the saturated mixture of vapor-liquid
(b) liquid in the saturated liquid
(c) solid in the saturated liquid
(d) solid in the saturated vapor

174. If any thermodynamic property is at the value greater than that of its equivalent saturated vapor at the given pressure or temperature, the matter is:

(a) superheated gas   (b) supercooled liquid
(c) saturated vapor   (d) dry vapor

175. Among the following equations, which represents the relation between the quality and property for a given pressure and temperature (s is entropy, h is enthalpy, u is internal energy, x is vapor quality, and f and g subscripts represent the liquid and gas states)?

(a) $h = h_f + x h_{fg}$ (b) $u = x u_f + (1-x) u_{fg}$
(c) $v = (1-x) v_f + v_g$ (d) $s = s_f + (1-x) s_g$

176. The conditions for water triple point are . . . and for critical point are . . .

(a) (0.01 °C, 0.6113 kPa) / (374.14 °C, 22.09 MPa)
(b) (0.01 °C, 0.6113 kPa) / (0.01 °C, 0.6113 MPa)
(c) (374.14 °C, 22.09 MPa) / (0.01 °C, 0.6113 kPa)
(d) (374.14 °C, 22.09 MPa) / (374.14 °C, 22.09 MPa)

177. Gibbs's phase rule is defined as . . . , where F, N, and π are the number of degrees of freedom, components, and phases, respectively:

(a) $F = 2 + N - \pi$ (b) $N = 2 + F - \pi$
(c) $\pi = 2 + N - F$ (d) $2 = F + N - \pi$

178. The number of degrees of freedom for the triple point is:

(a) 3   (b) 2   (c) 1   (d) 0

179. The number of degrees of freedom for the critical point is:

(a) 3   (b) 2   (c) 1   (d) 0

180. The difference of the saturated liquid and vapor properties of water at the critical point is:

(a) 1   (b) 0   (c) > 1   (d) < 1

181. The ratio of the saturated liquid to saturated vapor properties of water at the triple point is:

(a) 1   (b) 0   (c) > 1   (d) < 1

182. For a cycle that operates between two reservoirs at fixed temperatures (T_1 and T_2), where $T_1 > T_2$, the efficiency:

(a) cannot exceed that of the Carnot cycle
(b) varies depending on the temperatures
(c) is very large
(d) is very small

183. The efficiencies (η) of heat engine cycles presented in Figure 130, operating between the two heat sources, where $T_1 > T_2$, are (note that W is the work done and Q is the heat transferred):

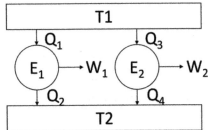

FIGURE 130. Heat engine cycle.

(a) $\eta_1 = \dfrac{W_1}{Q_1} = 1 - \dfrac{T_2}{T_1}, \eta_2 = \dfrac{W_2}{Q_3} = 1 - \dfrac{T_2}{T_1}$ (b) $\eta_2 = \dfrac{W_1}{Q_1} = 1 - \dfrac{T_2}{T_1}, \eta_1 = \dfrac{W_2}{Q_3} = 1 - \dfrac{T_2}{T_1}$

(c) $\eta_1 = \dfrac{W_1}{Q_1} = 1 - \dfrac{T_1}{T_2}, \eta_2 = \dfrac{W_2}{Q_3} = 1 - \dfrac{T_1}{T_2}$ (d) $\eta_2 = \dfrac{W_1}{Q_1} = 1 - \dfrac{T_1}{T_2}, \eta_1 = \dfrac{W_2}{Q_3} = 1 - \dfrac{T_1}{T_2}$

184. The efficiency of a reversible engine is:
(a) the highest among all possible engine types
(b) the lowest among all possible engine types
(c) 0
(d) 1

185. The Carnot relation, that shows the relation between the heat added to and rejected from a thermodynamic system, states that (Q is the heat trnaferred, T is temperature, and H and C subscripts represent hot and cold conditions):

(a) $\dfrac{Q_c}{T_c} = \dfrac{Q_H}{T_H}$ (b) $\dfrac{Q_c}{Q_H} = \dfrac{T_H}{T_C}$ (c) $\dfrac{Q_c}{T_c} = 1 - \dfrac{Q_H}{T_H}$ (d) $\dfrac{Q_c}{Q_H} = 1 - \dfrac{T_H}{T_C}$

186. In order to convert water from compressed liquid to superheated gas, the following processes may be employed:
(a) isothermal expansion and isobaric cooling
(b) isothermal contraction and isobaric heating
(c) isothermal expansion and isobaric cooling
(d) isothermal contraction and isobaric heating

187. Assuming that the enthalpy (h) at state 2 is twice that at state 1, pressures (P) are equal, and volume (V) at state 2 is twice that at state 1, the internal energy (u) of states 1 and 2 are related by:
(a) $u_2 = 1.5u_1$ (b) $u_2 = 2u_1$ (c) $u_2 = u_1$ (d) $u_2 = 0.5\,u_1$

188. If water at a constant pressure is cooled and as a result its vapor quality changes from x_1 to x_2, assuming that the process is adiabatic and reversible, the work done is (u is internal energy, h is enthalpy, and f and g subscripts represent the liquid and gas states):
(a) $(x_2 - x_1)u_{fg}$ (b) $(x_1 - x_2)h_{fg}$ (c) $(x_1 - x_2)u_{fg}$ (d) $(x_2 - x_1)h_{fg}$

189. Two bodies of water with equal mass, each having their own vapor quality, are mixed. If the vapor quality of the first body of water was 0.75 and the vapor quality of the mixture is 0.5, the vapor quality of the second body of water was:
(a) 0.25   (b) 0.5   (c) 0.75   (d) 1

190. A heat engine is made to operate between 100 °C and 300 °C. The efficiency of this heat engine is:

(a) 35% (b) 18% (c) 70% (d) 21%

191. A refrigerator is made to operate between 0 °C and 40 °C. The coefficient of performance of this refrigerator is:

(a) 0.15 (b) 0.68 (c) 6.8 (d) 1.5

192. A heat pump is made to operate between 40 °C and 140 °C. The coefficient of performance of this heat pump is:

(a) 1.04 (b) 4.13 (c) 2.07 (d) 8.26

193. The universal entropy for a system and its surroundings is always greater than zero, meaning that entropy is reduced. (T/F)

194. If a process is adiabatic and reversible, it must be:

(a) isentropic (b) isothermal (c) isobaric (d) isochoric

195. If a process is reversible and isentropic, it must be:

(a) Isothermal (b) isobaric (c) adiabatic (d) isochoric

196. Changes in entropy at a constant pressure (ds_p) and volume (ds_v), between two temperature levels, where $T_2 > T_1$, are gioven by (C_p and C_v are heat capacities at constant pressure and volume) :

(a) $ds_p = C_v \ln(T_1/T_2), ds_v = C_p \ln(T_1/T_2)$

(b) $ds_p = C_p \ln(T_1/T_2), ds_v = C_v \ln(T_1/T_2)$

(c) $ds_p = C_v \ln(T_2/T_1), ds_v = C_p \ln(T_2/T_1)$

(d) $ds_p = C_p \ln(T_2/T_1), ds_v = C_v \ln(T_2/T_1)$

197. Changes in entropy between 2 temperature levels, where $T_2 > T_1$, are given by (P is pressure, V is volume, C_p and C_v are heat capacities at constant pressure and volume, and $s = s(V, P)$):

(a) $ds = C_p \ln\left(\dfrac{P_2}{P_1}\right) + C_v \ln\left(\dfrac{V_2}{V_1}\right)$ (b) $ds = C_p \ln\left(\dfrac{V_2}{V_1}\right) + C_v \ln\left(\dfrac{P_2}{P_1}\right)$

(c) $ds_p = C_p \ln\left(\dfrac{T_2}{T_1}\right) + C_v \ln\left(\dfrac{T_2}{T_1}\right)$ (d) $ds_v = C_p \ln\left(\dfrac{T_1}{T_2}\right) + C_v \ln\left(\dfrac{T_1}{T_2}\right)$

198. The slope for the entropy changes for an isobaric process, between 2 temperature levels, with $T_2 > T_1$, is . . . that of the isochoric process.

(a) greater than (b) less than
(c) the same as (d) not comparable to

199. Isobaric and isochoric lines in an entropy-temperature diagram:

(a) cross one another once (b) cross one another twice
(c) do not cross one another (d) eventually converge

200. Changes of universal entropy for the system presented in Figure 131 are (assume there is no entropy generation; T is temperature and Q is the heat transferred):

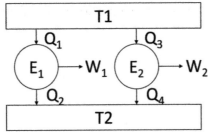

FIGURE 131. Heat engine cycle.

(a) $ds_p = -\dfrac{Q_1}{T_1} + \dfrac{Q_2}{T_2} - \dfrac{Q_3}{T_1} + \dfrac{Q_4}{T_2}$ 　　　　(b) $ds_p = \dfrac{Q_1}{T_1} + \dfrac{Q_2}{T_2} + \dfrac{Q_3}{T_1} + \dfrac{Q_4}{T_2}$

(c) $ds_p = -\dfrac{Q_1}{T_1} - \dfrac{Q_2}{T_2} - \dfrac{Q_3}{T_1} - \dfrac{Q_4}{T_2}$ 　　　　(d) $ds_p = \dfrac{Q_1}{T_1} - \dfrac{Q_2}{T_2} + \dfrac{Q_3}{T_1} - \dfrac{Q_4}{T_2}$

201. An isentropic process is always adiabatic. (T/F).

202. An isentropic process is always reversible. (T/F).

203. Isentropic and isothermal lines in a pressure-volume diagram:
    (a) cross one another once 　　　　(b) cross one another twice
    (c) do not cross one another 　　　　(d) eventually converge

204. An ideal gas in a closed system undergoes processes in which its pressure quadruples and its volume doubles. Change in entropy is (P is pressure, V is volume, C_p and C_v are heat capacities at constant pressure and volume; and $s = s(V, P)$):
    (a) $(2C_p + C_v)\ln(2)$ 　　　　(b) $(C_p + 2C_v)\ln(2)$
    (c) $(C_p + 0.5C_v)\ln(2)$ 　　　　(d) $(0.5C_p + C_v)\ln(2)$

205. An ideal gas in a closed system undergoes processes in which its pressure quadruples and its temperature doubles. Change in entropy is (P is pressure, V is volume, R is gas constant, C_p and C_v are heat capacities at constant pressure and volume, and $s = s(T, P)$):
    (a) $(C_p - 2R)\ln(2)$ 　　　　(b) $(C_p + 2R)\ln(2)$
    (c) $(C_p + 0.5C_v)\ln(2)$ 　　　　(d) $(0.5C_p + C_v)\ln(2)$

206. An ideal gas in a closed system undergoes processes in which its volume quadruples and its temperature doubles. Change in entropy is (P is pressure, V is volume, R is gas constant, C_p and C_v are heat capacities at constant pressure and volume, and $s = s(T, V)$):
    (a) $(C_p - 2R)\ln(2)$ 　　　　(b) $(C_p + 2R)\ln(2)$
    (c) $(C_p + 0.5C_v)\ln(2)$ 　　　　(d) $(0.5C_p + C_v)\ln(2)$

207. The difference between the Gibbs and Helmholtz energy changes is (P is pressure, V is volume, T is temperature, S is entropy, U is internal energy, and H is enthalpy):
    (a) $d(PV)$ 　　　(b) TS 　　　(c) $H - U$ 　　　(d) 0

208. The difference between the Gibbs and Helmholtz energy changes at a constant pressure (P) and volume (V) process is (T is temperature, S is entropy, U is internal energy, and H is enthalpy):
    (a) PV 　　　(b) TS 　　　(c) $H - U$ 　　　(d) 0

209. Select the correct Maxwell relation to describe the relationship between temperature (T), pressure (P), volume (V), and entropy (S).

(a) $\left(\dfrac{\partial T}{\partial V}\right)_S = -\left(\dfrac{\partial P}{\partial S}\right)_V$

(b) $\left(\dfrac{\partial T}{\partial P}\right)_S = -\left(\dfrac{\partial V}{\partial S}\right)_P$

(c) $\left(\dfrac{\partial V}{\partial T}\right)_P = \left(\dfrac{\partial S}{\partial P}\right)_T$

(d) $\left(\dfrac{\partial P}{\partial T}\right)_V = -\left(\dfrac{\partial S}{\partial V}\right)_T$

210. Select the correct reciprocal relationship often used in thermodynamics:

(a) $\left(\dfrac{\partial x}{\partial y}\right)_z \left(\dfrac{\partial y}{\partial z}\right)_x \left(\dfrac{\partial z}{\partial x}\right)_y = 1$

(b) $\left(\dfrac{\partial x}{\partial y}\right)_z \left(\dfrac{\partial y}{\partial z}\right)_x \left(\dfrac{\partial z}{\partial x}\right)_y = -1$

(c) $\left(\dfrac{\partial x}{\partial z}\right)_y \left(\dfrac{\partial z}{\partial y}\right)_x \left(\dfrac{\partial y}{\partial x}\right)_z = 1$

(c) $\left(\dfrac{\partial x}{\partial y}\right)_z \left(\dfrac{\partial z}{\partial y}\right)_x \left(\dfrac{\partial x}{\partial z}\right)_y = -1$

211. The Jacobean of (u, v) with respect to (x, y) is:

(a) $\dfrac{\partial(u,v)}{\partial(x,y)} \cdot \dfrac{\partial(x,y)}{\partial(p,q)} = \dfrac{\partial(p,q)}{\partial(u,v)}$

(b) $\dfrac{\partial(u,v)}{\partial(x,y)} \cdot \dfrac{\partial(x,y)}{\partial(p,q)} = -\dfrac{\partial(p,q)}{\partial(u,v)}$

(c) $\dfrac{\partial(u,v)}{\partial(x,y)} \cdot \dfrac{\partial(x,y)}{\partial(p,q)} = \dfrac{\partial(u,v)}{\partial(p,q)}$

(d) $\dfrac{\partial(u,v)}{\partial(x,y)} \cdot \dfrac{\partial(x,y)}{\partial(p,q)} = -\dfrac{\partial(u,v)}{\partial(p,q)}$

212. The difference between heat capacities at constant pressure and volume $(C_p - C_v)$ is (T is temperature, P is pressure, and V is volume):

(a) $\left(\dfrac{\partial P}{\partial T}\right)_V \left(\dfrac{\partial V}{\partial T}\right)_P$

(b) $-T\left(\dfrac{\partial P}{\partial T}\right)_V \left(\dfrac{\partial V}{\partial T}\right)_P$

(c) $T\left(\dfrac{\partial T}{\partial P}\right)_V \left(\dfrac{\partial T}{\partial V}\right)_P$

(d) $-T\left(\dfrac{\partial T}{\partial P}\right)_V \left(\dfrac{\partial T}{\partial V}\right)_P$

213. For a control volume, the relation between the rate of mass change $\left(\dfrac{dm_{cv}}{dt}\right)$, density (ρ), volume (V), mass rate $(\dot{m})$, and cross-sectional area (A) is:

(a) $\dfrac{dm_{cv}}{dt} = (\ A) = -\dot{m}_e - \dot{m}_i$

(b) $\dfrac{dm_{cv}}{dt} = (\rho V) = -\dot{m}_e + \dot{m}_i$

(b) $\dfrac{dm_{cv}}{dt} = (\rho A V) = \dot{m}_e + \dot{m}_i$

(d) $\dfrac{dm_{cv}}{dt} = (\rho A V) = \dot{m}_e - \dot{m}_i$

214. The relation between the rate of energy change $\left(\dfrac{dE_{cv}}{dt}\right)$, density (ρ), volume (V), cross-sectional area (A), enthalpy (h), heat rate $(\dot{Q})$, work rate $(\dot{W})$, mass rate $(\dot{m})$. height (z), and flow velocity (V) at inlet (i) and outlet (e) is:

(a) $\dfrac{dE_{cv}}{dt} = \dot{Q} - \dot{W} + \dot{m}_i\left(h_i + \dfrac{V_i^2}{2} + gz_i\right) - \dot{m}_e\left(h_e + \dfrac{V_e^2}{2} + gz_e\right)$

(b) $\dfrac{dE_{cv}}{dt} = \dot{Q} - \dot{W} + \dot{m}_i(e_i + p_i V_i) - \dot{m}_e(e_e + p_e V_e)$

(c) $\dfrac{dE_{cv}}{dt} = -\dot{Q} + \dot{W} - \dot{m}_i\left(h_i + \dfrac{V_i^2}{2} + gz_i\right) + \dot{m}_e\left(h_e + \dfrac{V_e^2}{2} + gz_e\right)$

(d) $\dfrac{dE_{cv}}{dt} = -\dot{Q} + \dot{W} + \dot{m}_i(e_i + p_iV_i) - \dot{m}_e(e_e + p_eV_e)$

215. For a turbine presented in Figure 132, the relation between inlet (i) and outlet (e) enthalpies (h_i, h_e) is (assume flow velocities entering and leaving the turbine are the same):

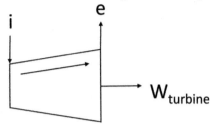

FIGURE 132. Turbine in an adiabatic process.

(a) $(h_e - h_i) \geq 0$ (b) $(-h_e - h_i) \geq 0$ (c) $(-h_e + h_i) \geq 0$ (d) $(h_e + h_i) \leq 0$

216. For a compressor presented in Figure 133, the relation between inlet (i) and outlet (e) enthalpies (h_i, h_e) is (assume flow velocities entering and leaving the compressor are the same):

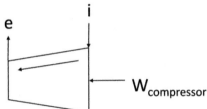

FIGURE 133. Compressor in an adiabatic process.

(a) $(h_e - h_i) \leq 0$ (b) $(-h_e - h_i) \geq 0$ (c) $(h_e - h_i) \geq 0$ (d) $(-h_e - h_i) \leq 0$

217. For the nozzle shown in Figure 134, the isentropic efficiency is:

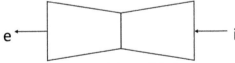

FIGURE 134. Nozzle in an adiabatic process.

(a) $\left(\dfrac{v_i^2}{v_e^2}\right) \geq 1$ (b) $\left(\dfrac{v_e^2}{v_i^2}\right) \geq 1$ (c) $\left(\dfrac{v_i^2}{v_e^2}\right) \leq 1$ (d) $\left(\dfrac{v_e^2}{v_i^2}\right) \leq 1$

218. The Clausius–Clapeyron relation for a constant pressure (P) and temperature (T) process is given by (V is volume and L is latent heat):

(a) $\dfrac{dP}{dT} = \dfrac{T}{L\Delta V}$ (b) $\dfrac{dP}{dT} = \dfrac{L}{T\Delta V}$

(c) $\dfrac{dP}{dT} = \dfrac{L\Delta V}{T}$ (d) $\dfrac{dP}{dT} = \dfrac{T\Delta V}{L}$

219. Enthalpy change at a constant temperature process, where the entropy change is 0.5 kJ/kgK and temperature is 200 °C, is:

(a) 261.2 kJ/kg   (b) 226.2 kJ/kg   (c) 208.7 kJ/kg   (d) 236.6 kJ/kg

220. Helium passes through an adiabatic environment where it exchanges heat with a liquid that also passes through it. Helium enters at room temperature (22 °C) and leaves at 100 °C. The heat capacity of helium is 5.2 kJ/kg K and that of the liquid is 4.2 kJ/kg K. The flow rate for helium is twice that of the liquid. If the liquid enters the environment at 1,200 °C, the liquid enthalpy change is:

(a) 811.2 kW   (b) −811.2 kW   (c) 405.6 kW   (d) −405.6 kW

221. Helium passes through an adiabatic environment where it exchanges heat with a liquid that also passes through it. Helium enters at room temperature (22 °C) and leaves at 100 °C. The heat capacity of helium is 5.2 kJ/kg K and that of the liquid is 4.2 kJ/kg K. The flow rate for helium is twice that of the liquid. If the liquid enters the environment at 1,200 °C, the liquid leaves the environment at:

(a) 1,006.9 °C   (b) 794.6 °C   (c) 1,393.1 °C   (d) 227.1 °C

222. Which of the following are correct statements?

(1) Fluids are substances that flow.
(2) Fluids are gases or liquids.
(3) Gases have non-interacting molecules that may collide.
(4) Solids have large compressibility ratios.

(a) (1), (2), and (3)   (b) (1), (3), and (4)
(c) (2) and (3)   (d) (1), (2), and (4)

223. If a compressible fluid is entrapped in a piston-cylinder system, with increasing pressure:

(a) volume decreases and temperature increases
(b) density decreases and temperature increases
(c) pressure increases and density decreases
(d) pressure increases and temperature decreases

224. Isothermal compressibility is the change of:

(a) density with temperature at a constant pressure
(b) volume with temperature at a constant pressure
(c) density with pressure at a constant temperature
(d) volume with pressure at a constant temperature

225. The ratio of the isothermal to isentropic compressibility is the:

(a) heat capacity ratio   (b) compressibility ratio
(c) pressure ratio   (d) thermal expansion coefficient

226. As the molecules in matter become more densely packed, the effect of the repulsive force between molecules:

(a) dominates   (b) decreases
(c) remains the same   (d) becomes negligible

227. Ideal gas molecules bounce back elastically when hitting walls. (T/F)

228. Surface orientation is described by a unit vector that is oriented:

(a) normal toward to the surface   (b) normal away from the surface
(c) parallel to the surface   (d) tangent to the surface

229. A perfect gas:

(a) has a constant viscosity   (b) has zero viscosity
(c) is incompressible   (d) has intermolecular forces that can be ignored

230. An instrument that is used to analyze a gas sample for its oxygen, carbon monoxide, and carbon dioxide content is a(an):

(a) Orsat   (b) hygrostat   (c) gas meter   (d) humidistat

231. A humidistat is sensitive to variations of . . . and . . .

(a) specific humidity / temperature
(b) humidity / not temperature
(c) humidity / temperature
(d) specific humidity / not temperature

232. A device that is used to convert mechanical energy into electrical energy is a:

(a) battery   (b) generator   (c) motor   (d) transformer

233. The engines on most trucks are:

(a) four-stroke steam engines with a turbocharger
(b) two-stroke gasoline engines with an intercooler
(c) four-stroke diesel engines with a turbocharger and intercooler
(d) four-stroke kerosene engines with a turbocharger and intercooler

234. Wind speed and direction are measured by a(an):

(a) altimeter   (b) tachometer   (c) psychrometer   (d) anemometer

235. A psychrometer is used to measure:

(a) dew point and wet bulb temperatures
(b) dry and wet bulb temperatures
(c) humidity and dew point temperatures
(d) humidity ratio and dry bulb temperature

236. When evaporation occurs, the . . . of the liquid . . . because the . . . is . . . it. This is known as . . .

(a) temperature / decreases / energy / removed from / evaporation cooling
(b) pressure / increases / heat / added to / condensation cooling
(c) temperature / decreases / energy / removed from / evaporation cooling
(d) pressure / decreases / heat / added to / condensation cooling

237. Entropy is related to a variation of . . . with . . .

(a) temperature / heat capacity   (b) pressure / pressure
(c) energy / temperature   (d) thermal conductivity / pressure variations

238. The ratio of the heat capacities at constant pressure to volume for monoatomic gases is about:

(a) 1.7   (b) 2.9   (c) 2.3   (d) 1.4

239. The ratio of the heat capacities at constant pressure to volume for diatomic gases is always greater than one. (T/F)

240. Ideal gas pressure is . . . than that for real gases at . . . pressures, based on the . . . equation.

(a) higher than / low / ideal gas law   (b) lower than / high / van der Waals
(c) lower than / high / van der Waals   (d) higher than / low / ideal gas law

241. The most efficient thermodynamic cycle is the . . . cycle.

(a) Stirling   (b) Diesel   (c) Carnot   (d) Otto

242. Which of the following statements concerning a mixture of gases is(are) true:

   (1) The total pressure of the ideal gas mixture is the sum of the partial pressures of the mixture components.

   (2) The partial pressure of the gas component is equal to the pressure it would have had if it occupied the entire volume of the mixture at the same temperature.

   (3) Gases react according to their partial pressures, and not according to their concentrations in gas mixtures or liquids.

   (a) (1), (2), and (3)   (b) (2) and (3)
   (c) (1) and (2)   (d) (1) and (3)

243. The gravimetric analysis of a gas results in the data presented in Table 72. The pressure of the mixture is:

TABLE 72. Gravimetric analysis results.

Components	Concentration(%)	Partial pressure (Pa)
A	10	0.25
B	20	0.35
C	30	0.45
D	40	0.55

   (a) 25 Pa   (b) 35 Pa   (c) 45 Pa   (d) 55 Pa

244. The engine performance is described by the:

   (a) power-to-voltage ratio   (b) mass-to-specific power ratio
   (c) power-to-volume ratio   (d) power-to-weight ratio

245. Speed of sound in an ideal gas . . . with . . .

   (a) decreases / pressure   (b) decreases / humidity
   (c) increases / density   (d) increases / temperature

246. The speed of sound in an ideal gas is equal to (γ is the ratio of the heat capacities, R is gas constant, T is temperature, K is bulk modulus, and ρ is density):

   (a) $c = \sqrt{\gamma R T}$ (b) $c = \gamma R T$ (c) $c = \sqrt{K/\rho}$ (d) $c = \sqrt{K\rho}$

247. The speed of sound in liquids is equal to (γ is the ratio of the heat capacities, R is gas constant, T is temperature, K is bulk modulus, and ρ is density):

   (a) $c = \sqrt{\gamma R T}$ (b) $c = \gamma R T$ (c) $c = \sqrt{K/\rho}$ (d) $c = \sqrt{K\rho}$

248. Someone claims that their heat engine achieves 100% thermal efficiency. Their claim:

   (a) must be false as it violates the zeroth law of thermodynamics
   (b) is possible as it agrees with the first law of thermodynamics
   (c) is possible as it agrees with the conservation of energy law
   (d) must be false as it violates the second law of thermodynamics

Answer Key

1. (c)	**2.** (b)	**3.** (a)	**4.** (b)	**5.** (b)	**6.** (c)	**7.** (b)	**8.** (a)	**9.** (b)	**10.** (a)
11. (d)	**12.** (a)	**13.** (b)	**14.** (c)	**15.** (d)	**16.** (a)	**17.** (d)	**18.** (a)	**19.** (a)	**20.** (a)
21. (c)	**22.** (b)	**23.** (d)	**24.** (a)	**25.** (b)	**26.** (a)	**27.** (d)	**28.** (b)	**29.** F	**30.** (a)
31. (d)	**32.** (a)	**33.** (b)	**34.** (d)	**35.** (c)	**36.** (a)	**37.** (b)	**38.** (c)	**39.** (d)	**40.** (a)
41. (d)	**42.** (a)	**43.** (a)	**44.** (b)	**45.** (c)	**46.** (b)	**47.** (a)	**48.** (a)	**49.** (a)	**50.** (b)
51. (c)	**52.** (a)	**53.** (b)	**54.** (d)	**55.** (b)	**56.** (a)	**57.** (a)	**58.** (d)	**59.** (a)	**60.** (b)
61. (b)	**62.** (a)	**63.** (b)	**64.** (a)	**65.** (c)	**66.** (d)	**67.** (a)	**68.** (c)	**69.** (d)	**70.** (b)
71. (d)	**72.** (a)	**73.** (a)	**74.** (a)	**75.** (d)	**76.** T	**77.** T	**78.** (d)	**79.** (c)	**80.** (d)
81. (d)	**82.** (a)	**83.** (b)	**84.** (d)	**85.** (b)	**86.** (a)	**87.** (d)	**88.** (b)	**89.** (a)	**90.** T
91. T	**92.** (d)	**93.** F	**94.** F	**95.** (a)	**96.** (a)	**97.** (d)	**98.** (d)	**99.** (c)	**100.** (d)
101. (b)	**102.** (a)	**103.** (a)	**104.** (b)	**105.** (b)	**106.** (a)	**107.** (a)	**108.** T	**109.** (a)	**110.** (b)
111. (c)	**112.** (a)	**113.** T	**114.** (a)	**115.** (b)	**116.** (d)	**117.** (d)	**118.** (a)	**119.** (a)	**120.** F
121. F	**122.** F	**123.** (a)	**124.** (d)	**125.** (c)	**126.** T	**127.** T	**128.** (b)	**129.** T	**130.** (a)
131. (c)	**132.** (b)	**133.** (a)	**134.** (c)	**135.** T	**136.** T	**137.** (a)	**138.** (b)	**139.** (a)	**140.** (a)
141. (d)	**142.** (a)	**143.** (c)	**144.** (b)	**145.** (a)	**146.** (a)	**147.** (b)	**148.** (a)	**149.** (d)	**150.** (a)
151. (b)	**152.** (a)	**153.** (b)	**154.** (b)	**155.** (c)	**156.** (d)	**157.** (b)	**158.** (b)	**159.** (c)	**160.** (a)
161. (a)	**162.** T	**163.** (a)	**164.** (b)	**165.** (d)	**166.** (d)	**167.** (b)	**168.** (c)	**169.** (b)	**170.** (d)
171. (b)	**172.** (c)	**173.** (a)	**174.** (a)	**175.** (a)	**176.** (a)	**177.** (a)	**178.** (d)	**179.** (c)	**180.** (b)
181. (b)	**182.** (a)	**183** (a)	**184.** (a)	**185.** (a)	**186.** (d)	**187.** (b)	**188.** (c)	**189.** (a)	**190.** (a)
191. (c)	**192.** (b)	**193.** T	**194.** (a)	**195.** (c)	**196.** (d)	**197.** (b)	**198.** (a)	**199.** (a)	**200.** (a)
201. F	**202.** F	**203.** (a)	**204.** (b)	**205.** (a)	**206.** (b)	**207.** (a)	**208.** (d)	**209.** (a)	**210.** (b)
211. (c)	**212.** (a)	**213.** (d)	**214.** (a)	**215.** (a)	**216.** (a)	**217.** (d)	**218.** (b)	**219.** (d)	**220.** (b)
221. (a)	**222.** (d)	**223.** (a)	**224.** (d)	**225.** (a)	**226.** (a)	**227.** T	**228.** (b)	**229.** (d)	**230.** (a)
231. (b)	**232.** (b)	**233.** (c)	**234.** (d)	**235.** (b)	**236.** (a)	**237.** (c)	**238.** (a)	**239.** T	**240.** (b)
241. (c)	**242.** (a)	**243.** (c)	**244.** (d)	**245.** (d)	**246.** (a)	**247.** (c)	**248** (d)		

CHAPTER 7

HEAT TRANSFER

1. The main modes of heat transfer are:
   (a) conduction, convection, and radiation
   (b) convection and electromagnetism
   (c) dynamism and electromagnetism
   (d) conduction and convection

2. Conduction heat transfer happens by means of:
   (a) electromagnetic waves   (b) contact between surfaces
   (c) currents in a fluid   (d) all of the above

3. Convection heat transfer happens by means of:
   (a) electromagnetic waves   (b) contact between surfaces
   (c) currents in a fluid   (d) all of the above

4. Radiation heat transfer happens by means of:
   (a) electromagnetic waves   (b) contact between surfaces
   (c) currents in a fluid   (d) all of the above

5. If A is area, e is emissivity, and s is the Stefan-Boltzmann constant, Stefan-Boltzmann's law states that $Q = \sigma \varepsilon A T^4$ Q is:
   (a) light intensity emitted by the body (b) irradiation from a body
   (c) power radiated from a black body (d) reflected energy from a black body

6. The convective heat transfer coefficient relates:
   (a) heat capacity to temperature variation
   (b) temperature to heat transfer flux
   (c) heat capacity to cross-sectional area variation
   (d) the difference between fluid and solid object temperature variations to heat transfer flux rate

7. A black body with a rough surface absorbs heat . . . the one with a smooth surface:
   (a) better than   (b) less than
   (c) the same as   (d) more or less than, depending on other factors

8. In convection heat transfer:
   (a) heat transfer occurs between solid bodies
   (b) heat transfer occurs among fluids
   (c) heat transfer occurs between solids and fluids
   (d) heat transfer occurs in the vacuum

9. Among the following, which one is the best radiation emitter?
   (a) a white object with rough surfaces   (b) a black object with rough surfaces
   (c) a black object with smooth surfaces (d) a white object with smooth surfaces

10. According to the Stefan-Boltzmann equation, the ratio of the radiated energy from a black body radiating at 600 K to that of a black body radiating at 500 K, given the ambient temperature of 22 °C, is:
    (a) 2.2   (b) 1.0   (c) 3.2   (d) 4.3

11. Objects A and B have the same interior temperatures and are exposed to the same environment. Object A is a 2-mm thick 1×2 m^2 plate with a thermal conductivity of 1 W/mK. Object B is a 3-mm thick 2×1 m^2 plate with a thermal conductivity of 1.5 W/mK. What is the ratio of the heat loss rate from object A relative to that from object B?
    (a) 0.5   (b) 1   (c) 1.5   (d) 2

12. A 1-cm thick 3×2 m^2 composite plate is being heated evenly across one face and maintains a 10 K temperature difference between its two faces. If heat is being conducted across the plate thickness at the rate of 6,000 W, the plate's thermal conductivity is:
    (a) 0.25 W/mK   (b) 0.5 W/mK   (c) 1 W/mK   (d) 1.25 W/mK

13. An object and the environment are at the same temperature. What can you observe about the heat convection coefficient of the object?
    (a) nothing   (b) it is infinite   (c) it is 0 W/m^{2}K   (d) it is 1 W/m^{2}K

14. A pipe has an interior diameter of D and an exterior diameter of $2D$, with a thermal conductivity of k. Conduction heat transfer from the interior to the exterior per unit length of the pipe per degree Celsius is given by:
    (a) $2\pi k / \ln(2)$ (b) $2\pi k$ (c) $\pi k / \ln(2)$ (d) πk

15. Thermal resistance of a pipe with a given heat transfer of $Q = 2\pi k L (T_i - T_s)/5$ is (k is thermal conductivity, T is temperature, L is length, and Q is heat transfer rate):
    (a) $(\pi k L)/5$ (b) $(2\pi k L)/5$ (c) $5/(\pi k L)$ (d) $5/(2\pi k L)$

16. The International System (SI) unit of temperature measurement is:
    (a) Celsius   (b) Kelvin   (c) Fahrenheit   (d) Rankine

17. Absolute zero is equal to . . . . It is the temperature at which the molecules . . .
    (a) –273.15 °C / cease moving   (b) 467 K / slow down
    (c) 273.16 K / cease moving   (d) 0 °C / speed up

18. The driving force for heat transfer is . . . ; molecules transfer heat from regions with . . . to that of the . . .
    (a) humidity variations / lower temperatures / higher temperatures
    (b) temperature change / lower temperatures / higher temperatures
    (c) humidity change / higher temperatures / lower temperatures
    (d) temperature change / higher temperatures / lower temperatures

19. Energy is a quantitative property that causes an object to . . . or . . .

(a) be heated / work to be conducted on it or by it

(b) be energized / work to be conducted by it

(c) be electrically charged / heated

(d) be heated or electrically charged / magnetized

20. Work is defined as force multiplied by distance or energy transferred by a system to its surroundings. (T/F)

21. Energy is defined as a property transferred to an object in order to heat it or do work on it. (T/F)

22. In a system, if a force is applied along the direction of the displacement, the work done by the system is negative. (T/F)

23. Energy is proportionate to the product of:

(a) force and area

(b) momentum and distance

(c) power and velocity

(d) pressure and volume

24. A system's internal energy is:

(a) the total molecular kinetic energy

(b) the total molecular potential energy

(c) the energy due to the changes in its internal state

(c) all of the above

25. A piston-cylinder system is filled with neon gas. If the neon moves the piston outward, the work done by the gas is positive. (T/F)

26. If the force applied and the resulting displacement are in the opposite direction, the work done is negative. (T/F)

27. Zero-degrees Fahrenheit in Celsius is:

(a) 32   (b) −17.8   (c) 17.8   (d) −32

28. The pressure-temperature curve on a pressure-temperature (*P-T*) diagram for pure materials, up to the triple point, is approximately:

(a) linear   (b) parabolic   (c) hyperbolic   (d) concave

29. The pressure-temperature curve for pure materials crosses the pressure axis at . . . and the temperature axis at . . .

(a) zero pressure / zero temperature

(b) zero pressure / absolute zero temperature

(c) absolute zero pressure / absolute zero temperature

(d) absolute zero pressure / zero temperature

30. 373 K in degrees Fahrenheit is:

(a) 212   (b) −212   (c) 100   (d) −100

31. Zero-degrees Fahrenheit in Kelvin is:

(a) −17.7   (b) 255.4   (c) 17.7   (d) −255.4

32. The average molecular kinetic energy is measured by a:

(a) thermometer   (b) manometer

(c) accelerometer   (d) power meter

33. As gas is heated, the molecules' kinetic energy . . . ; temperature . . . ; frequency of intermolecular collisions . . . ; and the number of gas molecules . . .
    (a) increases / increases / increases / remains the same
    (b) increases / decreases / increases / remains the same
    (c) increases / increases / decrease / remains the same
    (d) decreases / increases / increases / decreases

34. Thermal conductivity shows how:
    (a) well a substance lets heat be transmitted
    (b) poorly a substance allows energy to accumulate
    (c) disruptive a barrier is to energy
    (d) restrictive current is

35. A substance that does not allow heat to be easily transmitted is a(an):
    (a) thermal insulator   (b) electrical insulator
    (c) insulator   (d) energy insulator

36. A substance that allows heat to easily flow through it is a(an):
    (a) insulator   (b) combustor
    (c) conductor   (d) acoustic absorber

37. Radiation transfer can only take place between objects in a vacuum. (T/F)

38. In convection heat transfer, the medium of energy transfer is:
    (a) a fluid   (b) a solid   (c) the vacuum   (d) a gas

39. In conduction heat transfer, the medium of energy transfer is:
    (a) a fluid   (b) a solid   (c) the vacuum   (d) a gas

40. In convection heat transfer, warmer fluid . . . and cooler fluid . . .
    (a) sinks / rises   (b) rises / sinks   (c) sinks / sinks   (d) roses / rises

41. In conduction heat transfer, energy is transferred by:
    (a) inelastic collision of atoms
    (b) elastic collision of largely spaced atoms
    (c) inelastic collision of shortly spaced atoms
    (d) elastic collisions of molecules

42. If A is a cross-sectional area perpendicular to the heat flow direction, k is thermal conductivity, T is temperature, and x is thickness, conduction heat transfer (q) is described by the formula:

    (a) $q = -kA\dfrac{dT}{dx}$ (b) $q = kA\dfrac{dT}{dx}$ (c) $q = -k\dfrac{dT}{dx}$ (d) $q = k\dfrac{dT}{dx}$

43. If a material's thermal conductivity doubles, the heat transferred:
    (a) remains the same   (b) doubles
    (c) Halves   (d) quadruples

44. If solid object's area perpendicular to the heat flow direction triples, the heat transferred:
    (a) triples   (b) doubles   (c) halves   (d) becomes one-third

45. If a material's length along the heat flow direction quadruples, the heat transferred:
    (a) doubles   (b) halves   (c) quarters   (d) quadruples

46. If you feel cold after touching a surface, you conclude that the object temperature is:

    (a) higher than that of your skin   (b) lower than that of your skin

    (c) the same as that of your skin   (d) information is insufficient

47. If a surface feels neither cold nor warm to the touch, it can be concluded that the surface has:

    (a) the same temperature as your body

    (b) a lower temperature than your body

    (c) a higher temperature than your body

    (d) a very large thermal conductivity

48. At constant temperatures, if thermal conductivity of a material increases:

    (a) thermal resistivity decreases   (b) thermal resistivity increases

    (c) electrical resistivity increases   (d) electrical conductivity decreases

49. A 1×2 m^2 window is made of a single 5-cm thick layer of glass ($k = 1$ W/mK). The ambient temperature is 50 °C, and the interior temperature is 20 °C. Assume the window is coated by a thin layer that allows only 90% of the exterior radiative heat to be transmitted. The energy transferred to the interior from the window in a one-hour period is:

    (a) 4.32 MJ   (b) 5.61 MJ   (c) 5.61 MJ   (d) 3.89 MJ

50. A 2×3 m^2 window is made of a two-layer glass (with thermal conductivities of 0.8 W/mK for the exterior and 1 W/mK for the interior layer). Each layer is 2.5-cm thick. The ambient temperature is 40 °C, and the interior temperature is 22 °C. Assume the exterior surface is coated by a thin layer that allows only 80% of the exterior radiative heat to be transmitted. Energy transferred to the interior from the window in a 25-min period is:

    (a) 3.15 MJ   (b) 0.27 MJ   (c) 2.88 MJ   (d) 2.60 MJ

51. A 3×4 m^2 metal plate is made of three layers (with thermal conductivities of 400 W/mK, 350 W/mK, and 300 W/mK, from interior to exterior layers, respectively). The thicknesses of the layers, starting from the interior, varies linearly from 40 to 30 cm (5-cm increments). The exterior temperature is 45 °C, and the interior temperature is kept at 25 °C. Assume only 95% of the exterior radiative heat is transferred. Energy transferred from the window in a 15-min period is:

    (a) 80.3 MJ   (b) 72.2 MJ   (c) 0 MJ   (d) 72.2 MJ

52. A ball made of glass ($k = 1$ W/mK) takes 1 min to transfer 100 J of heat. If the ball's material is switched to aluminum ($k = 205$ W/mK), how long will it take to transfer the same quantity of heat?

    (a) 0.05 s   (b) 0.05 min   (c) 0.3 s   (d) 0.3 min

53. A 2-m diameter, 0.5-m long cylinder made of copper ($k = 385$ W/mK, $SG = 8.96$, $C_p = 385$ J/kgK) with its axis oriented horizontally. The cylinder comes in contact with a copper ball (2-m diameter) at 373 K. The equilibrium temperature is (assume cylinder initial temperature is 150 °C):

    (a) 386.7 K   (b) 423.2 K   (c) 193.4 K   (d) 211.6 K

54. A 2-m diameter, 0.5-m long cylinder made of aluminum ($k = 385$ W/mK, $SG = 8.96$, $C_p = 385$ J/kgK) is at the uniform temperature of 150 °C. A 2-m diameter copper ball ($k = 205$ W/mK, $SG = 2.7$, $C_p = 900$ J/kgK) is at 373 K. The two objects are brought into contact so that heat can flow between them while they are perfectly insulated from the environment. The equilibrium temperature of the cylinder-ball system is:

    (a) 390.4 °C   (b) 195.2 K   (c) 117.3 °C   (d) 58.6 °K

55. An aluminum cylinder with an internal diameter of 46 cm and 10-cm wall thickness ($k = 205$ W/mK) is heated on its interior surface. The heater is able to transfer 1 kJ energy to the cylinder in 2 min. The interior surface temperature reached after 2 min is (assume the exterior wall temperature is 22 °C):
    (a) 506 °C   (b) 506 K   (c) 453 °C   (d) 453 K

56. When heating up a bowl of rice in a microwave, the following heat transfer mechanism(s) is(are) dominant:
    (a) radiation   (b) convection and conduction
    (c) conduction and radiation   (d) radiation and convection

57. When not at thermodynamic equilibrium, there is:
    (a) exchange of work   (b) net flow of matter or energy
    (c) exchange of heat   (d) exchange of heat and work

58. One method to measure the average kinetic energy of molecules is to employ a:
    (a) speedometer   (b) barometer
    (c) hygrometer   (d) reversible liquid crystal temperature label

59. According to Wiedemann–Franz's law, thermal conductivity is . . . to electrical conductivity and . . .
    (a) directly related / temperature   (b) nonlinearly related / temperature
    (c) inversely related / temperature   (d) directly related / pressure

60. Thermal conductivity (k) is directly related to temperature (T). Electrical conductivity (σ) is inversely related to temperature by a constant known as . . . . The relationship between these three variables can be expressed as:
    (a) Franz number (Fr), $\sigma = kLT$ (b) Franz number (Fr), $T = \sigma kL$
    (c) Lorenz number (L), $L = \sigma kT$ (d) Lorenz number (L), $k = \sigma LT$

61. If the temperature of a copper wire is doubled, and its electrical conductivity is halved, thermal conductivity:
    (a) doubles   (b) remains the same
    (c) quadruples   (d) halves

62. If the temperature of a copper wire is halved, and its thermal conductivity doubles, electrical conductivity:
    (a) doubles   (b) remains the same
    (c) quadruples   (d) halves

63. If the temperature of a molecule doubles, its average kinetic energy:
    (a) doubles   (b) remains the same
    (c) quadruples   (d) halves

64. If the root mean square speed of a nitrogen molecule is 517 m/s and $\sqrt{\dfrac{3R}{M}} = 29.85$, the molecular temperature is:
    (a) 300 K   (b) 298 K   (c) 295 K   (d) 303 K

65. If the temperature of a nitrogen molecule is 500 K and $\sqrt{\dfrac{3R}{M}} = 29.85$, the molecular average speed is:
    (a) 667 m/s   (b) 615 m/s   (c) 595 m/s   (d) 520 m/s

66. There are three kinds of gas molecules (nitrogen, oxygen, and carbon dioxide) freely moving in a container at 1 atm and with root mean square speeds of 550, 525, and 500 m/s, respectively. The equivalent speed (V) and temperature (T) are (assume there are an equal number of molecules associated with each gas):
   (a) $V = 910$ m/s, $T = 105$ °C
   (b) $V = 525.4$ m/s, $T = 378$ °C
   (c) $V = 910$ m/s, $T = 378$ °C
   (d) $V = 525.4$ m/s, $T = 105$ °C

67. The Prandtl number is the ratio of the:
   (a) inertial to viscous forces
   (b) viscous forces to diffusion of heat
   (c) convection to conduction heat transfer
   (d) advective to diffusive transport rate

68. The Reynolds number is the ratio of the:
   (a) inertial to viscous forces
   (b) viscous forces to diffusion of heat
   (c) convection to conduction heat transfer
   (d) advective to diffusive transport rate

69. The Nusselt number is the ratio of the:
   (a) inertial to viscous forces
   (b) viscous forces to diffusion of heat
   (c) convection to conduction heat transfer
   (d) advective to diffusive transport rate

70. The Peclet number is the ratio of the:
   (a) inertial to viscous forces
   (b) viscous forces to diffusion of heat
   (c) convection to conduction heat transfer
   (d) advective to diffusive transport rate

71. The Stanton number is the ratio of the:
   (a) heat convection to capacity
   (b) speed of an object within a medium to speed of sound within that medium
   (c) viscous to mass diffusion
   (d) advective to diffusive transport rate

72. The Mach number is the ratio of the:
   (a) heat convection to capacity
   (b) speed of an object within a medium to speed of sound within that medium
   (c) viscous to mass diffusion
   (d) advective to diffusive transport rate

73. The Schmidt number is the ratio of the:
   (a) heat convection to capacity
   (b) speed of an object within a medium to speed of sound within that medium
   (c) viscous to mass diffusion
   (d) advective to diffusive transport rate

74. The Biot number is the ratio of the:
   (a) heat convection to capacity
   (b) speed of an object within a medium to speed of sound within that medium

(c) viscous to mass diffusion
(d) convection to conduction forces

75. The Grashof number is the ratio of the:
    (a) gravity to viscous forces   (b) gravity to diffusion of heat
    (c) diffusive to conductive forces   (d) viscous to mass diffusion

76. The Rayleigh number is the ratio of the:
    (a) gravity to viscous forces   (b) gravity to diffusion of heat
    (c) diffusive to conductive forces   (d) viscous to mass diffusion

77. The Fourier number is the ratio of the:
    (a) gravity to viscous forces
    (b) gravity to diffusion of heat
    (c) conduction forces to thermal energy storage
    (d) viscous to mass diffusion

78. The Eckert number is the ratio of the:
    (a) gravity to viscous forces
    (b) gravity to diffusion of heat
    (c) inertial forces to thermal energy storage
    (d) potential energy to heat dissipation

79. The Louis number is the ratio of the:
    (a) shear to inertial forces   (b) drag to inertial forces
    (c) diffusion of heat to mass   (d) sensible to latent energy

80. The Jakob number is the ratio of the:
    (a) shear to inertial forces   (b) drag to inertial forces
    (c) diffusive to conductive forces   (d) sensible to latent energy

81. The Knudsen number is the ratio of the:
    (a) mean free path to characteristic length
    (b) drag to inertial forces
    (c) diffusive to conductive forces
    (d) sensible to latent energy

82. A bimetallic strip is used to . . . the . . . into the . . . . The strip consists of two strips of different metals which expand at . . . as they are heated.
    (a) transform / pressure change / chemical composition variation / the same rate
    (b) harmonize / temperature change / mechanical displacement / different rates
    (c) transform / temperature change / mechanical displacement / different rates
    (d) harmonize / pressure change / chemical composition variation / the same rate

83. In continuous flows, the Knudsen number is:
    (a) > 1   (b) < 1   (c) 1   (d) 0

84. In slip flows, the Knudsen number is:
    (a) > 1   (b) < 1   (c) 1   (d) 0

85. In free molecular flows, the Knudsen number is:
    (a) > 1   (b) < 1   (c) 1   (d) 0

86. In slip flows, fluid velocity at the wall is . . . and in continuous flows, fluid velocity at the wall is . . .

(a) nonzero / zero   (b) zero / nonzero
(c) nonzero / nonzero   (d) zero / zero

87. Newton's law of cooling relates temperature difference between the surface and the surrounding ambient to the:

(a) convection coefficient   (b) conduction coefficient
(c) thermal conductivity   (d) thermal diffusivity

88. The units of the ratio of the convective to conductive heat transfer coefficients is:

(a) m   (b) m^2 (c) $1/m^2$ (d) $1/m$

89. The Navier-Stokes equation relates the inertial forces to the:

(a) viscous, pressure, and gravity forces   (b) viscous, pressure, and conductive forces
(c) pressure, gravity, and convective forces   (d) pressure, gravity, and conductive forces

90. The Navier-Stokes equation in a steady-state forced convection regime is simplified to (U, V, and W are velocities, P is pressure, t is time, x, y, and z are coordinates, g is gravity acceleration, and r is density):

(a) $\rho\left(\dfrac{\partial U}{\partial t}+U\dfrac{\partial U}{\partial x}+V\dfrac{\partial U}{\partial y}\right)=-\dfrac{\partial P}{\partial x}+\mu\left(\dfrac{\partial^2 U}{\partial x^2}+\dfrac{\partial^2 U}{\partial y^2}\right)+\rho g$

(b) $\left(U\dfrac{\partial U}{\partial x}\right)=-\dfrac{1}{\rho}\dfrac{\partial P}{\partial x}+v\left(\dfrac{\partial^2 U}{\partial x^2}\right)+g$

(c) $\left(\dfrac{\partial U}{\partial t}+U\dfrac{\partial U}{\partial x}+V\dfrac{\partial U}{\partial y}+W\dfrac{\partial U}{\partial z}\right)=-\dfrac{1}{\rho}\dfrac{\partial P}{\partial x}+v\left(\dfrac{\partial^2 U}{\partial x^2}+\dfrac{\partial^2 U}{\partial y^2}+\dfrac{\partial^2 U}{\partial z^2}\right)$

(d) $\rho\left(U\dfrac{\partial U}{\partial x}+V\dfrac{\partial U}{\partial y}+W\dfrac{\partial U}{\partial z}\right)=-\dfrac{\partial P}{\partial x}+\mu\left(\dfrac{\partial^2 U}{\partial x^2}+\dfrac{\partial^2 U}{\partial y^2}+\dfrac{\partial^2 U}{\partial z^2}\right)$

91. For an incompressible flow, at distance x from a flat surface edge, velocity profiles are defined by $U=x+y$, $V=x-y$; x is (assume flow is steady and the forced convection regime is applicable):

(a) 0   (b) 0.1   (c) 1   (d) 10

92. For an incompressible flow, at distance x from a flat surface edge, velocity profiles are defined by $U=V=x^2-y^2$. The relation between x and y is (assume flow is steady and the forced convection regime is applicable):

(a) $x-y=0$ (b) $x+y=0$ (c) $x=y=0$ (d) $x^2-y^2=0$

93. For an incompressible flow, at distance x from a flat surface edge, velocity profiles are defined by $U=V=x^2+y^2$. Surface temperature is 100 °C and potential flow temperature is 25 °C. If $x=y$, the boundary layer thickness and its temperature are (assume flow is steady and the free convection regime is applicable; g is gravity acceleration, ρ is density, v is dynamic viscosity, and $U\infty$ is free flow velocity):

(a) $y=\sqrt{1.2+0.5v}$, $T=-\dfrac{150}{U_\infty}(1.2+0.5v)^{2/3}+75$

(b) $y=\sqrt[3]{1.2+0.5v}$, $T=-\dfrac{150}{U_\infty}(1.2+0.5v)^{2/3}+100$

(c) $y = 1.2 + 0.5v - \rho g, T = \dfrac{150}{U_\infty}(1.2 + 0.5v - \rho g)^2 + 75$

(d) $y = (1.2 + 0.5v - \rho g)^2, T = -\dfrac{150}{U_\infty}(1.2 + 0.5v - \rho g)^4 + 100$

94. If a flat surface emits 100 kW radiative heat, and the surrounding temperature is 25 °C, the surface temperature is (assume surface area is 0.5 m² and its emissivity is 0.9):

(a) 1,407.7 K   (b) 1,153.7 K   (c) 1,371.2 K   (d) 1,335.6 K

95. If an object exposed to the Sun receives 0.5 kW of solar radiative heat, and its surface temperature stabilizes at 100 °C, assuming that the surrounding temperature is 30 °C and the surface area is 1 m², emissivity of the surface is:

(a) 0.19   (b) 0.81   (c) 0.5   (d) 1

96. The prism shell shown in Figure 135 is heated due to the exposure to an intense heat source, which causes the temperature of the internal surfaces to increase. Assuming that these surfaces are shiny and smooth, Surfaces 1 and 2 have equal areas, and the area of Surface 3 is 0.6 times that of Surface 1, the portions of energy of Surface 1, reaching Surfaces 2 and 3, are:

(a) $F_{1-2} = 0.7, F_{1-3} = 0.3$ (b) $F_{1-2} = 0.5, F_{1-3} = 0.5$
(c) $F_{1-2} = 0.6, F_{1-3} = 0.7$ (d) $F_{1-2} = 0.5, F_{1-3} = 0.3$

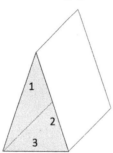

FIGURE 135. Prism exposed to an intense radiation.

97. The cylindrical shell shown in Figure 136 is exposed to intense radiation, which causes the temperature of the internal surfaces to increase. Assuming that the surfaces are shiny and smooth, the area of Surface 1 is double that of Surface 2, and 50% of the energy emitting from Surface 2 reaches Surface 3, the portions of energy that Surface 1 emits to itself and Surface 3 are:

(a) $F_{1-1} = 0.25, F_{1-3} = 0.25$ (b) $F_{1-1} = 0.25, F_{1-3} = 0.5$
(c) $F_{1-1} = 0.5, F_{1-3} = 0.5$ (c) $F_{1-1} = 0.5, F_{1-3} = 0.25$

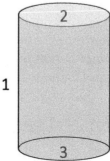

FIGURE 136. Cylindrical shell exposed to an intense radiation.

98. The cuboid shell shown in Figure 137 is exposed to an intense radiation, which causes the temperature of the internal surfaces to increase. These surfaces are shiny and smooth. The surface area of Surface 2 is twice that of Surface 1 and 0.6 times that of Surface 5. 50% of the energy emitting from Surface 5 reaches Surfaces 6 and 2 in equal portions. 25% of the energy emitting from Surface 1 reaches Surface 3. The portion of energy that Surface 1 emits to Surfaces 2 and 5 are:

(a) $F_{1-2} = 0.22$, $F_{1-5} = 0.17$ (b) $F_{1-2} = 0.11$, $F_{1-5} = 0.33$

(c) $F_{1-2} = 0.7$, $F_{1-5} = 0.25$ (d) $F_{1-2} = 0.125$, $F_{1-5} = 0.25$

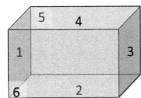

FIGURE 137. Cuboid shell exposed to intense radiation.

99. Emissivity shows the:

(a) ratio of the energy emitted from a body to the same for a gray body

(b) energy emitted from a black body

(c) ability of a body to emit infrared energy

(d) spectral directional energy emitted from a body

100. The space between the two hot plates is filled with hot gases. A thermal shield is employed to create two zones. If the shield is installed in the space in between the two surfaces, the portion of the energy radiated by the surfaces is:

(a) halved   (b) doubled   (c) quartered   (d) quadrupled

101. If T is temperature, r is radius, θ is angle, z is height, ρ is density, k is thermal conductivity, c_p is specific heat capacity at constant pressure, $\dot{q}$ is heat generation, and t is time, the conduction heat transfer equation for a heat-generating cylindrical isotropic object is expressed by:

(a) $\left(\dfrac{\partial}{\partial r}\left(r\dfrac{\partial T}{\partial r} + \dfrac{\partial^2 T}{\partial r^2} \right) + \dfrac{1}{r}\dfrac{\partial^2 T}{\partial \theta^2} + \dfrac{\partial^2 T}{\partial z^2} \right) + \dfrac{\dot{q}}{k} = \dfrac{\rho c_p}{k}\left(\dfrac{\partial T}{\partial t} \right)$

(b) $\left(\dfrac{\partial}{\partial r}\left(r\dfrac{\partial T}{\partial r} + \dfrac{\partial^2 T}{\partial r^2} \right) + \dfrac{1}{r^2}\dfrac{\partial^2 T}{\partial \theta^2} + \dfrac{\partial^2 T}{\partial z^2} \right) + \dfrac{\dot{q}}{k} = \dfrac{\rho c_p}{k}\left(\dfrac{\partial T}{\partial t} \right)$

(c) $k\left(\dfrac{1}{r}\dfrac{\partial T}{\partial r} + \dfrac{\partial^2 T}{\partial r^2} + \dfrac{1}{r}\dfrac{\partial^2 T}{\partial \theta^2} + \dfrac{\partial^2 T}{\partial z^2} \right) + \dot{q} = \rho c_p\left(\dfrac{\partial T}{\partial t} \right)$

(d) $k\left(\dfrac{\partial}{\partial r}\left(r^2\dfrac{\partial T}{\partial r} \right) - r\dfrac{\partial T}{\partial r} + \dfrac{\partial^2 T}{\partial \theta^2} + \dfrac{\partial^2 T}{\partial z^2} \right) - \dot{q} = \rho c_p\left(\dfrac{\partial T}{\partial t} \right)$

102. If T is temperature, x, y, and z are length, depth, and height, α is thermal diffusivity, $\dot{q}$ is heat generation, and t is time, the conduction heat transfer equation in the Cartesian coordinates for a heat-generating isotropic object is expressed by:

(a) $\left(\dfrac{\partial}{\partial x}\left(k\dfrac{\partial T}{\partial x} \right) + \dfrac{\partial}{\partial y}\left(k\dfrac{\partial T}{\partial y} \right) + \dfrac{\partial}{\partial z}\left(k\dfrac{\partial T}{\partial z} \right) \right) + \dot{q} = \alpha\left(\dfrac{\partial T}{\partial t} \right)$

(b) $\left(\dfrac{\partial}{\partial x}\left(\dfrac{\partial T}{\partial x}\right)+\dfrac{\partial}{\partial y}\left(\dfrac{\partial T}{\partial y}\right)+\dfrac{\partial}{\partial z}\left(\dfrac{\partial T}{\partial z}\right)\right)+\dfrac{\dot q}{k}=\alpha\left(\dfrac{\partial T}{\partial t}\right)$

(c) $\left(\dfrac{\partial}{\partial x}\left(k\dfrac{\partial T}{\partial x}\right)+\dfrac{\partial}{\partial y}\left(k\dfrac{\partial T}{\partial y}\right)+\dfrac{\partial}{\partial z}\left(k\dfrac{\partial T}{\partial z}\right)\right)+\dot q=\dfrac{1}{\alpha}\left(\dfrac{\partial T}{\partial t}\right)$

(d) $\left(\dfrac{\partial^2 T}{\partial x^2}+\dfrac{\partial^2 T}{\partial y^2}+\dfrac{\partial^2 T}{\partial z^2}\right)+\dfrac{\overset{y}{q}}{k}=\dfrac{1}{\alpha}\left(\dfrac{\partial T}{\partial t}\right)$

103. If T is temperature, r is radius, ρ is density, k is thermal conductivity, c_p is specific heat capacity at constant pressure, $\dot q$ is heat generation, t is time, and temperature variation along the angular coordinates can be ignored, the conduction heat transfer equation in spherical coordinates for a heat-generating isotropic object is expressed by:

(a) $\dfrac{1}{r}\dfrac{\partial}{\partial r}\left(kr\dfrac{\partial T}{\partial r}\right)+\dfrac{\dot q}{k}=\dfrac{\rho c_p}{k}\left(\dfrac{\partial T}{\partial t}\right)$ (b) $\dfrac{1}{r^2}\left(\dfrac{\partial^2 T}{\partial r^2}+\dfrac{2}{r}\dfrac{\partial T}{\partial r}\right)+\dfrac{\dot q}{k}=\dfrac{\rho c_p}{k}\left(\dfrac{\partial T}{\partial t}\right)$

(c) $\left(\dfrac{\partial^2 T}{\partial r^2}+\dfrac{2}{r^2}\dfrac{\partial T}{\partial r}\right)+\dfrac{\dot q}{k}=\dfrac{\rho c_p}{k}\left(\dfrac{\partial T}{\partial t}\right)$ (d) $k\left(\dfrac{\partial^2 T}{\partial r^2}+\dfrac{2}{r}\dfrac{\partial T}{\partial r}\right)+\dot q=\rho c_p\left(\dfrac{\partial T}{\partial t}\right)$

104. A solid sphere is at steady-state conditions, with the temperature of 100 K at the radius of 0.25 m and 200 K at the radius of 0.5 m. If r is the radial distance from sphere's center, the temperature profile can be expressed as (r is distance from the center):

(a) $-\dfrac{50}{r}+300$ (b) $\dfrac{50}{r}-300$ (c) $50r+88$ (d) $-50r+225$

105. A 3-m long solid rectangular aluminum fin (10×20 cm^2) is used to dissipate heat from the base of a column with a rectangular cross section. If the temperature at the base is 400 °C, fin effectiveness (e_f) is (assume the ambient temperature is 22 °C, heat convection coefficient is 10 W/m^{2}K, and fin thermal conductivity is 205 W/mK):

(a) $e_f=24.8$ (b) $e_f=72.4$ (c) $e_f=27.6$ (d) $e_f=75.2$

106. A 3-m long rectangular aluminum fin (10×20 cm^2) is used to dissipate heat from the base of a column with a rectangular cross section. If the temperature at the base is 400 °C, fin efficiency (η_f) is (assume the ambient temperature is 22°C, heat convection coefficient is 10 W/m^{2}K, and fin thermal conductivity is 205 W/mK):

(a) $\eta_f=24.8\%$ (b) $\eta_f=72.4\%$ (c) $\eta_f=27.6\%$ (d) $\eta_f=75.2\%$

107. A 3-m long rectangular aluminum fin (10×20 cm^2) is used to dissipate heat from the base of a column with a rectangular cross section. If the temperature at the base is 400 °C, the temperature of the fin at 0.5 m from the base is (assume the ambient temperature is 22 °C, heat convection coefficient is 10 W/m^{2}K, and fin thermal conductivity is 205 W/mK):

(a) 379 °C   (b) 362 °C   (c) 206 °C   (d) 228 °C

108. A 3-m long rectangular aluminum fin (10×20 m^2) is used to dissipate heat from the base of a column with a rectangular cross section. If the entire fin temperature is 400 °C, the energy transfer from the fin after 30 min is (assume the ambient temperature is 22 °C, heat convection coefficient is 10 W/m^{2}K, and fin thermal conductivity is 205 W/mK):

(a) 12.4 MJ   (b) 15.1 MJ   (c) 6.8 MJ   (d) 24.4 MJ

109. A 1-m long aluminum pipe (0.5-m internal radius) is filled with mercury at 400 °C. The thickness of the wall is 10 cm. Total thermal resistance of the system is (assume the ambient temperature is 22 °C, heat convection coefficients for the internal and external fluids are 10 W/m²K, and pipe thermal conductivity is 205 W/mK):

(a) 0.058 K/W   (b) 0.032 K/W   (c) 0.026 K/W   (d) 0.003 K/W

110. A 1-m long aluminum pipe (0.5-m internal radius) is filled with mercury at 400 °C. The thickness of the wall is 10 cm. The total heat transfer rate from the system, ignoring the end, is (assume the ambient temperature is 22 °C, heat convection coefficients for the internal and external fluids are 10 W/m²K, and pipe thermal conductivity is 205 W/mK):

(a) 7.4 kW   (b) 6.5 kW   (c) 3.3 kW   (d) 4.5 kW

111. A 1-m long aluminum pipe (0.5-m internal radius) is filled with water at 400 °C. The thickness of the wall is 10 cm. If the heat transfer rate from the internal fluid and pipe is 5 kW, the temperature of the exterior surface is (assume the ambient temperature is 22 °C, heat convection coefficient for the external fluid is 10 W/m²K and for the internal fluid is 20 W/m²K, and pipe thermal conductivity is 205 W/mK):

(a) 592 K   (b) 592 °C   (c) 319 K   (d) 319 °C

112. A 2-m long, 3-m high wall is made of 0.5-m thick brick coated with a 10 cm gypsum layer on one side. The environment temperature is 30 °C and the room temperature is 20 °C. Wall thermal resistance (R_w) and the total heat transfer rate (Q) are (assume heat convection coefficients for the internal and external fluids are 15 W/m²K, brick thermal conductivity is 0.6 W/mK, and gypsum thermal conductivity is 0.18 W/mK):

(a) R_w = 0.56 K/W, Q = 18 W   (b) R_w = 0.28 K/W, Q = 5.4 W
(c) R_w = 0.14 K/W, Q = 9.2 W   (d) R_w = 0.23 K/W, Q = 36.2 W

113. An aluminum sphere (0.4-m internal radius and 0.5-m external radius) is enclosed in a 0.1-m thick steel layer. The temperature of the aluminum sphere's interior surface is 600 °C. Total thermal resistance (R_t) and the total heat transfer rate (Q) are (assume the ambient temperature is 22 °C, heat convection coefficients for the external and internal fluids are 15 W/m²K, steel thermal conductivity is 50.2 W/mK, and aluminum thermal conductivity is 205 W/mK):

(a) R_t = 0.002 K/W, Q = 126.2 kW   (b) R_t = 0.005 K/W, Q = 37.7 kW
(c) R_t = 0.048 K/W, Q = 11.9 kW   (d) R_t = 0.010 K/W, Q = 19.1 kW

114. Heat is generated inside a 0.5-m thick wall made of steel. Exterior surfaces are kept at a constant temperature (400 K). Heat generation inside the wall is 100 kW/m³. The temperature at the center of the wall is (assume heat convection coefficient for the external fluid is 10 W/m²K, and steel thermal conductivity is 50.2 W/mK):

(a) 462 K   (b) 425 K   (c) 649 K   (d) 400 K

115. Heat is generated inside a 1.2-m thick wall made of aluminum. Exterior surfaces are kept at a constant temperature (T_w). Heat generation per unit volume is 100 kW/m³. The temperature at the center of the wall is 700 K. If the temperature at the distance 0.2 m from the wall's exterior surface is 661 K, the wall's exterior surface temperature is (assume heat convection coefficient for the external fluid is 15 W/m²K, and aluminum thermal conductivity is 205 W/mK):

(a) 330 K   (b) 661 K   (c) 612 K   (d) 306 K

116. Heat is generated inside a 2.4-m thick insulated wall made of aluminum. The exterior surfaces are kept at a constant temperature (T_w). Heat generation per unit volume is 215 kW/m^3. The temperature at the center of the wall is 802 K. If the wall exterior surface temperature is 100 K, temperature at the distance 0.5 m from the wall center is (assume heat convection coefficient for the external fluid is 10 W/m^{2}K, and aluminum thermal conductivity is 205 W/mK):

   (a) 125 K   (b) 283 K   (c) 598 K   (d) 175 K

117. A square element $(dx \times dy)$, where $dx = dy = 0.1$ m, is used to calculate temperature distribution within a body using the finite element technique. Temperatures of four points surrounding the point located at (1,1) are 100, 110, 120, and 130 °C. The temperature at point (1,1) is (assume heat convection coefficient for the surrounding fluids are 10 W/m^{2}K, and aluminum thermal conductivity is 205 W/mK):

   (a) 115 °C   (b) 114 °C   (c) 113 °C   (d) 112 °C

118. A square element $(dx \times dy)$, where $dx = dy = 0.2$ m, is used to calculate temperature distribution within a body using the finite element technique. Temperatures of four points surrounding the point located at (1,1) are 120, 140, 160, and 180°C. If the heat generation rate is 1000 kW/m^3, the temperature at point (1,1) is (assume heat convection coefficients for the surrounding fluids are 10 W/m^{2}K, and aluminum thermal conductivity is 205 W/mK):

   (a) 164 °C   (b) 388 °C   (c) 198 °C   (d) 236 °C

119. For 2D problems to be steady, the Fourier number should be:

   (a) smaller than 0.25   (b) greater than 0.25
   (c) smaller than 0.5   (d) greater than 0.5

120. Assume heat transfer when cooking hamburgers can be described using the lumped capacity method. Assume meat's total surface area is 251 cm^2, density is 1,000 kg/m^3, heat capacity is 3,800 J/kgK, and heat convection coefficient to the environment is 15 W/m^{2}K. The hamburger's internal temperatures after 15, 30, and 45 min are:

   (a) 75.4 °C, 58.6 °C, 47.0 °C   (b) 90.7 °C, 75.4 °C, 58.6 °C
   (c) 75.4 °C, 47.0 °C, 50.4 °C   (d) 90.7 °C, 58.6 °C, 41.4 °C

121. A hot semi-infinite steel plate at the initial temperature of 100 °C is placed in an environment at 25 °C. Steel heat capacity is 511 J/kgK, thermal conductivity is 50.2 W/mK, and density is 8,050 kg/m^3. The plate's temperatures at the distance 10 cm from the base after 15, 30, and 45 min are:

   (a) 90.7 °C, 75.4 °C, 58.6 °C   (b) 62.5 °C, 52.5 °C, 47.7 °C
   (c) 75.4 °C, 47.1 °C, 50.4 °C   (d) 90.7 °C, 58.6 °C, 41.4 °C

122. A hot semi-infinite steel plate at the initial temperature of 100 °C is placed in an environment at 25 °C. Steel heat capacity is 511 J/kgK, thermal conductivity is 50.2 W/mK, and density is 8,050 kg/m^3. The plate's heat flux rates at the distance of 20 cm from the base after 15, 30, and 45 min are:

   (a) 20.3 kW/m^2, 14.3 kW/m^2, 11.7 kW/m^2
   (b) 14.3 kW/m^2, 11.7 kW/m^2, 10.6 kW/m^2
   (c) 11.7 kW/m^2, 10.6 kW/m^2, 9.7 kW/m^2
   (d) 10.6 kW/m^2, 9.7 kW/m^2, 8.5 kW/m^2

123. Thermal conductivity of a 1-m radius aluminum sphere is 205 W/mK. The heat convection coefficient of the surrounding fluid is 20 W/m²K. The sphere's Biot number is:

(a) 0.006   (b) 0.012   (c) 0.024   (d) 0.033

124. The convection heat transfer rate from a (2×4 m²) polyethylene surface is 10 W, the radiation heat transfer rate is 20 W, and the total heat transfer rate is 100 W. The conduction heat transfer rate (Q) and surface temperature (T) are (environment temperature is 25 °C and polyethylene thermal conductivity is 0.33 W/mK):

(a) $Q = 80$ W, $T = 55.3$ °C   (b) $Q = 70$ W, $T = 51.5$ °C

(c) $Q = 130$ W, $T = 74.2$ °C   (d) $Q = 90$ W, $T = 59.1$ °C

Answer Key									
1. (a)	**2.** (b)	**3.** (c)	**4.** (a)	**5.** (c)	**6.** (d)	**7.** (a)	**8.** (c)	**9.** (b)	**10.** (a)
11. (b)	**12.** (c)	**13.** (a)	**14.** (a)	**15.** (d)	**16.** (b)	**17.** (a)	**18.** (d)	**19.** (a)	**20.** T
21. T	**22.** F	**23.** (d)	**24.** (d)	**25.** T	**26.** T	**27.** (b)	**28.** (a)	**29.** (b)	**30.** (a)
31. (b)	**32.** (a)	**33.** (a)	**34.** (a)	**35.** (a)	**36.** (c)	**37.** F	**38.** (a)	**39.** (b)	**40.** (b)
41. (d)	**42.** (a)	**43.** (b)	**44.** (a)	**45.** (c)	**46.** (b)	**47.** (a)	**48.** (a)	**49.** (b)	**50.** (a)
51. (d)	**52.** (c)	**53.** (a)	**54.** (c)	**55.** (b)	**56.** (a)	**57.** (c)	**58.** (a)	**59.** (a)	**60** (d)
61. (b)	**62.** (c)	**63.** (a)	**64.** (a)	**65.** (b)	**66.** (d)	**67.** (b)	**68.** (a)	**69.** (c)	**70.** (d)
71. (a)	**72.** (b)	**73.** (c)	**74.** (d)	**75.** (a)	**76.** (b)	**77.** (c)	**78.** (c)	**79.** (c)	**80.** (d)
81. (c)	**82.** (a)	**83.** (b)	**84.** (c)	**85.** (a)	**86.** (a)	**87.** (a)	**88.** (d)	**89.** (a)	**90.** (d)
91. (a)	**92.** (a)	**93.** (b)	**94.** (a)	**95.** (b)	**96.** (a)	**97.** (d)	**98.** (d)	**99.** (c)	**100.** (a)
101. (c)	**102.** (d)	**103.** (d)	**104.** (a)	**105.** (a)	**106.** (c)	**107.** (d)	**108.** (a)	**109.** (a)	**110.** (b)
111. (a)	**112.** (d)	**113.** (c)	**114.** (a)	**115.** (c)	**116.** (b)	**117.** (a)	**118.** (c)	**119.** (a)	**120.** (a)
121. (b)	**122.** (a)	**123.** (d)	**124.** (b)						

CHAPTER 8

FLUID MECHANICS

1. Specific gravity represents the:
   (a) ratio of the density of the matter to that of water
   (b) density of the matter
   (c) ratio of mass of the matter to its volume
   (d) ratio of mass of the matter to that of water

2. If the specific gravity of a matter is greater than that of a liquid, the matter:
   (a) floats over the liquid   (b) sinks into the liquid
   (c) mixes with the liquid   (d) all of the above

3. Pressure is a . . . property and is the ratio of the . . . applied . . . to the surface of an object per . . . over which the force is distributed.
   (a) scalar / force / perpendicular / unit area
   (b) vector / acceleration / parallel / area
   (c) scalar / acceleration / perpendicular / unit surface
   (d) vector / force / parallel / surface

4. You are standing on the ground. The average pressure you apply to the ground is equal to the:
   (a) weight
   (b) weight divided by your foot area
   (c) weight divided by your contact surface area
   (d) weight by body area

5. Standing on one bare foot . . . the pressure by a factor of . . .. compared to standing on both bare feet.
   (a) decreases / two   (b) increases / four
   (c) decreases / four   (d) increases / two

6. A free diver looking for pearls descends from the surface to a depth of 10 m. The pressure increase that he experiences at that depth is:
   (a) 0.01 MPa   (b) 0.1 MPa   (c) 1 MPa   (d) 10 MPa

7. A scuba diver is observing the ocean wildlife at the 10-m depth. He then decides to go deeper, to the 30-m depth, where the best corals are located. When he reaches the new depth, the pressure . . . and the pressure difference from his previous depth is . . .

   (a) increases / 0.2 MPa
   (b) decreases / 2 MPa
   (c) increases / 2 MPa
   (d) decreases / 0.2 MPa

8. Total pressure is also known as . . . and is the . . . of the . . . pressure and the . . . pressure.

   (a) absolute pressure / sum / gauge / atmospheric
   (b) dynamic pressure / pressure difference / atmospheric / relative
   (c) static pressure / pressure difference / relative / atmospheric
   (d) relative pressure / sum / gauge / atmospheric

9. Pressure on an object submerged in a fluid and the height of the fluid above an object are . . . related.

   (a) directly   (b) inversely   (c) quadratically   (d) logarithmically

10. Gauge pressure shows the deviation from the:

   (a) relative pressure
   (b) atmospheric pressure
   (c) absolute pressure
   (d) static pressure

11. An air pressure sensor installed to measure pressure inside a hyperbaric chamber shows the:

   (a) gauge pressure
   (b) total pressure
   (c) atmospheric pressure
   (d) static pressure

12. The difference between the air pressure inside the tire and in the surrounding air is called the:

   (a) absolute pressure
   (b) relative pressure
   (c) gauge pressure
   (d) static pressure

13. Apparent weight is an object weight in the:

   (a) air   (b) fluid   (c) gas   (d) vacuum

14. The buoyant force is the:

   (a) net upward force exerted by fluid on the immersed object
   (b) object weight in a fluid
   (c) difference between an object weight and displaced fluid weight
   (d) downward force applied to the floating object

15. The buoyant force on a submerged object is the:

   (a) tendency of an object to float
   (b) weight of an object
   (c) weight of fluid displaced by an object
   (d) pressure of fluid

16. The buoyant force is calculated by using the:

   (a) tendency of an object to accelerate (b) floating principle
   (c) Archimedes principle   (d) equilibrium principle

17. The buoyant force can be expressed by the formula (ρ_{fluid} is fluid density, V_{fluid} and V_{displac} are fluid volume and displaced volume, and g is acceleration gravity):

   (a) $\rho_{\text{fluid}} V_{\text{displac}} g$
   (b) $\rho_{\text{fluid}} V_{\text{fluid}} g$
   (c) $r_{\text{displaced}} V_{\text{displaced}} g$
   (d) $\rho_{\text{displaced}} V_{\text{fluid}} g$

18. A $4 \times 5 \times 2$-m^3 brick-shaped lead block is kept submerged in a mercury bath of sufficient size to contain the block. If the specific gravity of lead is 11.34 while the specific gravity of mercury is 13.59, the force required to keep it below the liquid surface is:

(a) 883 kN   (b) 4,449.8 kN   (c) 88.3 kN   (d) 5,332 kN

19. A solid object of volume V is submerged in water. If the net downward force is 20,000 N, the specific gravity of the object (SG) is:

(a) $1 - \dfrac{1}{2V}$ (b) $1 + \dfrac{1}{2V}$ (c) $1 - \dfrac{2}{V}$ (d) $1 + \dfrac{2}{V}$

20. A solid object is sinking in a lake. As the object sinks deeper, the buoyant force:

(a) increases   (b) decreases

(c) remains the same   (d) approaches zero

21. A solid object of 1 m^3 volume is kept submerged by force in a pool of water. If the force required to keep it submerged is 5,000 N, the specific gravity of the object (SG) is:

(a) 1   (b) 2   (c) 0.5   (d) 1.5

22. A solid object of 1 m^3 volume is floating in a pool of water. 50% of the object's volume is below the water surface. The density of the object (ρ) is:

(a) 1,010 kg/m^3 (b) 500 kg/m^3 (c) 550 kg/m^3 (d) 2,020 kg/m^3

23. A solid object ($\rho = 2,000$ kg/m^3) is floating in a pool of mercury ($\rho = 11,340$ kg/m^3) while displacing 0.1 m^3 of the fluid. The buoyant force it experiences is:

(a) 11.1 kN   (b) 2.0 kN   (c) 9.2 kN   (d) 13.1 kN

24. One-meter high and 2-m diameter cylindrical balloons are filled with helium ($\rho = 0.178$ kg/m^3) and are used to lift a solid spherical mass with the 1-m diameter and 150 kg/m^3 density. The mass is located on the Earth's surface with the air density of 1.225 kg/m^3. If the balloon shell weighs 1 kg, the minimum number of balloons needed is:

(a) 121   (b) 61   (c) 170   (d) 85

25. A 4-m high and 2-m diameter cylindrical balloon is filled with helium ($\rho = 0.178$ kg/m^3) and is used to lift a solid spherical mass with the 1-m diameter and 350 kg/m^3 density. The mass is located on the Earth's surface with the air density of 1.225 kg/m^3. If the balloon shell weighs m kg, assuming that the minimum of 131 balloons are needed to lift the ball, the balloon shell mass (m) and density (ρ) are:

(a) $m = 4.3$ kg, $\rho = 0.34$ kg/m^3 (b) $m = 2.0$ kg, $\rho = 0.16$ kg/m^3

(c) $m = 4.3$ kg, $\rho = 0.16$ kg/m^3 (d) $m = 2.0$ kg, $\rho = 0.34$ kg/m^3

26. A 4-m high and 2-m diameter cylindrical balloon is filled with helium ($\rho = 0.178$ kg/m^3) and is used to lift a solid mass. The balloon is located on the Earth's surface with air ($\rho = 1.225$ kg/m^3). If the balloon shell weighs m kg and it is trying to lift a spherical mass with the 2-m diameter and 350 kg/m^3 density, assuming that the minimum number of balloons needed is 132, the balloon shell mass (m) and density (ρ), if the weight of the helium is ignored, is:

(a) $m = 4.3$ kg, $\rho = 0.34$ kg/m^3 (b) $m = 2.1$ kg, $\rho = 0.16$ kg/m^3

(c) $m = 4.3$ kg, $\rho = 0.16$ kg/m^3 (d) $m = 2.1$ kg, $\rho = 0.34$ kg/m^3

27. A cubic balloon, with 1-m sides, is filled with helium ($\rho = 0.178$ kg/m^3) and weighs 280 g. At least 6 of these balloons are required to make a sphere with the 1-m diameter and 10 kg/m^3 density float in the air ($\rho = 1.225$ kg/m^3). The buoyant force per balloon is:

(a) 39.2 N   (b) 78.3 N   (c) 13.1 N   (d) 26.2 N

28. A block is being slowly lowered into a pool of water. From the point it makes initial contact with the water surface until it is submerged under its surface and released, the buoyant force . . .

(a) decreases   (b) remains the same
(c) doubles   (d) increases

29. Fluid flow rate can be used to determine the . . . of flow as it moves through a constriction.

(a) velocity   (b) acceleration   (c) volume   (d) density

30. In an incompressible fluid, the change of the:

(a) density with pressure is negligible
(b) temperature with pressure is not negligible
(c) pressure with temperature is significant
(d) Mach number is greater than 3

31. In an incompressible fluid, fluid . . . is zero.

(a) velocity   (b) divergence   (c) acceleration   (d) all of the above

32. A laminar flow is associated with . . . flow following an . . . path.

(a) smooth / uninterrupted   (b) disturbed / interrupted
(c) smooth / interrupted   (d) disturbed / uninterrupted

33. A turbulent flow is associated with . . . flow following an . . . path.

(a) smooth / uninterrupted   (b) swirling / interrupted
(c) smooth / interrupted   (d) swirling / uninterrupted

34. Viscosity determines how . . . a fluid moves, given a specific . . .

(a) compliant with bend / time
(b) resistive to deformation / deformation rate
(c) compliant with deformation / time
(d) resistive to pressure / deformation rate

35. For continuity to be valid, flow must be:

(a) laminar, low viscosity, and incompressible
(b) turbulent, low viscosity, and compressible
(c) laminar, high viscosity, and incompressible
(d) turbulent, high viscosity, and incompressible

36. A stream of fluid moves from location 1 to 2 at a constant density. If the inlet area is half that of the outlet, fluid velocity at the outlet is:

(a) half that of the inlet   (b) double that of the inlet
(c) quarter that of the inlet   (d) equal to that of the inlet

37. A stream of fluid moves from location 1 to 2 at a constant density. If the outlet diameter is half that of the inlet, velocity at the outlet is:

(a) double that of the inlet   (b) half that of the inlet
(c) four times that of the inlet   (d) a quarter of that of the inlet

38. Fluid at the rate of 0.3 m³/s flows from a single 30-cm-diameter pipe inlet into 3 smaller pipes, each having the 10-cm diameter. Assuming outlet velocities for the 3 smaller pipes are the same, the inlet (V_{inlet}) and outlet (V_{outlet}) velocities are:

(a) V_{inlet} = 4.24 m/s, V_{outlet} = 12.73 m/s (b) V_{inlet} = 8.48 m/s, V_{outlet} = 25.44 m/s
(c) Vi_{nlet} = 2.12 m/s, V_{outlet} = 6.36 m/s (d) V_{inlet} = 1.41 m/s, V_{outlet} = 4.23 m/s

39. Fluid at the rate of 0.3 m³/s flows from a single 10-cm-diameter pipe inlet into 3 larger 30-cm-diameter pipes. Assuming outlet velocities are the same, outlet flow rates (Q_{outlet}) are:

(a) Q_{outlet} = 0.3 m³/s (b) Q_{outlet} = 0.1 m³/s
(c) Q_{outlet} = 0.03 m³/s (d) Q_{outlet} = 0.01 m³/s

40. A water irrigation system requires a number of pumps to be working while water is transferred from 2 m underground to the ground level. The water flow rate needs to be at least 2.5 m³/s for sufficient irrigation. Assuming that the diameter of the pipe at the pump outlet is 20 cm and the flow velocity is 15 m/s, the minimum number of pumps required is:

(a) 5 (b) 4 (c) 6 (d) 7

41. Pumps with the 30-cm outlet diameter and 5-m/s flow velocity are used to fill a $12 \times 14 \times 8$ m³ swimming pool. The minimum number of pumps required to fill the pool in 20 min or less is:

(a) 2 (b) 3 (c) 5 (d) 4

42. Pumps with the 20-cm outlet diameter and 12-m/s flow velocity are used to fill a $12 \times 14 \times 8$ m³ swimming pool. The minimum number of pumps required, considering that they can only operate up to 10 min, is:

(a) 7 (b) 6 (c) 5 (d) 4

43. Pumps with the 5-cm outlet diameter and 5 m/s flow velocity are used to fill a $2 \times 4 \times 8$ m³ swimming pool. There are 5 pumps working for t s. The time it takes to fill up the pool is:

(a) 5 min (b) 11 min (c) 22 min (d) 44 min

44. For a flow in a channel, the product of the flow velocity and channel cross-sectional area is constant. (T/F)

45. The product of velocity by area is constant for . . . flows and is known as the. . . equation.

(a) laminar / continuity (b) turbulent / mass continuity
(c) laminar / volume continuity (d) turbulent / volume continuity

46. One rectangular and one circular channel merge into a single circular channel. The inlet circular channel diameter is 0.5 m and the outlet circular channel diameter is 1 m. The flow velocity of the rectangular inlet is half that of the circular inlet, where it is 1 m/s. The flow velocity at the outlet is the summation of those of inlets. The total inlet channel area is:

(a) 2.15 m² (b) 0.20 m² (c) 1.96 m² (d) 1.37 m²

47. The conservation of energy law states that:

(a) energy can be created but not destroyed
(b) energy cannot be created but can be destroyed
(c) energy cannot change its form
(d) enegy can neither be created nor destroyed

48. Energy is the ability to . . . work.

(a) generate (b) transform (c) recreate (d) minimize

49. Water inside a rain barrel sitting on the ground flows from a horizontally oriented tap located 20 cm above the barrel's bottom. The barrel's height is 2 m and it is half full. Assume the water leaves the barrel at the atmospheric conditions. The instantaneous velocity of the water coming out of the tap is:

(a) 4.3 m/s (b) 4.6 m/s (c) 3.7 m/s (d) 4 m/s

50. The Bernoulli equation is derived from the conservation of . . . law.

(a) mass   (b) energy   (c) momentum   (d) velocity

51. A person has an upset stomach. He vomits and fluid leaves his throat at the 4 m/s velocity. Flow transitions from a narrow channel to a wide channel (mouth). It follows contours of the channel. If the length of his esophagus is 25 cm, his stomach's pressure is (assume the atmospheric pressure is 101.325 kPa):

(a) 111.8 KPa   (b) 111.8 Pa   (c) 431.3 KPa   (d) 431.3 Pa

52. A person has an upset stomach. He vomits and fluid leaves his throat at the 2 m/s velocity. Flow transitions from a narrow channel to a wide channel (mouth). It follows contours of the channel. The mouth is 12 cm higher than the throat. If the length of his esophagus is 25 cm, and the diameter of his throat is 2 cm, the velocity of liquid at his throat (V_{throat}) before reaching his mouth and his mouth diameter (d_{mouth}) are:

(a) $V_{throat} = 4$ m/s, $d_{mouth} = 2.8$ cm   (b) $V_{throat} = 9$ m/s, $d_{mouth} = 4.2$ cm

(c) $V_{throat} = 1.3$ m/s, $d_{mouth} = 2.5$ cm   (d) $V_{throat} = 6$ m/s, $d_{mouth} = 3.5$ cm

53. A pump at the ground level is used to transfer water uphill, 2 m above the ground level, at the 10 m/s velocity to the atmospheric conditions (101.325 k Pa). If the same pump is located 3 m below the ground level, the flow velocity up the hill is:

(a) 14.1 m/s   (b) 7.2 m/s   (c) 12.3 m/s   (d) 6.4 m/s

54. A pump is used to transfer water uphill, 2 m above the ground level, at the 20 m/s velocity to the atmospheric conditions (101.325 k Pa). If the same pump is located 3 m below the ground level and the pump pressure is 300 kPa, the velocity of water at 3 m below the ground level is:

(a) 17.3 m/s   (b) 10 m/s   (c) 24.5 m/s   (d) 20.2 m/s

55. A pump is used to transfer water uphill, 3 m above the ground level, at the 15 m/s velocity to the atmospheric conditions. The pump is located 2 m below the ground level. The pump pressure is 233 kPa and the flow velocity is 10.3 m/s. The pressure of water at 1 m below the ground level is:

(a) 110 kPa   (b) 200 kPa   (c) 300 kPa   (d) 150 kPa

56. Water is transferred from below the ground level to above the ground level. Select the most accurate:

(a) downstream water is more energetic
(b) upstream water is less energetic
(c) downstream water has more potential
(d) any of the above is possible

57. Water is transferred from a location below the ground level to a location above the ground level. The fluid parcels with more energy are the ones that are:

(a) faster   (b) at the higher elevation
(c) faster and elevated   (d) any of the above

58. For an airfoil that is generating an upward lift force, the air traveling over its upper surface is . . . the air traveling under its lower surface.

(a) faster than   (b) slower than
(c) at the same speed   (d) either of (a) or (b)

59. For a thrown baseball traveling horizontally through the air while it is spinning about the axis perpendicular to the direction of motion and clockwise, when looking from the right side of the person who throws it, air pressure over its upper surface is . . . that under its lower surface.
   (a) more than   (b) less than   (c) the same as   (d) either (a) or (b)

60. To comply with the Bernoulli relation with increasing pressure, what can you say about the flow velocity?
   (a) it will decrease   (b) it will increase
   (c) it will remain the same   (d) the information provided is not sufficient

61. A very large tank contains fluid with its surface exposed to the atmospheric conditions. Surface velocity is:
   (a) very large   (b) very small
   (c) zero   (d) the same as that of the outlet

62. Liquid fuel is stored in a large tank with its surface exposed to the atmospheric conditions. The outlet at the bottom of the tank discharges fluid to the atmosphere at 10 m/s. Pressure at the inlet is . . . that of the outlet.
   (a) the same as   (b) lower than
   (c) higher than   (d) the information is not sufficient

63. Two pieces of paper are suspended by strings and are hanging parallel to one another. If you direct a fast-moving air stream so that it passes between these two pieces of paper, you would expect that the papers:
   (a) get away from one another   (b) come closer to one another
   (c) do not move   (d) any of the above conditions may happen

64. Two pieces of paper are suspended by strings and are hanging parallel to one another. If you direct a fast-moving air stream so that it passes the exterior surfaces of papers, you would expect that the papers:
   (a) get away from one another   (b) come closer to one another
   (c) do not move   (d) any of the above conditions may happen

65. A water pond, which is located 20 m above the sea level, is filled with water at the rate of 1,000 kg/s. The pond releases water to the surrounding land at the rate of 800 kg/s. The rate at which the volume of water in the pond changes is:
   (a) 0.2 m³/s   (b) 0.1 m³/s   (c) 0.8 m³/s   (d) 1 m³/s

66. A water pond, which is located 20 m above the sea level and has the capacity of 500 m³, is filled with water at the rate of 1,500 kg/s. The pond releases water to the surrounding land at the rate of 600 kg/s. The pond is initially partially filled with water (15% of capacity). The pond needs to be fully filled with water in time for the flower show. If the show starts in 15 min, the time difference between the start of the show and when the pond is fully filled is:
   (a) 7.1 min before the show starts   (b) 0.71 min before the show starts
   (c) 7.1 min after the show starts   (d) 0.71 min after the show starts

67. The operation of which of these devices does not involve the Bernoulli principle?
   (a) Parachute   (b) Airplane   (c) Race car   (d) Clock

68. A water-filled barometer whose column rises to the height of 10.3 m on the Earth at the standard atmospheric conditions is taken to another planet, where the liquid column rises to the same height. However, we know that the atmospheric pressure on this new planet is 50 kPa. The gravitational acceleration on the planet's surface is:

(a) 0.3 m/s^2 (b) 2 m/s^2 (c) 1 m/s^2 (d) 0.5 m/s^2

69. A 4-m-diameter and 10-m-high cylindrical water tank is supported by 20-m steel legs that extend from the ground to the bottom of the tank. If water fills 60% of the tank, the pump pressure required to maintain the current water level is (assume the atmospheric pressure is 101.325 kPa and the flow pump velocity is $V = -0.1P^2 - 0.1P + 50$, V is in m/s and P is in Pa):

(a) 26.5 kPa   (b) 14.8 kPa   (c) 36.1 kPa   (d) 10.1 kPa

70. A 5-m-diameter and 20-m-high cylindrical water tank is filled up to 75% and is supported by 10-m steel legs that extend from the ground to the bottom of the tank. The reservoir has 100 holes located at the bottom of the tank, each having the diameter of about 20 mm. If 20% of the holes are closed, the flow velocity from each hole is (assume the water pressure at the bottom of the tank is 175 kPa):

(a) 15.5 m/s   (b) 14.3 m/s   (c) 13.8 m/s   (d) 18.4 m/s

71. A hemispherical colander of the 40-cm diameter has 100 holes that are uniformly distributed, each having the diameter of 2 mm. The colander is submerged in liquid and is then quickly lifted out. It is filled with 80% liquid and it takes about 1 min for the liquid to drain completely. The speed of the liquid leaving each hole right after it is lifted is (assume a constant flow rate):

(a) 1.4 m/s   (b) 0.7 m/s   (c) 0.9 m/s   (d) 1.8 m/s

72. A hemispherical colander with 15 cm radius has 75 holes that are uniformly distributed, each having the diameter of 4 mm. 70% of the colander volume is filled with liquid. The colander is originally filled with a liquid and it takes about 5 s for the liquid to drain completely. If 20 of the holes are blocked, the speed of the liquid leaving each hole is (assume a constant flow rate):

(a) 1.4 m/s   (b) 1.2 m/s   (c) 1.0 m/s   (d) 1.6 m/s

73. Passengers are to board a ship that is to sail in salt water ($\rho = 1{,}029$ kg/m^3). The weight of the ship without cargo, crew, and passengers is 2,100 tons. The ship's horizontal cross-sectional area at the water level is 1,065 m^2 and it has the maximum draft of 2.13 m. 100 passengers and crew have boarded the ship, with the average weight of 75 kg. The weight of the cargo is 20 tons. The actual ship's draft after boarding and loading is (assume the ship has vertical hull walls above water):

(a) 1.94 m   (b) 3.75 m   (c) 0.97 m   (d) 2.5 m

74. Passengers are to board a ship that is to sail in salt water ($\rho = 1{,}029$ kg/m^3). The weight of the ship without cargo, crew, and passengers is 2,100 tons. The ship's horizontal cross-sectional area at the water level is 1,065 m^2 and it has the maximum draft of 2 m. Assume a person's average weight is 75 kg and the weight of the cargo is 20 tons. The maximum allowed number of passengers and crew is:

(a) 970   (b) 928   (c) 956   (d) 914

75. A residential water flow rate ranges between 1.5 and 1.75 gal/min (1 gal/min = 6.309 × 10^{-5} m^3/s). Assume in this case that the supplied water flow rate is the average of the

2 range limits. The shower head has 72 holes and has the diameter of 25 cm. If the size of the holes is 0.5 mm, the velocity of the water leaving the holes (V) and the water pressure when leaving the nozzles (P) are (assume the length of the pipe from the water tap to the head is about 1.5 m and the atmospheric pressure is 101.325 kPa):

(a) $V = 7.3$ m/s, $P = 142$ kPa   (b) $V = 0.73$ m/s, $P = 142$ kPa

(c) $V = 7.3$ m/s, $P = 71$ kPa   (d) $V = 0.73$ m/s, $P = 7.1$ kPa

76. A solid block of titanium (ρ = 4,500 kg/m^3), weighing 50 kg, is suspended in water (ρ = 1,000 kg/m^3) by means of 2 air-filled balls attached to it by strings (ignore the weight of the air). The volume of one ball is double the other one. The volume of the block is one-tenth that of the two balls combined. The weight of the smaller ball is 3 kg. The submerged volume (V) and its ratio to that of the total volume (x) are:

(a) $V = 0.06$ m^3, $x = 0.5$ (b) $V = 0.02$ m^3, $x = 0.3$

(c) $V = 0.1$ m^3, $x = 0.6$ (d) $V = 0.6$ m^3, $x = 0.2$

77. A 1-m-diameter cylindrical rain barrel of 1,000-liter capacity is discharging water to the backyard. At the moment, the water level in the barrel is 90% full. The velocity of the water leaving the barrel from a tap located at the bottom of the barrel is:

(a) 1.51 m/s   (b) 0.53 m/s   (c) 0.15 m/s   (d) 4.74 m/s

78. A 1-m-diameter cylindrical rain barrel of 1,000-liter capacity is discharging water to the backyard. At the moment, the water level in the barrel is 90% full. If the barrel is placed on supports that raise it by 25 cm above the ground level, water escapes from the tap placed at the bottom of the barrel, hitting the ground at a distance from the tap. The time it takes for the water to reach the ground is (assume water leaves the barrel horizontally from a tap located at the barrel's bottom):

(a) 0.12 s   (b) 0.23 s   (c) 0.46 s   (d) 0.29 s

79. A 1-m-diameter cylindrical rain barrel of 1,000-liter capacity is discharging water to the backyard. The barrel is 90% full. Water leaves the barrel horizontally from a tap located at the barrel's bottom. If the barrel is placed on supports that raise it by 25 cm above the ground level, water escapes from the tap placed at the bottom of the barrel, hitting the ground at a distance from the tap that is equal to (assume water leaves the barrel horizontally):

(a) 0.3 m   (b) 0.1 m   (c) 1.1 m   (d) 3.4 m

80. If a person (SG = 1.03) tries to float in salty water, would she float or sink? If she floats, what percentage of her body remains above water surface?

(a) Floats, 2.9%   (b) Floats, 97.1%

(c) First floats and then sinks   (d) Sinks

81. A person swims in salt water. Which of the following is not a force?

(a) Weight   (b) Buoyant force

(c) Pressure by area   (d) Gravity acceleration

82. A 50-kg person is riding on a 10-kg bike. She wishes to fly and contemplates attaching helium balloons to her bike to lift her into the air. How many 60-cm-diameter balloons does she require to achieve her goal? If the balloon's diameter is tripled, the minimum number of balloons required to keep the person aloft will be (assume air and helium density at 20 °C are 1.2041 kg/m^3 and 0.1634 kg/m^3, respectively).

(a) 510, 19   (b) 365, 51   (c) 1150, 38   (d) 249, 100

83. The continuity equation is valid for flows of constant or variable velocities. (T/F)

84. Air is supplied by a main air-conditioning duct at the rate of 450 CFM. The main duct is split into 4 smaller branches of circular ducts. Their diameters start at 5 inches and increase progressively in 3-inch increments. The flow velocity at the main branch (V_{inlet}) and outlets (V_{outlet}) are (assume the inlet diameter is the summation of those of outlets and flow velocities at the outlets are the same):
    (a) V_{inlet} = 0.32 m/s, V_{outlet} = 3.41 m/s (b) V_{inlet} = 0.13 m/s, V_{outlet} = 1.13 m/s
    (c) V_{inlet} = 0.32 m/s, V_{outlet} = 1.13 m/s (d) V_{inlet} = 0.13 m/s, V_{outlet} = 3.41 m/s

85. Pascal's law states that pressure . . . at any point in a(an) . . . fluid is transmitted through a fluid so that this change is . . . everywhere.
    (a) level / unconfined compressible / not the same
    (b) variation / confined incompressible / the same
    (c) level / unconfined incompressible / not the same
    (d) variation / confined incompressible / the same

86. When conducting weight-balance calculations inside denser-than-air fluids such as water, . . . becomes important.
    (a) apparent weight
    (b) weight
    (c) gravity by mass
    (d) mass per gravity

87. An aluminum can with a total surface area A filled with liquid nitrogen (-195.79 °C) at P_{in} is suddenly submerged in a water basin, which is placed inside a room at the atmospheric conditions (P_{atm}). The pressure difference between the interior and exterior of the can is . . . and the can . . .
    (a) $P_{atm} - P_{in}$, collapses
    (b) $P_{atm} - P_{in}$, expands
    (c) P_{atm}, collapses
    (d) P_{in}, extends

88. An open-top cylindrical container filled with liquid with density ρ is exposed to the atmospheric pressure (P_{atm}). The container's bottom surface has area A. Pressure P at the depth h within this cylinder can be expressed as (g is gravitational acceleration):
    (a) $P = P_{atm} - \rho g h A$
    (b) $P = P_{atm} + \rho g h P$
    (c) $P = P_{atm} + \rho g h$
    (d) $P = P_{atm} - \rho g A$

89. Hydrostatic conditions imply that fluid is . . . moving . . . flowing. Furthermore, pressure is transferred . . . to all points at . . .
    (a) either / or / nonuniformly / different heights
    (b) neither / nor / uniformly / the same height
    (c) neither / nor / nonuniformly / different heights
    (d) either / or / uniformly / the same height

90. An open container is exposed to the surrounding atmosphere and is being filled with water. The pressure above the water surface is . . . (that) of the atmosphere.
    (a) the same as   (b) lower than   (c) higher than   (d) zero

91. The weight of a column of air per unit area at an elevation that is at height d from the sea level is:
    (a) pressure altitude
    (b) gauge pressure
    (c) pressure
    (d) density altitude

92. At the standard atmospheric conditions, the weight of a column of air above a surface area of 1 cm² located at the mean sea level is:
    (a) 10.13 N   (b) 10.13 kN   (c) 101.3 N   (d) 101.3 kN

93. Consider a column of height b within a fluid, with the column's bottom at depth a from the fluid's surface. If fluid density is ρ, pressure difference (P) between the top and bottom ends of this column is:

(a) $P = \rho g(a - b)$ (b) $P = \rho g(a + b)$ (c) $P = \rho g b$ (d) $P = \rho g a$

94. Pressure is the same along any horizontal line in a single continuous hydrostatic fluid. (T/F)

95. Fluid flow is due to the . . . , and flow continues until . . .

(a) change in pressure / pressures become distinctly different
(b) equalized pressure / pressure differences become zero
(c) change in pressure / pressures equalize
(d) change in pressure / flows reverse

96. For the flow channel presented in Figure 138, calculate the pressure at the selected points (A to C). Assume the liquid is salt water ($SG = 1.029$).

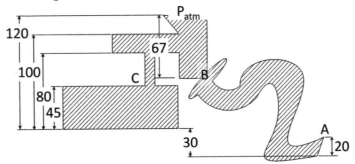

FIGURE 138. Flow channel (dimensions are in cm).

(a) $P_A = 14$ kPa, $P_B = 7$ kPa, $P_C = 8$ kPa
(b) $P_A = 115$ kPa, $P_B = 8$ kPa, $P_C = 7$ kPa
(c) $P_A = 14$ kPa, $P_B = 109$ kPa, $P_C = 108$ kPa
(d) $P_A = 115$ kPa, $P_B = 108$ kPa, $P_C = 109$ kPa

97. For the flow channel presented in Figure 138, calculate the average force imposed by the fluid at the selected cross sections (A to C). Assume the liquid is mercury ($SG = 13.6$). The diameters of the circular pipe at the cross sections A, B, and C, are, respectively, 15, 10, and 5 cm.

(a) $F_A = 0.6$ kN, $F_B = 3.4$ kN, $F_C = 1.6$ kN
(b) $F_A = 1.6$ kN, $F_B = 3.4$ kN, $F_C = 0.6$ kN
(c) $F_A = 0.4$ kN, $F_B = 1.6$ kN, $F_C = 5.1$ kN
(d) $F_A = 3.4$ kN, $F_B = 0.6$ kN, $F_C = 1.6$ kN

98. P is gas pressure in the grayed area. The hatched area is filled with a hydrostatic fluid with density ρ. The diagram shown in Figure 139 represents a physically possible flow channel. (T/F)

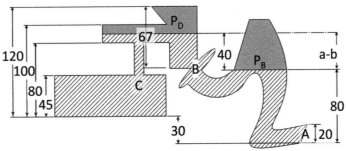

FIGURE 139. Flow channel (dimensions are in cm)

99. Calculate the instantaneous flow velocity at the cross section A in Figure 139. The liquid (hatched area) is salt water ($SG = 1.029$) and the gas (solid gray area) is air ($SG = 0.001225$). P_D is equal to the atmospheric pressure, and the flow at the cross section A is discharged to the atmosphere. The cross section at the top of the flow channel is large enough for the flow to be considered stationary.

(a) 17.5 m/s   (b) 4.6 m/s   (c) 13.7 m/s   (d) 4.5 m/s

100. Calculate pressure (P_D) in Figure 139 if the flow velocity at the cross section A (V_A) is 5 m/s. The liquid (hatched area) is mercury ($SG = 13.6$) and the liquid (solid gray area) is salt water ($SG = 1.029$). The cross section at the top of the flow channel is large enough for the flow to be stationary. The flow at the cross section A is discharged to the atmosphere.

(a) 106 kPa   (b) 101 kPa   (c) 106 MPa   (d) 101 MPa

101. The flow channel in Figure 139 is filled with liquid. The liquid (hatched area) is salt water ($SG = 1.029$) and the gas (solid gray area) is air ($SG = 0.001225$). The pressure difference between the cross sections B and C is 50 kPa. If the flow velocity at the cross section B is 0.5 m/s, the flow velocity at the cross section C is:

(a) 10.8 m/s   (b) 1.1 m/s   (c) 0.5 m/s   (d) 5.4 m/s

102. In an incompressible fluid, volume . . . with . . .

(a) changes / pressure   (b) does not change / pressure
(c) changes / temperature   (d) does not change / temperature

103. The gauge pressure (P) in Figure 140 shows 10 Pa. Assume the liquid (hatched area) is salt water ($SG = 1.029$) and the gas (solid green area above the liquid) is air ($SG = 0.001225$). A cylinder located at the right side of the tank has the initial internal volume of 0.1 m³ and is filled with 1 mole of helium, $R = 2,077.1$ J/kgK. Three springs with the 10 N/m constant are connected to the piston (shown in blue on the left of the cylinder). The temperature of the gas inside the cylinder is increased by 50 °C and the volume of the cylinder under the piston is expanded. New gauge pressure (P) and percent of gas expansion under the piston (x) are:

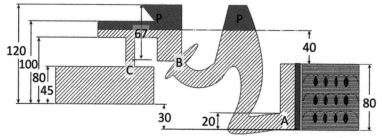

FIGURE 140. Flow channel (dimensions are in cm).

(a) $P = 8$ kPa, $x = 10\%$ (b) $P = 8$ Pa, $x = 51\%$
(c) $P = 16$ kPa, $x = 103\%$ (d) $P = 16$ kPa, $x = 5\%$

104. Total pressure for a very small column of gas is . . . the . . .

(a) approximately the same as / atmospheric pressure
(b) the same as / gauge pressure
(c) higher than / atmospheric pressure
(d) lower than / gauge pressure

105. Gauge pressure is:

(a) the difference between the absolute and the atmospheric pressure
(b) the same as that of the barometer

(c) the addition of the absolute and the atmospheric pressure

(d) lower than that of the barometer

106. For a sufficiently tall column of gas, gauge pressure (P_{guage}) will always be (P_{atm} is the atmospheric pressure):

(a) positive if $P_{absolute} > P_{atm}$

(b) zero if $P_{absolute} < P_{atm}$

(c) negative if $P_{absolute} = P_{atm}$

(d) very large if $P_{absolute} > P_{atm}$

107. A mercury-filled barometer (mercury density is ρ) is at the atmospheric pressure (P_{atm}). The height of the mercury column within the barometer is given by:

(a) $\rho g/P_{atm}$

(b) $P_{atm} + \rho g / P_{atm}$

(c) $P_{atm}/\rho g$

(d) $P_{atm} + P_{atm}/\rho g$

108. The minimum gauge pressure over a very short column of gas is approximately the same as that of the:

(a) atmosphere   (b) gauge   (c) barometric   (d) vacuum

109. A barometer is filled with salt water ($SG = 1.029$). It is exposed to the atmospheric pressure (101.325 kPa). Water column height is:

(a) 9.7 m   (b) 10.0 m   (c) 1.5 m   (d) 97 m

110. A barometer is filled with mercury ($SG = 13.6$). It is exposed to the pressure of 200 kPa. Mercury column height is:

(a) 9.7 m   (b) 10.0 m   (c) 1.5 m   (d) 97 m

111. Gravity on the Venus is 2.4 times greater than that on the Mars ($g_{Mars} = 3.711$ m/s^2); the atmospheric pressure on the Venus is 15,246 times that on the Mars ($P_{Mars} = 610$ Pa). The column height ratio of a barometer filled with mercury ($SG = 13.6$) on the Venus compared to the Mars is:

(a) 1,210   (b) 8,200   (c) 6,000   (d) 6,353

112. A sphygmomanometer is also known as a blood:

(a) pressure meter

(b) temperature meter

(c) density meter

(d) flowrate meter

113. A blood pressure meter consists of an inflatable cuff that first:

(a) constricts and then releases pressure imposed on the artery in a controlled manner

(d) releases and then constricts pressure imposed on the artery in a controlled manner

(c) constricts and then releases pressure imposed on the artery in an uncontrolled manner

(d) releases and then constricts pressure imposed on the artery in an uncontrolled manner

114. If the density of ice is 916.8 kg/m^3 and the density of water is 1,000 kg/m^3, what volume does the tip of the iceberg (i.e., volume above the water surface) have as a fraction of the total volume of the iceberg?

(a) 0.9   (b) 0.08   (c) 0.04   (d) 0

115. For the flow to be continuous, the . . . of the moving fluid should . . .

(a) mass / change

(b) volume / change

(c) mass / remain the same

(d) volume / remain the same

116. For the continuity law to be valid in an incompressible flow, the following relationship is applicable, where A is the cross-sectional area, dx is the length of the flow element, and V is the flow velocity. Subscripts 1 and 2 represent two flow regions.

(a) $(Adx/dt)_1 = (Adx/dt)_2$

(b) $(Adx)_1 = 0.5(Adx)_2$

(c) $(AV)_1 = 1/(AV)_2$

(d) $(AV)_1 = 0.5/(AV)_2$

117. A water tap is opened by a curious student. He observes that the water stream diameter gets smaller as it gets farther away from the tap. The reason for this phenomenon is:
(a) air pressure makes water stream progressively narrower
(b) water flow speeds up due to the gravity and therefore its cross-sectional area gets smaller, based on the continuity equation
(c) only an optical illusion and the water stream is actually the same in diameter
(d) a capillary force causes the water stream to become narrower

118. Hydraulics to pneumatics is like:
(a) solid to fluid   (b) gas to liquid   (c) liquid to gas   (d) fluid to solid

119. A Cessna 150 is a small two-seat aircraft that weighs 680 kg. It has been placed on a hydraulically operating lift platform for servicing. The platform and other mechanism parts being lifted weigh 100 kg. The cylinder that lifts the platform has the cross-sectional area of 7 m^2. The cylinder where the actuating force is applied has an area of 0.25 m^2. The viscous fluid within the system has the density of 868 kg/m^3. What actuating force needs to be applied in order to lift the plane by 0.1 m?
(a) 46 N   (b) 485 N   (c) 242 N   (d) 24 N

120. According to Pascal's law, a change in pressure applied to a bounded fluid is spread:
(a) uniformly to all parts of the fluid as well as the container walls
(b) nonuniformly to all parts of the fluid as well as the container walls
(c) uniformly to some parts of the fluid or the container walls
(d) nonuniformly to some parts of the fluid or the container walls

121. A block of fluid with $1 \times 2 \times 3$ m^3 dimensions flows through a rectangular tube of the 1×2 m^3 cross-section. It then moves to another tube with the 2×3 m^3 cross section. The length of the same block of the fluid is: (assume continuity applies):
(a) 1 m   (b) 2 m   (c) 3 m   (d) 4 m

122. Work generated by a gas inside a cylinder is positive if the force exerted by the cylinder and its:
(a) displacement are in the same direction
(b) displacement are in the opposite direction
(c) volume increase simultaneously
(d) volume decrease simultaneously

123. Given vectors of force, $\vec{F}$, and displacement, $\vec{D}$, as well as the angle between these two vectors, q, work can be calculated by:
(a) $W = \vec{F} \cdot \vec{D} = |\vec{F}||\vec{D}|\cos\theta$ (b) $W = -\vec{F} \cdot \vec{D} = |\vec{F}||\vec{D}|\cos\theta$
(c) $W = \vec{F} \cdot \vec{D} = |\vec{F}||\vec{D}|\sin\theta$ (d) $W = -\vec{F} \cdot \vec{D} = |\vec{F}||\vec{D}|\sin\theta$

124. A load moves within a frictionless cylinder. A 10 N force is applied at an angle of 45 degrees to the cylinder axis and causes the load to move by 1 m. The work done is:
(a) 3.5 N   (b) 7.1 N   (c) 10.6 N   (d) 14.1 N

125. 10 J work is performed on a load by a 10 N force applied at an angle θ relative to the direction of motion, causing the load to move by 10 m. The angle (θ) is:
(a) 1.6 degrees   (b) 0 degrees   (c) 84.3 degrees   (d) 89.4 degrees

126. If an object moves from point 1 to point 2 ($h_2 > h_1$) with a height difference of h, and constant velocity (V), change in energy (assuming a constant temperature) is:
(a) $-mgh$ (b) mgh (c) $-gh$ (d) gh

127. Water moves through a Venturi tube, causing a pressure difference of 981 Pa between points A and B on either side of the constricted area. The 2 points are located at different heights with respect to the Venturi centerline; the difference in their height is (assume the tube cross section does not vary):

(a) 0.01 m   (b) 0.1 m   (c) 1 m   (d) 10 m

128. Water moves through a Venturi tube, causing a pressure difference between points A and B on either side of the constricted area. The 2 points are located at different heights with respect to the Venturi centerline; the difference in their height is 5 cm. The pressure difference between the 2 sides is (assume the tube cross section does not vary):

(a) 491 Pa   (b) 245 Pa   (c) 122 Pa   (d) 600 Pa

129. Force F_1 is applied to Piston 1, causing it to move by x (Figure 141). This causes force F_2 to be transmitted to Piston 2. If Spring 1 compresses by 0.25 m, forces F_1 and F_2 are (pistons are cylindrical, spring constant of each of the three springs is 20 N/m, and enclosed space is filled with a fluid with $SG = 1.225$):

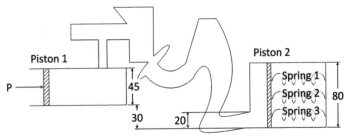

FIGURE 141. Flow channel (dimensions are in cm).

(a) $F_1 = 15$ N, $F_2 = 244$ N   (b) $F_1 = 244$ N, $F_2 = 15$ N
(c) $F_1 = 19$ N, $F_2 = 65$ N   (d) $F_1 = 65$ N, $F_2 = 19$ N

130. Under the standard atmospheric conditions on the Earth's surface, the maximum height (in meters) of a fluid, with the specific gravity x, which may be siphoned is:

(a) $x/10$ (b) $(1 - x)/10$ (c) $10/(1 - x)$ (d) $10/x$

131. A model airplane is being tested in a wind tunnel. The model's scale is 1/6 that of the full-size prototype. If the full-size plane is expected to fly at about 100 km/h, at what wind speed will the wind tunnel need to operate?

(a) 2 times   (b) 1/2 times   (c) 1/6 times   (d) 6 times

132. One Newton is the force required to cause a(an):

(a) acceleration of 1 m/s² for a 1 kg mass
(b) velocity of 1 m/s for a 1 kg mass
(c) acceleration of 10 m/s² for a 100 kg mass
(d) acceleration of 1 m/s² for any object

133. The relationship between absolute viscosity (μ) kinematic viscosity (n) and density (ρ) is:

(a) $v = \mu\rho$ (b) $\mu = v\rho$ (c) $v^2 = \mu\rho$ (d) $\mu^2 = \rho v$

134. For a thin-walled cylinder of mean internal diameter d, wall thickness t, containing fluid at pressure P, the circumferential or hoop stress is given by:

(a) Pd/t (b) $Pd/2t$ (c) $2t/Pd$ (d) $3t/Pd$

135. Flow at pressure 20 N/cm² is flowing through a 180-cm-diameter pipe (inside dimension). The pipe wall is 12-mm thick. Hoop stress in the pipe is:

(a) 112.5 N/cm²   (b) 150.0 N/cm²   (c) 1,125 N/cm²   (d) 1,500 N/cm²

136. With an increase in temperature, the viscosity of lubricating oil:

(a) decreases

(b) increases

(c) does not change

(d) becomes zero

137. If the atmospheric pressure is P_{atm} and liquid density is ρ, the maximum height of a siphon is given by:

(a) $\dfrac{\rho g}{P_{atm}} - \dfrac{2g}{v_B^2}$

(b) $\dfrac{P_{atm}}{\rho g}$

(c) $\dfrac{P_{atm}}{\rho g} - \dfrac{v_B^2}{2g}$

(d) $\dfrac{\rho g}{P_{atm}}$

138. Among the following, which one is the most accurate?

(a) A fluid can flow and conform to the shape of its container

(b) A fluid deforms continuously when exposed to shear stress

(c) A solid can form eddies

(d) Both (a) and (b)

139. For a non-Newtonian fluid, shear stress and deformation are linearly dependent. (T/F)

140. An ideal fluid is both frictionless and compressible. (T/F)

141. A small-diameter open-ended tube is partially inserted into an open-surface liquid. Liquid rises within the tube above the liquid general surface level. This happens due to the:

(a) air pressure   (b) gravity   (c) capillarity   (d) osmosis

142. The separation of two different liquids, such as water and oil, is due to the:

(a) interface tension

(b) tension

(c) capillarity

(d) osmosis

143. Oil droplets suspended in water agglomerate spontaneously and form larger droplets due to the:

(a) tension

(b) surface tension

(c) capillarity

(d) osmosis

144. The phenomenon in which oil droplets suspended in water agglomerate spontaneously and form larger droplets is called:

(a) tension

(b) surface tension

(c) emulsion

(d) osmosis

145. Surface tension between a liquid and gas is not affected by gas properties. (T/F)

146. Surface tension between water and air is less than that between water and a solid container. (T/F)

147. Forces acting on the molecules of the same matter are . . . forces, while those acting on the molecules of different matter are . . . forces.

(a) adhesive / cohesive

(b) cohesive / adhesive

(c) adhesive / tensive

(d) cohesive / tensive

148. A smaller surface tension for the liquid adhering to its container wall compared to that of the liquid-air causes the liquid level at the tube center to be higher than that at the wall. (T/F)

149. Fluid is a substance that:

(a) cannot be exposed to shear forces

(b) expands until it fills its container

(c) has the same shear stress at any point

(d) does not remain at rest if exposed to shear forces

150. In a static fluid:
   (a) resistance to shear stress is small   (b) fluid pressure and viscosity are zero
   (c) linear deformation is small   (d) only normal stresses exist

151. A fluid is ideal if:
   (a) incompressible and non-viscous   (b) viscous and incompressible
   (c) non-viscous and compressible   (d) viscous and compressible

152. If a fluid flow is incompressible, the mass continuity equation changes to the volume continuity equation. (T/F)

153. If no resistance is encountered when the matter is displaced, the matter is a(n):
   (a) ideal fluid   (b) liquid   (c) gas   (d) perfect solid

154. The volumetric change of fluid caused by a resistive force is known as:
   (a) volumetric strain   (b) volumetric index
   (c) compressibility   (d) adhesion

155. The density of water is maximum at:
   (a) 0 °C   (b) 20 °C   (c) 4 °C   (d) 100 °C

156. The density of water at 0°C, expressed in g/cm^3, is:
   (a) 0.9987   (b) 0.9997   (c) 1.0000   (d) 0.9998

157. Molecules within a fluid are attracted to one another by means of:
   (a) adhesion   (b) cohesion   (c) viscosity   (d) compressibility

158. Mercury does not wet glass because of its:
   (a) low adhesion   (b) high cohesion
   (c) high surface tension   (d) low viscosity

159. The property of a fluid which enables it to resist tensile stress is known as:
   (a) compressibility   (b) surface tension
   (c) cohesion   (d) adhesion

160. The specific mass of water is 1,000 kg/m^3 at:
   (a) normal pressure of 760 in Hg   (b) 4 °C temperature and 1 atm
   (c) 0.1 °C temperature and 100 Pa   (d) 25 °C temperature and 1 kPa

161. The specific weight of water in SI units is:
   (a) 1 kN/m^3 (b) 10 kN/m^3 (c) 9.81 kN/m^3 (d) 9,810 kN/m^3

162. When flow parameters at any given instant do not vary over time, flow is:
   (a) quasi-static   (b) steady state   (c) laminar   (d) uniform

163. Among the following, the dimensionless property is the specific:
   (a) weight   (b) volume   (c) density   (d) gravity

164. Pressure at a point within a fluid is not the same in all directions if fluid is:
   (a) moving and viscous   (b) viscous
   (c) stationary and viscous   (d) non-viscous and moving

165. A non-accelerating object of mass 10 kg is placed on a spring balance and its weight reads as 16 N. The value of gravitational acceleration at this location is:
   (a) 9.8 m/s^2 (b) 1.6 m/s^2 (c) 3.2 m/s^2 (d) 4.8 m/s^2

166. The tendency of a liquid to shrink in order to minimize its surface area is the cause of:
    (a) cohesion
    (b) adhesion
    (c) viscosity
    (d) surface tension

167. Surface tension of mercury at room temperature compared to that of water is:
    (a) greater by 6.7 times
    (b) smaller by 6.7 times
    (c) the same
    (d) dependent upon the size of the glass tube

168. At high pressures, viscosity of most gases . . . as temperature . . .
    (a) remains the same / decreases
    (b) decreases / decreases
    (c) increases / decreases
    (d) increases / increases

169. Viscosity of water in comparison to mercury is:
    (a) higher
    (b) lower
    (c) the same
    (d) higher or lower, depending on the temperature

170. Bulk modulus of elasticity decreases with either decrease in pressure or with increase in volume. (T/F)

171. Bulk modulus of elasticity has the units of:
    (a) 1/pressure
    (b) pressure
    (c) pressure per volume
    (d) volume

172. A helium balloon rises in air due to the:
    (a) continuity law
    (b) conservation of energy
    (c) buoyancy
    (d) conservation of momentum

173. Compressibility has the units of:
    (a) 1/pressure
    (b) pressure
    (c) pressure per volume
    (d) volume

174. Surface tension has the units and dimensions of:
    (a) N/m^2, $ML^{-1}T^{-2}$
    (b) N/m^3, $ML^{-2}T^{-2}$
    (c) N/m, MT^{-2}
    (d) Nm, ML^2T^{-2}

175. Surface tension is applied:
    (a) normal to the plane interface
    (b) at an angle to the plane surface
    (c) tangent to the plane interface
    (d) none of the above

176. The stress and strain rate tensors in a Newtonian fluid are related:
    (a) linearly, with viscosity being the proportionality constant
    (b) nonlinearly, with viscosity being the proportionality constant
    (c) linearly, with the inverse of viscosity being the proportionality constant
    (d) nonlinearly, with the inverse of viscosity being the proportionality constant

177. A deep-sea diver is more compressed as she dives deeper. If she dives 10 m below the sea surface, and the bulk modulus of bone is $1.5 \times 10^{10}\,N/m^2$, her bone's volume change as a fraction of the initial volume is (assume the atmospheric pressure is 101.325 kPa):
    (a) 3.3×10^{-6}
    (b) 0
    (c) -6.5×10^{-6}
    (d) 1.6×10^{-6}

178. The units of dynamic and kinematic viscosity are, respectively:
    (a) kgm/s and m/s
    (b) kg/m s and m^2/s
    (c) kgm/s^2 and m^2/s
    (d) $kg/m\ s^2$ and m/s

179. Dynamic and kinematic viscosity depend, respectively, on the:
(a) strain rate and density
(b) density and strain
(c) density and pressure
(d) strain and pressure

180. Viscosity can be measured by:
(a) a viscometer
(b) a rheometer
(c) a barometer
(d) either (a) or (b)

181. If a water barometer is replaced with an alcohol barometer, if the column height for water was 100 cm, the new height is ($SG = 0.8$, assume the environmental conditions have not changed):
(a) 125 cm
(b) 12.5 cm
(c) 65 cm
(d) 6.5 cm

182. Pressure at the outer edge of a liquid (density ρ) inside a drum (radius r), rotating (ω angular velocity), is (interior and exterior radius of drum are r_i and r_o):

(a) $P = -\dfrac{\rho}{2}\omega^2\left(r_i^2 - r_0^2\right) + P_{atm}$

(b) $P = \dfrac{\rho}{2}\omega^2\left(r_i^2 - r_0^2\right)$

(c) $P = -2\rho\omega^2\left(r_i^2 - r_0^2\right)$

(d) $P = 4\rho\omega^2\left(r_i^2 - r_0^2\right)$

183. In a one-dimensional flow, fluid characteristics can change perpendicular to the fluid flow direction. (T/F)

184. Alcohol is often used instead of water in manometers because it has a longer column height due to its higher density relative to water. (T/F)

185. The pressure of the 1 m of head of water is equal to:
(a) 98.1 kPa
(b) 9.8 kPa
(c) 98.1 Pa
(d) 9.8 Pa

186. The specific weight of sea water is less than that of pure water due to the dissolved salt. (T/F)

187. If 500 g of fluid occupies a volume of 1 m³, 4.9 represents its specific:
(a) gravity
(b) mass
(c) weight
(d) density

188. Water drops tend to form spheres due to the property of:
(a) adhesion
(b) polarity
(c) surface tension
(d) viscosity

189. A liquid would wet a solid if adhesion forces are . . . compared to cohesion forces.
(a) lower
(b) greater
(c) equal
(d) lower at low and greater at high temperatures

190. If the cohesion between molecules of a fluid is greater than the adhesion between fluid and glass, the free level of fluid in a glass tube is . . . that of the liquid surface.
(a) higher than
(b) the same as
(c) lower than
(d) unpredictable

191. The . . . is a point through which the resultant pressure field acts on a body.
(a) meta center
(b) center of pressure
(c) center of buoyancy
(d) center of gravity

192. A 2 × 3 m² rectangular sluice gate is immersed vertically in water with the long side being vertical and with the water level rising to 2 m above the gate's bottom edge. The total pressure on the gate is:
(a) 98.1 kPa
(b) 98.1 Pa
(c) 9.81 Pa
(d) 9.81 kPa

193. Conditions for stable equilibrium of a floating body are:
(a) the metacenter should be above the center of gravity
(b) the center of buoyancy and gravity must be on the same vertical line
(c) a righting couple should be formed
(d) the center of buoyancy is between the meta center and the center of gravity

194. Poise is the unit of:
(a) surface tension   (b) capillarity
(c) dynamic viscosity   (d) shear stress

195. Metacentric height is the distance between the center of:
(a) gravity and the metacenter   (b) gravity and buoyancy
(c) gravity and pressure   (d) buoyancy and the metacenter

196. Buoyancy depends on the displaced liquid:
(a) height   (b) viscosity   (c) pressure   (d) volume

197. The center of gravity of the volume of a displaced liquid with an immersed body is the:
(a) metacenter   (b) center of pressure
(c) center of buoyancy   (d) center of gravity

198. A solid block of cast iron ($SG = 8$) is placed in mercury ($SG = 13.6$). What fraction of the block's volume will remain above the mercury surface?
(a) 1   (b) 0.6   (c) 0.4   (d) 0

199. The contact angle for a liquid on a solid surface depends on the:
(a) net forces between the liquid and solid
(b) the gas surrounding the liquid and solid
(c) surface tension
(d) all of the above

200. Surface tension units are:
(a) force per unit length   (b) energy per unit area
(c) force per unit area   (d) energy per unit length

201. The rise or fall of the head h in a capillary tube of radius r, liquid surface tension σ, density ρ, and acceleration gravity g is equal to (θ is contact angle):
(a) $2\sigma\cos\theta/\rho g r$ (b) $\gamma r/\sigma\cos\theta$ (c) $\sigma\cos\theta/\gamma r$ (d) $\gamma r/2\sigma\cos\theta$

202. A 2-cm-radius ceramic tube is filled with water at 20 °C. The water-ceramic surface tension is 0.0728 N/m. The capillary rise is approximately (assume the contact angle is zero degrees):
(a) 0.74 mm   (b) 7.4 mm   (c) 74 mm   (d) 0.074 mm

203. A 20-cm-radius soda-lime glass tube is filled with water at 20 °C. The water-glass surface tension is 0.0728 N/m. The capillary rise is 0.65 mm. The contact angle is:
(a) 35 degrees   (b) 29 degrees   (c) 60 degrees   (d) 45 degrees

204. The pressure inside a liquid drop is (T is surface tension and r is radius):
(a) Tr (b) T/r (c) $4T/r$ (d) $2T/r$

205. If the surface of a liquid becomes more convex, then the pressure due to the cohesion:
(a) is unaffected
(b) is decreased
(c) is increased
(d) may or may not be affected, depending on the temperature

206. An air vessel is provided at . . . in order to avoid an interruption in the flow of a siphon.

(a) the inlet   (b) the outlet

(c) the summit   (d) any point between inlet and outlet

207. The vapor pressure over a concave surface is . . . that of the saturated vapor pressure of a . . . surface.

(a) less than / flat   (b) the same as / convex

(c) greater than / flat   (d) zero / convex

208. The resistance to relative motion between liquid layers is directly proportional to the:

(a) surface tension   (b) cohesion

(c) viscosity   (d) osmosis

209. The diffusion of the solvent through a semi-permeable membrane that separates cells of fluid compartments can be described as:

(a) viscosity   (b) osmosis

(c) surface tension   (d) cohesion

210. The atmospheric pressure . . . with an increase in altitude:

(a) increases linearly   (b) decreases exponentially

(c) decreases linearly   (d) increases exponentially

211. An object weighing 100 N and occupying 0.06 m³ volume will be completely submerged in a fluid having the specific gravity of . . . or less.

(a) 0.51   (b) 0.75   (c) 1.25   (d) 0.17

212. A(An) . . . is used to measure a submarine's speed relative to water.

(a) pitot tube   (b) pitometer log

(c) Venturi meter   (d) orifice plate

213. A(An) . . . is used to measure an aircraft's speed relative to air.

(a) pitot tube   (b) pitometer log

(c) Venturi meter   (d) orifice plate

214. A piezometer is used to measure:

(a) the atmospheric pressure   (b) very low pressures

(c) static pressure   (d) absolute pressure

215. . . . uses the Bernoulli principle to measure the fluid flow velocity.

(a) A Venturi meter   (b) An orifice plate

(c) A pitot tube   (d) all of the above

216. From the following list of manometers, the most accurate is a(n):

(a) U-tube with water   (b) inclined U-tube

(c) displacement type   (d) micro-manometer with water

217. The ratio of the pressure indicated by a manometer inclined by angle θ (from horizontal) to that of a straight U-tube (at 90 degrees) is:

(a) $\sin\theta$ (b) $1/\sin\theta$ (c) $\cos\theta$ (d) $1/\cos\theta$

218. Fluid statics problems are influenced by . . . forces:

(a) gravity and viscous   (b) gravity and pressure

(c) surface tension and viscous   (d) surface tension and gravity

219. A matter unable to resist any shear without deformation is a(an):

(a) fluid
(b) Newtonian fluid
(c) viscous fluid
(d) inviscid fluid

220. A real fluid having internal friction, in which the rate of deformation is directly proportional to applied shear stress, is a(an):

(a) Newtonian fluid
(b) non viscous fluid
(c) inviscid fluid
(d) viscous fluid

221. The ratio of the pressure stress to volumetric strain is the:

(a) bulk modulus of elasticity
(b) dynamic viscosity
(c) kinematic viscosity
(d) shear stress

222. A device that senses absolute pressure is a:

(a) manometer
(b) pitot tube
(c) barometer
(d) Venturi meter

223. A device that senses differential pressure is a:

(a) barometer
(b) manometer
(c) pitot tube
(d) Venturi meter

224. According to . . . , any force applied to a confined fluid is transmitted uniformly in all directions throughout the fluid, irrespective of the container's shape.

(a) Bernoulli law
(b) Stokes's law
(c) Blasius's boundary layer relations
(d) Pascal's law

225. The expression $(P/\rho + gz + 0.5v^2)$ is commonly used to express the Bernoulli equation, where P is pressure, ρ is density, z is height, v is velocity, and g is the acceleration due to the gravity. This expression has the units of energy per unit . . .

(a) volume
(b) weight
(c) mass
(d) cross-sectional area of flow

226. The Bernoulli theorem is applicable to:

(a) compressible fluids flowing in continuous streams
(b) incompressible fluids flowing in continuous streams
(c) steady compressible fluids up to approximately Mach 0.3
(d) both (b) and (c)

227. Irrotational flow is described by the . . . equation.

(a) Cauchy-Riemann
(b) Reynolds
(c) Laplace
(d) Bernoulli

228. Any non-nuclear continuum mechanics flow analysis obeys the:

(a) Bernoulli equation
(b) Newton's law of viscosity
(c) Darcy's equation
(d) continuity equation

229. The continuity equation:

(a) is based on the Bernoulli theorem
(b) expresses the relation between work and energy
(c) describes mass flow rate along a streamline
(d) determines flow by pitot tube

230. The Bernoulli theorem is based on the . . . law.
  (a) conservation of mass
  (b) conservation of force
  (c) conservation of energy
  (d) conservation of momentum

231. If the Bernoulli equation terms are expressed as energy per unit weight, this is equivalent to expressing them in the units of:
  (a) energy
  (b) work
  (c) mass
  (d) length

232. According to the Bernoulli equation for steady ideal fluid flow, the:
  (a) velocity and pressure are inversely proportional
  (b) total energy is constant throughout
  (c) the resultant of all forms of energy in a fluid along a streamline is the same for all points on that streamline
  (d) fluid particles are under fluid pressure and weight

233. Eulerian method describes flow fields as functions of:
  (a) space and time (x, y, z, t), where a control volume is used
  (b) time (t), where fluid particles are tracked
  (c) space (x, y, z), where a control volume is used
  (d) space and time (x, y, z, t), where fluid particles are tracked

234. Lagrangian method describes flow fields as functions of:
  (a) space and time (x, y, z, t) , where a control volume is used
  (b) time (t), where fluid particles are tracked
  (c) space (x, y, z), where a control volume is used
  (d) space and time (x, y, z, t) , where fluid particles are tracked

235. The Navier-Stokes equation accounts for the following forces on a fluid:
  (a) gravity, pressure, and turbulent
  (b) pressure, viscous, and turbulent
  (c) gravity, pressure, and viscous
  (d) pressure, viscous, and laminar

236. Which of the following conditions will linearize the Navier-Stokes equation to make it amenable for an analytical solution?
  (a) high Reynolds numbers $(Re >> 1)$
  (b) low Mach numbers $(M << 1)$
  (c) low Reynolds numbers $(Re << 1)$
  (d) high Mach numbers $(M >> 1)$

237. Based on the conservation of momentum law:
  (a) the balance of linear momentum of a system remains constant
  (b) the balance of angular momentum of a system remains constant
  (c) the sum of linear and angular momenta of a system remains constant
  (d) both (a) and (b)

238. Across a normal shock wave in a converging-diverging nozzle for adiabatic flow, the following equations apply:
  (a) equations of state, continuity, energy, and isentropic equations
  (b) equations of state, energy, momentum, and isentropic equations
  (c) equations of state, continuity, energy, and momentum equations
  (d) equations of state, isentropic, momentum, and mass-conservation equations

239. In an irrotational flow, the . . . equation is applicable, requiring the . . .

  (a) Laplace / stream function to be zero, $\dfrac{\partial^2 \psi}{\partial x^2} + \dfrac{\partial^2 \psi}{\partial y^2} = 0$

(b) Laplace / velocity potential to be zero, $\dfrac{\partial^2 \psi}{\partial x^2} + \dfrac{\partial^2 \varphi}{\partial y^2} = 0$

(c) Poisson / stream function to be zero, $\dfrac{\partial^2 \psi}{\partial x^2} + \dfrac{\partial^2 \psi}{\partial y^2} = 0$

(d) Poisson / velocity potential to be zero, $\dfrac{\partial^2 \varphi}{\partial x^2} + \dfrac{\partial^2 \varphi}{\partial y^2} = 0$

240. In a boundary layer, flow is:
   (a) viscous and rotational
   (b) inviscid and irrotational
   (c) inviscid and rotational
   (d) viscous and irrotational

241. The Euler equation for the motion of liquids is based on the assumption that flow is:
   (a) streamline and isobaric
   (b) continuous and isochoric
   (c) inviscid and adiabatic
   (d) turbulent or laminar

242. The Froude number is a . . . number in the . . . equation, representing the ratio of the . . .
   (a) dimensionless / Euler / inertial forces to gravity
   (b) dimensioned / Lagrange / viscous to inertial forces
   (c) dimensionless / Lagrange / inertial forces to gravity
   (d) dimensioned / Euler / viscous to inertial forces

243. In a steady, inviscid, and incompressible flow, the Euler equation is transformed to the:
   (a) continuity equation   (b) variable flow equation
   (c) Bernoulli equation   (d) Navier-Stokes equation

244. The Reynolds number in a laminar flow regime is approximately less than:
   (a) 10,000   (b) 29,000   (c) 500,000   (d) 2,300

245. The dynamic viscosity of most gases . . . with a rise in temperature:
   (a) increases   (b) decreases
   (c) remains unaffected   (d) is unpredictable

246. In Figure 142, curve . . . applies to a high-viscosity and curve . . . applies to a low-viscosity Newtonian fluid.

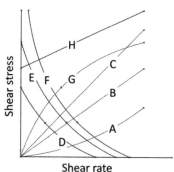

FIGURE 142. Shear stress versus shear rate.

   (a) A / D   (b) E / F   (c) C / B   (d) H / B

247. Figure 143 shows capillarity action as a function of circular glass tube diameter. For water, mercury, and ethyl alcohol, the following curves are applicable, respectively:

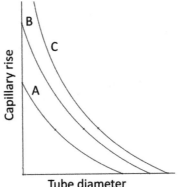

FIGURE 143. Capillary rise versus tube diameter.

(a) A / C / B   (b) A / B / C   (c) C / A / B   (d) C / B / A

248. The working principle of a dead weight pressure gauge tester is based on:

(a) Pascal's law   (b) Dalton's law
(c) Newton's law   (d) Gay-Lussac's law

249. For a submerged sluice, the center of pressure is . . . the center of gravity.

(a) above   (b) below
(c) at the same point as   (d) above or below, depending on the sluice height

250. The horizontal component of the buoyant force on a vertical sluice is:

(a) negligible   (b) the same as the buoyant force
(c) zero   (d) weight

251. The horizontal component of resultant force on an inclined sluice (θ degrees from horizontal) is:

(a) negligible   (b) the same as the buoyant force
(c) the resultant force $\times \sin\theta$ (d) the resultant force $\times \cos\theta$

252. A solid object weighing 8 kg in air was found to weigh 1 kg when submerged in water; its specific gravity is:

(a) 1.14   (b) 1.33   (c) 0.88   (d) 0.75

253. A vertical rectangular gate of the width w and height h is placed at the bottom of a tank containing a liquid with density ρ. If the liquid height is $2h$, the hydrostatic force on the gate's surface is:

(a) $\dfrac{3}{2}\rho gwh$ (b) ρgwh (c) $\dfrac{3}{2}\rho gwh^2$ (d) ρgwh^2

254. A triangular gate of the width w and height h is placed at the bottom of a tank so that the wall's side of the length w is at the bottom and the wall is inclined by the angle θ from horizontal. If liquid density is ρ and its height is $2h$, the hydrostatic force on the gate's surface is:

(a) $\dfrac{5}{6}\rho gwh^2 \sin\theta$ (b) ρgwh (c) $\dfrac{5}{6}\rho gwh^2$ (d) ρgwh^2

255. A rectangular surface stands vertically, with its longer edge at the water surface. h_G is the depth of centroid of its surface, A is the surface area, and I_G is the moment of inertia with respect to the horizontal centroidal axis, parallel to the free surface. The location of the resultant force acting on the surface is:

(a) $h_G + I_G/Ah_G$ (b) $h_G + AI_G/h_G$

(c) $h_G + I_G/h_G$ (d) $I_G h_G + h_G/AI_G$

256. A fully submerged 1×2 m^2 gate is oriented vertically with its longer side on the water surface. The hydrostatic force on the gate's surface is:

(a) 98 kN   (b) 9.8 kN   (c) 9,810 N   (d) 981 N

257. A 6×1 m^2 gate is oriented vertically with its longer side on the water surface. The water surface is 2 m above the gate's top. The depth of the center of pressure is:

(a) 3.0 m   (b) 2.5 m   (c) 3.5 m   (d) 2.0 m

258. Oil ($SG = 0.8$) sits in a tank exposed to the atmospheric pressure of 101.325 kPa. The pressure at the 3-m depth below the oil surface is:

(a) 125 Pa   (b) 125 kPa   (c) 12.5 kPa   (d) 12.5 Pa

259. For an object floating in liquid, normal pressure is applied to:

(a) the bottom-facing surfaces   (b) the center of gravity

(c) the metacenter   (d) all points on the object's submerged surface

260. The ratio of the drag force to cohesion force is the:

(a) Mach number   (b) Froude number

(c) Reynolds number   (d) Weber number

261. A full-scale ship has the hull length of 30 m and typically travels at 2 m/s. A scaled down model of the ship is made with the length reduced by a factor of 2.5. At what velocity should the model be tested to provide valid results?

(a) 5 m/s   (b) 2.5 m/s   (c) 7.5 m/s   (d) 10 m/s

262. A tank of the constant horizontal cross-section contains a volume of water that drains through an outlet at its bottom in 10 min. If the tank's water volume is tripled, the time it takes to drain it is:

(a) 5 min   (b) 17.3 min   (c) 15 min   (d) 30 min

263. A model of a submarine, whose original velocity is 5 m/s, is scaled and retested in a sufficiently big tank. The new prototype's velocity is 25 m/s, and the scale factor is:

(a) 20   (b) 0.2   (c) 2   (d) 10

264. The flow whose field properties, such as velocity, are a function of time and space is:

(a) a one-dimensional flow   (b) a uniform flow

(c) a turbulent flow   (d) a steady flow

265. The length of the converging part of a Venturi meter is shorter than that of the diverging part. (T/F)

266. Streamlines are a family of curves that are instantaneously . . . to the flow . . . vector.

(a) perpendicular / acceleration   (b) tangent / acceleration

(c) perpendicular / velocity   (d) tangent / velocity

267. Streamlines represent a . . . fluid element that travels . . . to the central axis flow.

(a) massless / parallel   (b) massive / perpendicular

(c) massless / perpendicular   (d) massive / parallel

268. A solid piece of wood, weighing 2 kg, floats in water with 25% of its volume submerged; the specific gravity of wood is:

    (a) 0.35   (b) 0.25   (c) 0.15   (d) 0.45

269. A tank filled with water has an opening at depth h below the tank's water surface. If the water is freely flowing from the opening, the velocity of the water stream is:

    (a) $\sqrt{gh}$ (b) $\sqrt{gh}/2$ (c) $\sqrt{2gh}$ (d) $\sqrt{gh}/4$

270. When there are no external forces, a . . . evolves quickly toward a(n) . . . pattern, and the . . . is . . . related to the . . . ; an irrotational vortex is known as . . .

    (a) vortex / irrotational flow / flow velocity / inversely / distance / free vortex
    (b) free vortex / vortex / irrotational flow / flow velocity / inversely / distance
    (c) vortex / free vortex / flow velocity / inversely / distance / irrotational flow
    (d) free vortex / vortex / flow velocity / inversely / distance / irrotational flow

271. The recirculating flow is the region of the . . . immediately behind a moving or stationary body, caused by . . . , which may be accompanied by flow separation and turbulence.

    (a) recirculating flow / viscosity   (b) wake / viscosity
    (c) wake / viscosity   (d) viscosity / recirculating flow

272. A wake is the . . . on a liquid surface, . . . of an object in a flow, or produced by a moving object, caused by the . . . of fluids above and below the free surface.

    (a) density differences / free surface / wake
    (b) free surface / wake / wave pattern
    (c) wave pattern / downstream / density differences
    (d) wake / wave pattern / downstream

273. When there is a relative motion between an object and the surrounding fluid, the flow separation or boundary layer separation is the detachment of the boundary layer from the . . . into the . . .

    (a) fluid surface / recirculating flow   (b) object surface / recirculating flow
    (c) fluid surface / wake   (d) object surface / wake

274. In hypersonic flows, the Mach number is:

    (a) less than 1   (b) greater than 0.5
    (c) greater than 1   (d) greater than 5

275. Velocity distribution in a turbulent boundary layer follows the:

    (a) linear law   (b) square law   (c) power law   (d) parabola law

276. In order for the flow to take place between two points in a pipeline, the differential pressure between these points must be greater than the:

    (a) friction and viscous forces   (b) viscous forces
    (c) friction forces   (d) shear stress

277. Two pipe systems can be said to be equivalent when the following quantities are identical:

    (a) head loss and discharge flow rate   (b) length and diameter
    (c) flow loss and length   (d) friction coefficient and diameter

278. A hydraulic press has a circular ram of the 0.5-m diameter and a circular plunger of the 0.25-m diameter. It is required to lift a weight of 100 kg. The force required on the plunger, equivalent to the weight of a mass, is equal to:

    (a) 25 kg   (b) 250 kg   (c) 2.5 kg   (d) 2500 kg

279. The Buckingham π theorem is a formalization of the Rayleigh's method of . . . analysis. The theorem states that if there is a meaningful equation involving n physical variables, the equation can be rewritten in terms of a set of the . . . dimensionless parameters, constructed from the . . . variables, where k is the number of physical dimensions involved.

(a) dimensional / original / $p = n - k$
(b) dimensionless / transformed / $p = n - k$
(c) dimensional / transformed / $p = n + k$
(d) dimensionless / original / $p = n + k$

280. The Buckingham π theorem indicates that for laws of physics to be valid, the specific unit system is irrelevant. (T/F)

281. The Buckingham π theorem indicates that any physical law can be expressed as an identity involving the:

(a) dimensioned combinations of variables linked by the law
(b) dimensionless combinations of variables not linked by the law
(c) dimensionless combinations of variables linked by the law
(d) dimensioned combinations of variables not linked by the law

282. The Buckingham π theorem would not be valid if the:

(a) dimensioned combinations of the variables do not vary with the systems of units
(b) dimensioned combinations of the variables vary with the systems of units
(c) dimensionless combinations of the variables do not vary with the systems of units
(d) dimensionless combinations of the variables vary with the systems of units

283. The Darcy–Weisbach equation is a(an) . . . equation, which relates the . . . loss due to the . . . along a given length of pipe to the . . . velocity of fluid flow for a(an) . . . fluid.

(a) analytical / pressure / drag / average / incompressible
(b) numerical / head / friction / instantaneous / compressible
(c) empirical / pressure or head / drag / instantaneous / compressible
(d) empirical / head or pressure / friction / the average / incompressible

284. The Darcy–Weisbach equation relates pressure loss to the Darcy friction coefficient, density of fluid, and:

(a) mean flow pressure
(b) mean flow velocity squared
(c) flow temperature
(d) flow volume

285. A flow separation or . . . is the . . . of a boundary layer from a surface into a . . . , and occurs in a flow that is . . . , with . . . pressure.

(a) wake / diverging / boundary layer / slowing down / increasing
(b) boundary layer convergence / separation / wake / speeding up / decreasing
(c) boundary layer separation / separation / wake / slowing down / increasing
(d) wake / converging / boundary layer / speeding up / decreasing

286. An orifice plate is a restriction used to measure . . . or to control . . .

(a) pressure / flow
(b) temperature / pressure
(c) velocity / velocity
(d) flow rate / pressure or flow

287. Coefficients of discharge, velocity, and contraction (C_D, C_V, and C_C) are related as

(a) $C_c = C_V + C_D$
(b) $C_c = C_V - C_D$
(c) $C_D = C_V/C_c$
(d) $C_D = C_V C_c$

288. A fluid jet discharging from a 9-cm-diameter orifice has the diameter of 4.5 cm at the vena contracta; the coefficient of contraction is:

    (a) 1.12   (b) 0.25   (c) 0.75   (d) 0.52

289. To avoid the formation of shock waves anywhere within a converging-diverging nozzle, the Mach number should be:

    (a) < 1   (b) > 1   (c) = 1   (d) 0

290. A channel geometry is the most efficient if it allows for the maximum discharge. The highest efficiency is achieved if the velocity of the discharge is . . . , the hydraulic diameter is . . . , and the wetted perimeter is . . .

    (a) minimum / minimum / maximum   (b) maximum / maximum / maximum
    (c) minimum / minimum / minimum   (d) maximum / maximum / minimum

291. In a vortex, the fluid flow velocity is the . . . next to its axis and . . . with . . . distance from the axis.

    (a) smallest / decreases directly / decreasing
    (b) largest / increases directly / increasing
    (c) largest / decreases inversely / increasing
    (d) smallest / increases inversely / decreasing

292. The coefficient of discharge for flow through an orifice depends on:

    (a) the mass flow rate of fluid through constriction
    (b) the orifice area
    (c) the gravitational acceleration
    (d) all of the above

293. The coefficient of velocity for flow through an orifice is the ratio of the actual velocity of the jet at the vena contracta to that of the theoretical one. (T/F)

294. The coefficient of contraction for flow through an orifice depends on:

    (a) the available liquid head   (b) the orifice size and shape
    (c) the jet velocity   (d) both (a) and (b)

295. If some of the water inside a tank freezes, the water level will:

    (a) decrease
    (b) increase
    (c) increase or decrease, depending on the tank's shape
    (d) remain the same

296. The Manning formula is an . . . formula estimating the average . . . of a liquid flowing in an open or partially open conduit.

    (a) empirical / pressure   (b) analytical / velocity
    (c) analytical / pressure   (d) empirical / velocity

297. When normal water depth in an open channel is larger than the critical depth, the flow is:

    (a) critical   (b) steep   (c) adverse   (d) mild

298. When normal water depth in an open channel is smaller than the critical depth, the flow is:

    (a) critical   (b) steep   (c) adverse   (d) mild

299. The hydraulic grade line is . . . the energy grade line by . . . (V is the flow velocity and g is gravity acceleration).

    (a) smaller than / $V_1^2/2g$ (b) larger than / $V_1^2/2g$
    (c) smaller than / V_1^2/g (d) larger than / V_1^2/g

300. A surge tank is a . . . at the downstream end of a closed . . . to absorb a sudden . . . of pressure, as well as to quickly provide . . . during a short . . . in . . .

(a) storage reservoir / barrage pipe / decrease / water / increase / flow
(b) storage reservoir / aqueduct / increase / excess water / decrease / pressure
(c) standpipe / feeder / increase / excess water / decrease / pressure
(d) standpipe / damp / decrease / water / increase / flow

301. Choked flow is where . . . flow does not . . . with further . . . in downstream . . . for a fixed upstream . . .

(a) volumetric / decrease / increase / flow / temperature
(b) mass / increase / decrease / pressure / pressure
(c) volumetric / decrease / increase / flow / pressure and temperature
(d) mass / increase / decrease / pressure / pressure and temperature

302. In parallel pipes, the:

(a) flow rates are the same
(b) head losses are the same
(c) head loss depends upon flow rate
(d) total head loss is the sum of head losses in all pipe branches

303. Every jet, even if not leaving the nozzle horizontally, has a . . . path.

(a) parabolic   (b) hyperbolic   (c) sinusoidal   (d) convex

304. A hydraulic ram is a cyclic water pump powered by hydropower. Water enters the ram at the specific pressure and flow rate, . . . , and outputs water at the higher . . . and lower . . . The device uses the . . . effect.

(a) pressure / pressure / mass flow rate / fluid hammer
(b) hydraulic head / pressure / volumetric flow rate / pressure surge
(c) hydraulic head / pressure / flow rate / water hammer
(d) pressure / pressure / flow / momentum

305. Critical velocity is the maximum speed at which fluid can flow through a conduit without becoming:

(a) laminar   (b) constricted   (c) blocked   (d) turbulent

306. Laminar flow is defined as a flow of fluid in . . . with . . . of layers.

(a) parallel layers / no disruption   (b) transverse layers / disruption
(c) parallel layers / no disruption   (d) transverse layers / disruption

307. If pipe inlet pressure is 100 Pa and the pressure drop over the pipeline is 30 Pa, the pressure transmission efficiency is:

(a) 60%   (b) 70%   (c) 40%   (d) 30%

308. Hydraulic gradient is equal to the ratio of the:

(a) head loss to the total flow rate
(b) head loss to the flow path length
(c) wetted perimeter to the conduit length
(d) conduit length to the cross-sectional area

309. Head loss between points A and B along a pipe is equal to 1 m. If the flow path length between these two points is 5 m, the hydraulic gradient is equal to:

(a) 0.2   (b) 5   (c) 1   (d) 0.5

310. If the side of a square channel is equal to the diameter of a circular one, the ratio of the hydraulic diameter of the square channel to the circular one is:

(a) 0.7   (b) 0.5   (c) 0.3   (d) 1

311. The Chézy formula describes the:

(a) relation between the average flow velocity of turbulent open channel flow, hydraulic radius, and hydraulic gradient
(b) average flow velocity of turbulent open channel flow
(c) relation between flow hydraulic radius and hydraulic gradient
(d) average flow velocity and hydraulic radius

312. The magnitude of the water hammer depends on:

(a) the fluid density   (b) the speed of sound in a fluid
(c) the fluid velocity variation   (d) all of the above

313. Parasitic drag is the:

(a) combination of form and skin friction drags
(b) form drag
(c) skin friction drag
(d) ram drag

314. Total drag components include:

(a) wave, ram, and parasite drags
(b) skin friction, ram, and parasite drags
(c) lift–induced, wave, ram, and parasite drags
(d) form, lift–induced, wave, and ram drags

315. When a hydraulic jump occurs, liquid at the . . . velocity region discharges into the lower velocity region. Fluid then . . . and its height . . .

(a) lower / speeds up / increases   (b) higher / slows down / decreases
(c) lower / speeds up / decreases   (d) higher / slows down / increases

316. When a hydraulic jump occurs, some of the flow's . . . energy is converted into a(an) . . . in . . . energy; this is a(an) . . . process with some loss through turbulence to . . .

(a) potential / increase / kinetic / reversible / energy
(b) kinetic / increase / potential / irreversible / heat
(c) kinetic / decrease / potential / irreversible / heat
(d) potential / decrease / kinetic / reversible / energy

317. Water depths before and after a hydraulic jump are 1 m and y_2, respectively; y_2, the Froude number (Fr), jump efficiency h, and energy loss expressed as a head loss (ΔE), are:

(a) $y_2 = 6.4$ m, $Fr = 1.8$, $\eta = 93.4\%$, $\Delta E = 0.5$ m
(b) $y_2 = 0.2$ m, $Fr = 0.3$, $\eta = 177\%$, $\Delta E = 0.8$ m
(c) $y_2 = 4$ m, $Fr = 3.2$, $\eta = 71.4\%$, $\Delta E = 1.7$ m
(d) $y_2 = 5.5$ m, $Fr = 2.3$, $\eta = 86.6\%$, $\Delta E = 0.9$ m

318. The thickness of a laminar boundary layer at the 1-mm distance from the leading edge of a smooth plate is 2 mm. The boundary layer thickness at the 5-cm distance from the leading edge is:

(a) 10 mm   (b) 1 cm   (c) 1 mm   (d) 10 cm

319. For a flow with the Reynolds number of 10,000, the thickness of a laminar boundary layer at the 10-mm distance from the leading edge of a smooth plate is:

(a) 5 mm   (b) 0.5 mm   (c) 0.5 cm   (d) 5 cm

320. The pressure difference between the interior and exterior of an 8-mm diameter soap bubble, if the surface tension is 0.05 N/m, based on Laplace's law for a spherical membrane, is:

(a) 2.5 Pa   (b) 25 Pa   (c) 5 Pa   (d) 50 Pa

321. Based on the recommended flow calculation by the Bureau of Reclamation for a fully contracted 90-degree V-notch Weir with free flow conditions, if flow depth over the V-notch is quadrupled, the volumetric flow rate will increase by a factor of (assume the head correction factor is zero):

(a) 45   (b) 8   (c) 16   (d) 31

322. A cylindrical tank with the 2-m diameter and 5-m height is filled with oil ($\rho = 847$ kg/m^3), is mounted horizontally on a ship, and is being delivered to an oil reservoir. To remove the oil, a hose of the 20-cm diameter is used. The outlet of the hose is located 0.5 m below the bottom of the tank. If the oil is drained by gravity only, the instantaneous speed of oil at the hose exit (V) and time that it takes for half of the oil to drain (t) the tank is:

(a) $V = 7.7$ m/s, $t = 0.8$ min   (b) $V = 5.4$ m/s, $t = 0.4$ min
(c) $V = 7.1$ m/s, $t = 0.6$ min   (d) $V = 9.4$ m/s, $t = 1.2$ min

323. Skin friction coefficient is the ratio of the:

(a) shear to inertial forces   (b) drag to inertial forces
(c) diffusion of heat to mass   (d) sensible to latent energy

324. Drag coefficient is the ratio of the:

(a) shear to inertial forces   (b) drag to inertial forces
(c) diffusion of heat to mass   (d) sensible to latent energy

Answer Key									
1. (a)	**2.** (b)	**3.** (a)	**4.** (c)	**5.** (d)	**6.** (b)	**7.** (a)	**8.** (a)	**9.** (a)	**10.** (b)
11. (b)	**12.** (c)	**13.** (b)	**14.** (a)	**15.** (c)	**16.** (c)	**17.** (a)	**18.** (a)	**19.** (d)	**20.** (c)
21. (c)	**22.** (b)	**23.** (a)	**24.** (a)	**25.** (b)	**26.** (a)	**27.** (c)	**28.** (d)	**29.** (a)	**30.** (a)
31. (b)	**32.** (a)	**33.** (b)	**34.** (b)	**35.** (a)	**36.** (a)	**37.** (c)	**38.** (a)	**39.** (b)	**40.** (c)
41. (d)	**42.** (b)	**43.** (c)	**44.** F	**45.** (a)	**46.** (a)	**47.** (d)	**48.** (a)	**49.** (d)	**50.** (b)
51. (a)	**52.** (c)	**53.** (a)	**54.** (b)	**55.** (b)	**56.** (d)	**57.** (d)	**58.** (a)	**59** (b)	**60.** (d)
61. (c)	**62.** (a)	**63.** (b)	**64.** (a)	**65.** (a)	**66.** (a)	**67.** (d)	**68.** (d)	**69.** (b)	**70.** (a)
71. (c)	**72.** (a)	**73.** (a)	**74.** (c)	**75.** (a)	**76.** (a)	**77.** (d)	**78.** (b)	**79.** (c)	**80.** (a)
81. (d)	**82.** (a)	**83.** T	**84.** (c)	**85.** (b)	**86.** (a)	**87.** (a)	**88.** (c)	**89.** (b)	**90.** (a)
91. (a)	**92.** (a)	**93.** (c)	**94.** T	**95.** (c)	**96.** (d)	**97.** (c)	**98.** F	**99.** (b)	**100.** (a)
101. (b)	**102.** (b)	**103.** (c)	**104.** (a)	**105.** (a)	**106.** (a)	**107.** (c)	**108.** (d)	**109.** (b)	**110.** (c)
111. (d)	**112.** (a)	**113.** (a)	**114.** (b)	**115.** (d)	**116.** (a)	**117.** (b)	**118.** (c)	**119.** (b)	**120.** (a)
121. (a)	**122.** (a)	**123.** (a)	**124.** (b)	**125** (c)	**126.** (b)	**127.** (b)	**128.** (a)	**129.** (b)	**130.** (d)
131. (d)	**132.** (a)	**133.** (b)	**134.** (b)	**135.** (d)	**136.** (a)	**137.** (b)	**138.** (d)	**139.** F	**140.** F
141. (c)	**142.** (a)	**143.** (b)	**144.** (c)	**145.** F	**146.** T	**147.** (b)	**148.** T	**149.** (d)	**150.** (d)
151. (a)	**152.** T	**153.** (c)	**154** (c)	**155.** (c)	**156.** (a)	**157.** (b)	**158.** (c)	**159.** (c)	**160.** (b)

161. (c)	**162.** (b)	**163.** (d)	**164.** (a)	**165.** (b)	**166.** (d)	**167.** (a)	**168.** (c)	**169.** (b)	**170.** T
171. (b)	**172.** (c)	**173.** (a)	**174.** (c)	**175.** (c)	**176.** (a)	**177.** (c)	**178.** (b)	**179.** (a)	**180.** (d)
181. (a)	**182.** (a)	**183.** F	**184.** F	**185.** (b)	**186.** F	**187.** (c)	**188.** (c)	**189.** (b)	**190.** (a)
191. (b)	**192.** (d)	**193.** (a)	**194.** (c)	**195.** (a)	**196.** (d)	**197.** (c)	**198.** (c)	**199.** (d)	**200.** (b)
201. (a)	**202.** (a)	**203.** (b)	**204.** (d)	**205.** (d)	**206.** (c)	**207.** (a)	**208.** (c)	**209.** (b)	**210.** (b)
211. (d)	**212.** (b)	**213.** (a)	**214.** (c)	**215.** (a)	**216.** (d)	**217.** (a)	**218.** (b)	**219.** (a)	**220.** (a)
221. (a)	**222.** (c)	**223.** (b)	**224.** (d)	**225.** (c)	**226.** (d)	**227.** (c)	**228.** (d)	**229.** (c)	**230.** (c)
231. (d)	**232.** (c)	**233.** (a)	**234.** (b)	**235.** (c)	**236.** (c)	**237.** (d)	**238.** (c)	**239.** (b)	**240.** (a)
241. (c)	**242.** (a)	**243.** (c)	**244.** (d)	**245.** (a)	**246.** (c)	**247.** (d)	**248.** (a)	**249.** (b)	**250.** (a)
251. (d)	**252.** (a)	**253.** (c)	**254.** (a)	**255.** (a)	**256.** (b)	**257.** (b)	**258.** (b)	**259.** (d)	**260.** (d)
261. (a)	**262.** (b)	**263.** (b)	**264.** (c)	**265.** T	**266.** (d)	**267.** (a)	**268.** (b)	**269.** (c)	**270.** (a)
271. (b)	**272.** (c)	**273.** (d)	**274.** (d)	**275.** (c)	**276.** (a)	**277.** (a)	**278.** (a)	**279.** (a)	**280.** T
281. (c)	**282.** (d)	**283.** (d)	**284.** (b)	**285.** (c)	**286.** (d)	**287.** (d)	**288.** (b)	**289.** (a)	**290.** (d)
291. (c)	**292.** (d)	**293.** T	**294.** (d)	**295.** (c)	**296.** (d)	**297.** (d)	**298.** (b)	**299.** (a)	**300.** (b)
301. (d)	**302.** (b)	**303.** (a)	**304.** (c)	**305.** (d)	**306.** (a)	**307.** (b)	**308.** (a)	**309.** (a)	**310.** (d)
311. (a)	**312.** (d)	**313.** (a)	**314.** (c)	**315.** (d)	**316.** (b)	**317.** (c)	**318.** (d)	**319.** (b)	**320.** (d)
321. (d)	**322.** (c)	**323.** (a)	**324.** (b)						

CHAPTER 9

MECHANICAL ENGINEERING DRAWING

Orthographic Views

1. The system of presenting the three principal views of an object generally used in engineering drawings is:
   (a) trimetric projection
   (b) oblique projection
   (c) isometric projection
   (d) orthographic projection

2. For a sphere:
   (a) any section is an ellipse
   (b) auxiliary views are necessary to fully describe the shape
   (c) the bottom view is necessary to adequately represent an object
   (d) all principal views are the same

3. On a drawing sheet, the heaviest lines would usually be the:
   (a) hidden lines
   (b) dimension lines
   (c) visible object edge lines
   (d) centerlines

4. If the front and top views of a solid object are circles of the same diameter, then the object is a(an):
   (a) torus
   (b) ellipsoid
   (c) cylinder
   (d) sphere

5. On a drawing showing front, rear, and side elevations and plans, the projected views are most likely to be:
   (a) isogonic
   (b) orthographic
   (c) isographic
   (d) isometric

6. An engineering drawing is:
   (a) a technical drawing to represent the engineering product requirements
   (b) an artistic method of presenting engineering components
   (c) a drawing representing only the main geometrical features
   (d) an approximate way to communicate the overall part shape

7. Perspective projections have the following characteristics:
   (a) converge at focal point
   (b) present a realistic view

(c) can be ambiguous because the view depends on the observer's position

(d) all of the above

8. Parallel projections have the following characteristics:

(a) focal point is at infinity

(b) orthographic projection is a type of parallel projection

(c) both (a) and (b)

(d) none of the above

9. Orthographic projections have the following characteristics:

(a) they are used for engineering drawings

(b) projection lines are parallel to the projection plane

(c) projection lines are perpendicular to the projection plane

(d) both (a) and (c)

10. Orthographic projections show:

(a) all features (visible and hidden)   (b) visible features

(c) hidden features   (d) only the most important features

11. Orthographic projections show edges and discontinuities only. (T/F)

12. The maximum number of principal views that can be used to represent an object by orthographic projection is:

(a) six   (b) three   (c) two   (d) four

13. View selection criteria are:

(a) the most informative

(b) the stable part orientation

(c) the minimum necessary number of views

(d) all of the above

14. When creating engineering drawings, line types that can be used are as follows:

(a) visible   (b) hidden

(c) both (a) and (b)   (d) neither (a) nor (b)

15. When creating engineering drawings, line characteristics are as follows:

(a) solid and continuous lines for visible edges; dashed and hidden lines for invisible edges

(b) invisible lines for axis of symmetries

(c) centerlines for axis of symmetries

(d) both (a) and (c)

16. A drawing cycle consists of:

(a) draw, observe, correct, and compare

(b) observe, correct, and draw

(c) correct and compare

(d) none of the above

17. The best sketching techniques consist of:

(a) arm constrained, paper fixed, darker lines as the first step

(b) paper rotation for comfort, boxing in dimensions and small features

(c) arm free movements, paper rotation for comfort, practicing, boxing in dimensions and small features, creating darker lines as the last step

(d) none of the above

18. Regarding systems used for projecting and unfolding the views, one can say that . . . projection is used in . . .

(a) third-angle / the United States and Canada
(b) third-angle / Europe and Asia
(c) second-angle / the United States and Canada
(d) second-angle / Europe and Asia

19. Regarding systems used for projecting and unfolding the views, one can say that . . . projection is used in . . .

(a) first-angle / the United States and Canada
(b) first-angle / Europe and Asia
(c) second-angle / the United States and Canada
(d) second-angle / Europe and Asia

20. How many sides does an N-side surface have when it is seen edge-on?

(a) N sides
(b) N − 1 sides
(c) N + 1 sides
(d) Depends on the viewing direction, either N + 1 or N-1 sides

21. If a surface is seen as a surface only in two of three standard views, it is called a(an):

(a) inclined surface
(b) oblique surface
(c) straight surface
(d) horizontal surface

22. If a surface is seen as a surface in all three standard views, it is called a(an):

(a) inclined surface
(b) oblique surface
(c) straight surface
(d) vertical surface

23. Parallel part edges:

(a) remain parallel in any view
(b) change alignment depending on the view
(c) are only seen as parallel in one view
(d) none of the above

24. Fillets and rounds are:

(a) rounded corners used to minimize stress concentrations
(b) rounded corners used to add strength
(c) rounded corners used to improve appearance and add strength
(d) both (a) and (c)

25. Fillets, rounds, and chamfers have the following characteristics, respectively:

(a) add, remove, and remove material
(b) remove, add, and remove material
(c) remove, remove, and remove material
(d) add, add, and remove material

26. Fillets, rounds, and chamfers are specified by (in the order presented):

(a) radius, height, and radius
(b) radius, radius, height, and height
(c) radius, radius, angle, and offset
(d) height, radius, radius, and offset

Isometric Views

27. Pictorial sketches are used to:
    (a) represent all three dimensions in a single image
    (b) illustrate the main features of a product or structure
    (c) communicate technical information in a form that is easy to visualize
    (d) all of the above

28. A pictorial drawing in which two of the three axes are at 30 degrees with respect to the horizontal is known as:
    (a) diametric drawing
    (b) isometric drawing
    (c) perspective drawing
    (d) oblique drawing

29. The three angles between the isometric axes are:
    (a) 120, 120, and 120 degrees
    (b) 90, 135, and 135 degrees
    (c) 30, 30, and 90 degrees
    (d) chosen to best represent the object, as long as the three angles add up to 360 degrees

30. An isometric view is:
    (a) a type of a pictorial sketch
    (b) formed by an orthographic projection
    (c) both (a) and (b)
    (d) neither (a) nor (b)

31. When drawing an isometric sketch:
    (a) principal axes are drawn at equal angles
    (b) height is laid out vertically, "up-and-down" the page
    (c) width and depth are at 30 degrees from horizontal, straight across the page line
    (d) all of the above

32. An isometric drawing:
    (a) is about 20% larger than true isometric projection along each dimension
    (b) still appears proportionate
    (c) (a), (b), and (d)
    (d) is easier to sketch compared to the true isometric projection

33. What are the properties of isometric projections?
    (a) Diagonals of a square are not of equal length
    (b) Right-angle corners have 90-degree angles
    (c) Angles are distorted
    (d) Both (a) and (c)

34. The correct technique for starting the drawing of an isometric sketch is to draw:
    (a) three principal axes (height being vertical), and transfer height, width, and depth to define the enclosing box
    (b) two principal axes, and transfer height and depth to define the enclosing box
    (c) height as the principal axis and transfer height to define the enclosing box
    (d) three principal axes (height being vertical), and transfer width and depth to define the enclosing box

35. When drawing inclined and oblique surfaces on an isometric sketch, their vertices:
    (a) both (b) and (c) are acceptable
    (b) are located by measuring horizontally and vertically on the drawing

(c) are located by measuring along the surface sides and transferring these distances directly

(d) are located by measuring only parallel to the isometric axes

36. Regarding including hidden lines on isometric sketches, the following applies:

(a) hidden lines are generally not to be used on isometric sketches

(b) hidden lines may be shown to define a feature that would be unclear otherwise

(c) (a), (b), and (d)

(d) the best approach is to choose an orientation to eliminate the need for the use of hidden lines

Auxiliary Views

37. Three principal views of an object are shown, with surfaces labeled in the top view (Figure 144). Which of the following is a correct statement?

(a) Surface C is true size in the top view

(b) Surface A is seen as an edge in the right-side view

(c) Surface D is true size in the top view

(d) Surface A is seen as an edge in the front view

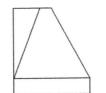

FIGURE 144. Three principal views of an object.

38. For the drawing shown in Question 37, to draw the true size view of surface A, the *fold line* must be drawn:

(a) perpendicular to the edge view of surface A, as seen in the right-side view

(b) parallel to the edge view of surface A, as seen in the right-side view

(c) perpendicular to the edge view of surface A, as seen in the front view

(d) parallel to the edge view of surface A, as seen in the front view

39. For the drawing shown in Question 37, to draw the true size view of surface D, the *projection lines* must be drawn:

(a) parallel to the edge view of surface D, as seen in the top view

(b) perpendicular to the edge view of surface D, as seen in the top view

(c) parallel to the edge view of surface D, as seen in the right-side view

(d) perpendicular to the edge view of surface D, as seen in the right-side view

40. Three principal views of an object are shown, with surface B and its three vertices labeled (Figure 145). The edge view of surface B:

(a) can be seen in the front view

(b) is not visible in any of the given views

(c) can be seen in the top view

(d) can be seen in the right-side view

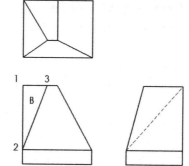

FIGURE 145. Three principal views of an object.

41. For the drawing shown in Question 40, the edge labeled 1–3 appears as true length in:

(a) the front view   (b) none of the views

(c) the top view   (d) the right-side view

42. For the drawing shown in Question 40, in order to draw the edge view of surface B using the top view as the parent of this edge view, the fold line must be drawn:

(a) parallel to edge 1–3 as seen in the top view

(b) perpendicular to edge 1-3 as seen in the top view

(c) perpendicular to edge 1-3 as seen in the front view

(d) parallel to edge 1-2 as seen in the front view

43. Figure 146 shows front and top views of an object . A primary auxiliary view is partially constructed; it is supposed to show the true size view of surface A only. If the view is completed correctly, distance 7–8 will be equal to:

(a) distance 2–3   (b) distance 1–2   (c) distance 1–4   (d) distance 6–5

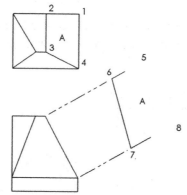

FIGURE 146. Front and top views of an object.

Section Views

44. Section views should be used in order to:

(a) show objects that consist of different sections

(b) provide a clearer representation of an object's interior features

(c) provide some variety to object views

(d) show how an object looks from different directions

45. The device that can be used for measuring cross-sectional areas on a printed or sketched engineering drawing is a:

(a) pantograph   (b) French curve   (c) geodimeter   (d) planimeter

46. Based on the provided top and isometric views of an object (Figure 147), select the section view (Figure 148) which is correct and adheres to all section view conventions.

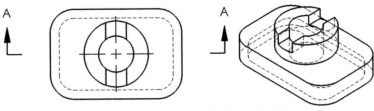

FIGURE 147. Top and isometric views of an object.

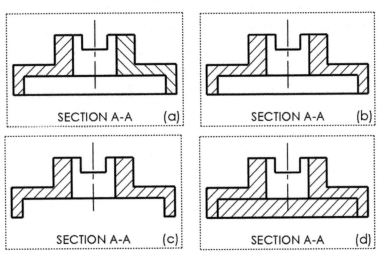

FIGURE 148. Section views of an object.

47. Hidden lines are normally not shown on a section view. (T/F)

48. Given the top view and the section view in place of the front view (Figure 149), the correctly drawn right-side view, adjacent to the top view of the given geometry (Figure 150), is:

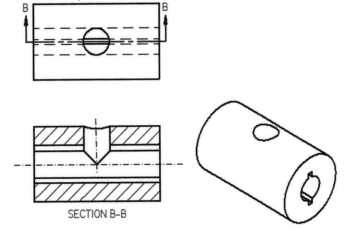

FIGURE 149. Top view and section view in place of the front view of an object.

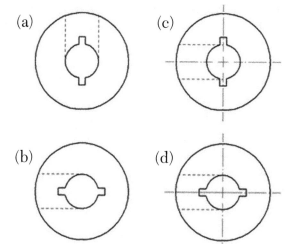

FIGURE 150. Right-side views, adjacent to the top view of an object.

49. Top and right-side views are given (Figure 151). The section view that replaces the front view (Figure 152) is:

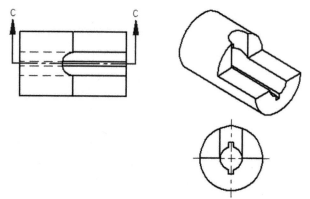

FIGURE 151. Top and right-side views of an object.

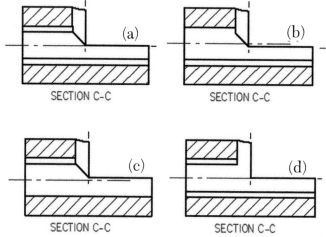

FIGURE 152. Section views that replace the front view of an object.

50. Given as follows are a top view and a section view that replace the front view (Figure 153). Select the correct right-side view (Figure 154) to be placed adjacent to the section view.

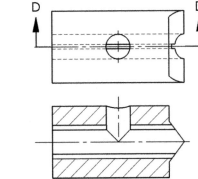

FIGURE 153. Top view and section view that replace the front view of an object.

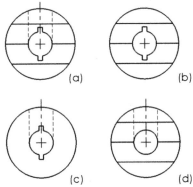

FIGURE 154. Right-side views to be placed adjacent to the section view of an object.

51. In the following four views, a top view is shown together with a section view which replaces the front view (Figure 155). Assuming that the third-angle projection is used, which of these view pairs is correctly arranged?

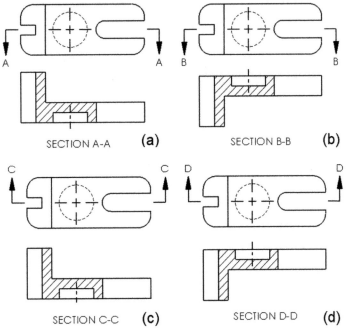

FIGURE 155. A top view together with a section view which replaces the front view of an object.

52. According to the section view drawing convection, ribs are:
   (a) not hatched when cutting plane passes flatwise through them
   (b) are not hatched when cutting plane passes thickness-wise through them
   (c) hatched when cutting plane passes flatwise through them
   (d) to be treated the same as any other part of an object

53. Based on the provided top view, with the cutting plane indicated on it, and the isometric view (Figure 156), identify which of the following four section views (Figure 157) is correct and adheres to section view conventions.

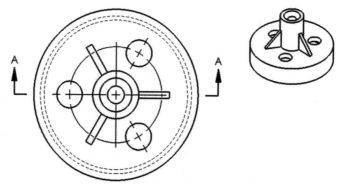

FIGURE 156. Top view including the cutting plane and isometric view of an object.

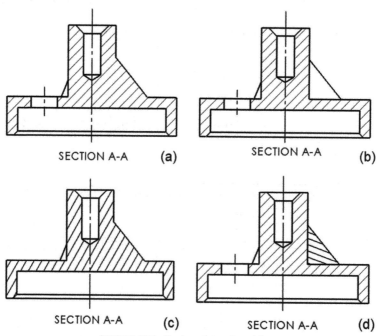

FIGURE 157. Section views of an object.

54. It is possible to have a visible line separating two adjacent hatched areas in a section view of a single object. (T/F)

55. Based on the provided top and isometric views (Figure 158), identify which of the following is a *full section* view (Figure 159):

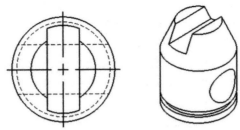

FIGURE 158. Top and isometric views of an object.

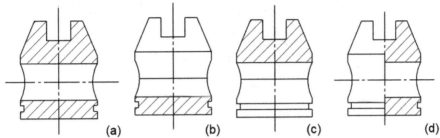

(a)   (b)   (c)   (d)

FIGURE 159. Full section views of an object.

56. Among the choices provided for Question 55, which represents the *half section* view?

57. Based on the provided top and isometric views (Figure 160), identify which of the following is the correct *offset section* view (Figure 161):

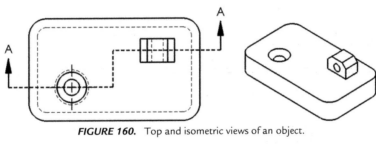

FIGURE 160. Top and isometric views of an object.

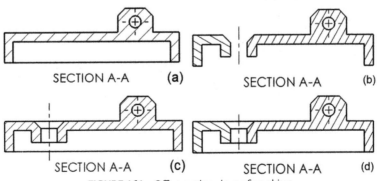

SECTION A-A   **(a)**   SECTION A-A   **(b)**

SECTION A-A   **(c)**   SECTION A-A   **(d)**

FIGURE 161. Offset section views of an object.

58. Based on the provided top and isometric views (Figure 162), identify which of the following is the correct *aligned section* view (Figure 163):

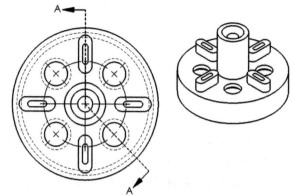

FIGURE 162. Top and isometric views of an object.

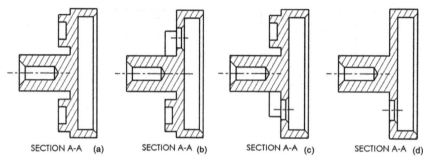

SECTION A-A (a)   SECTION A-A (b)   SECTION A-A (c)   SECTION A-A (d)

FIGURE 163. Aligned section views of an object.

59. Orthographic views are never used in assembly drawings. (T/F)

60. Assembly drawings must be fully dimensioned. (T/F)

61. Hidden lines are generally omitted in assembly drawings. (T/F)

62. All components in assembly sections must be hatched using the same pattern. (T/F)

63. Solid shafts, bolts, screws, nuts, and other hardware items are normally not hatched in assembly sections. (T/F)

Dimensioning and Working Drawings

64. Adding dimensions on a drawing for manufacture is optional because one can easily measure distances on a scaled drawing. (T/F)

65. Dimension lines can be placed as close to an object's outlines as possible, as long as the dimension number is visible. (T/F)

66. The units of measurement must be always added to the dimension number. (T/F)

67. A small gap must be always left between the extension line and the point on an object being measured. (T/F)

68. When using the unidirectional system, the dimension numbers are always aligned with the dimension lines. (T/F)

69. For the following orthographic views of a cylindrical part (Figure 164), choose the answer that correctly applies the centerlines and center marks.

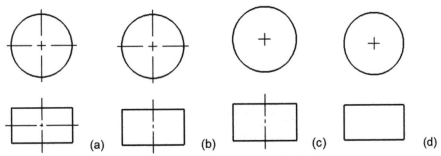

FIGURE 164. Orthographic views of a cylindrical part.

70. Which answer contains all the necessary dimensions while avoiding redundant dimensions (Figure 165)?

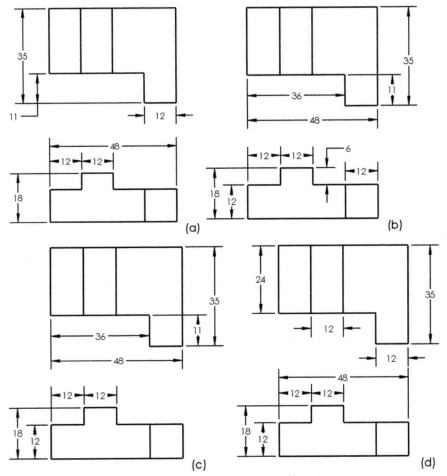

FIGURE 165. Drawings with necessary dimensions.

71. Select the drawing which follows all the recommended dimensioning practices (Figure 166).

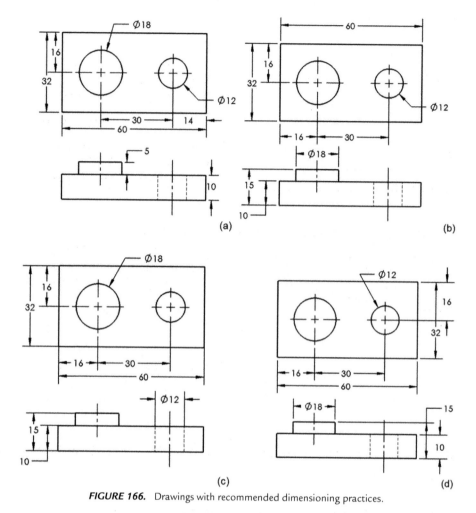

FIGURE 166. Drawings with recommended dimensioning practices.

72. Which of the provided drawings dimensions and locates the cylindrical shapes correctly (Figure 167)?

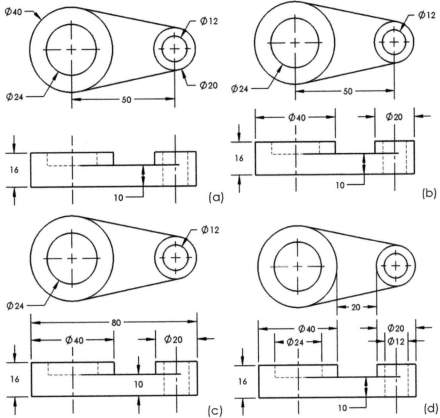

FIGURE 167. Dimensions and locations of the cylindrical shapes.

73. If a Metric thread is specified on a drawing as M20 × 1.0, the number "1.0" represents:
(a) athread depth in mm
(b) thread pitch in mm
(c) wire diameter in mm (to measure the thread characteristics)
(d) thread diameter in cm

74. If a thread on a screw is described by 1.75-16 UN-2A, then its *pitch* is:
(a) 2″   (b) 16″   (c) 1.75″   (d) 0.0625″

75. If a thread on a screw is described by 1.75-16 UN-2A, then its *major diameter* is:
(a) 1.75″   (b) 16″   (c) 2″   (d) 0.0625"

76. Tolerances can only be specified by adding the necessary information explicitly for each dimension. (T/F)

77. For the given two-component assembly (Figure 168), what is *the hole tolerance*?
(a) 0.021   (b) 23.000   (c) 0.007   (d) 0.020

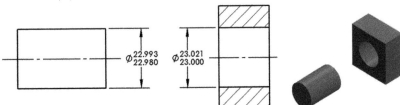

FIGURE 168. A two-component assembly.

78. For the two-component assembly in Question 77, what is *the shaft tolerance?*
   (a) 22.993   (b) 0.013   (c) 0.021   (d) 0.007

79. For the two-component assembly in Question 77, what is *the minimum clearance* (allowance)?
   (a) 23.000   (b) 0.021   (c) 0.007   (d) 0.020

80. For the two-component assembly in Question 77, what is *the maximum clearance?*
   (a) 22.993   (b) 0.007   (c) 23.021   (d) 0.041

81. What type of fit is specified for the two-component assembly in Question 77?
   (a) Transition fit
   (b) Fit type cannot be determined based on the provided information
   (c) Clearance fit
   (d) Interference fit

82. For the given two-component assembly (Figure 169), what is *the minimum clearance* (allowance)?
   (a) 23.048   (b) −0.048   (c) 23.000   (d) −0.021

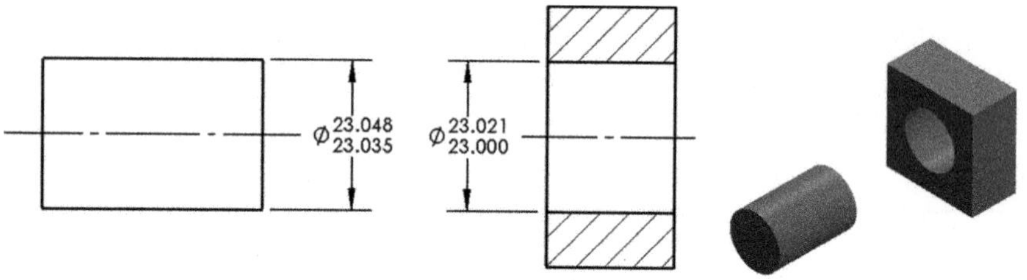

FIGURE 169. A two-component assembly.

83. For the two-component assembly in Question 82, what is *the maximum clearance?*
   (a) 0.048   (b) −0.021   (c) −0.014   (d) 23.048

84. What type of fit is specified for the two-component assembly in Question 82?
   (a) Clearance fit
   (b) Transition fit
   (c) Fit type cannot be determined based on the provided information
   (d) Interference fit

85. For the two-component assembly in Question 82, what is *the basic size?*
   (a) 23.035   (b) 23.000   (c) 23.048   (d) 20.000

86. A drawing based on which a part can be fabricated is known as:
   (a) perspective drawing   (b) contour drawing
   (c) trilinear drawing   (d) detail or shop drawing

87. The item least likely to be found in the title block of an engineering drawing is:
   (a) number of hours required to complete the drawing
   (b) designer name
   (c) drawing scale
   (d) completion date

88. On a detail/shop drawing of a part, a V- or -shaped symbol that touches a visible edge with its vertex indicates:
   (a) fit class
   (b) surface finish

(c) knurl type to be machined on that surface

(d) steel class to be used for the component

89. From four given drawings of front and right-side orthographic views (Figure 170), select the drawing which applies the finish marks correctly.

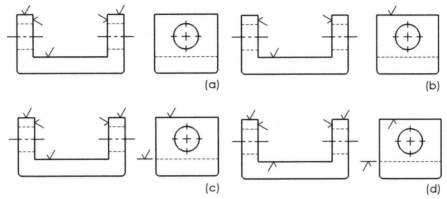

FIGURE 170. Front and right-side orthographic views and finish marks.

90. Finish marks:

(a) are used together with limit dimensions for accurate location of features

(b) should be omitted if all part surfaces are finished and instead a note "finish all over" should be added

(c) must be placed everywhere, even if all surfaces are finished

(d) both (a) and (b)

91. Surface roughness on a drawing is represented by a:

(a) circle   (b) triangle or V shape

(c) square   (d) parallelogram

92. Working drawings consist of:

(a) only detail/shop drawings

(b) only assembly drawings

(c) both detail/shop and assembly drawings

(d) only pictorial drawings

93. Drawings of standard (off-the-shelf) components, such as screws, must be included as part of the working drawings. (T/F)

94. When describing surface roughness, RMS value means:

(a) root minimum square value   (b) root mean square value

(c) root maximum square value   (d) the rate of mean square value

95. CLA value is used to describe:

(a) surface dimensions   (b) surface hardness

(c) surface roughness   (d) surface brittleness

CAD with SolidEdge®

96. To create a solid model of an individual component, one must use a(an):

(a) draft module   (b) assembly module

(c) part module   (d) design module

97. To create drawings, one must use a(an):

    (a) draft module   (b) assembly module

    (c) part module   (d) design module

98. To create solid models with multiple components, one must use a(an):

    (a) draft module   (b) assembly module

    (c) part module   (d) design module

99. When creating a solid model of a part, its *Base* feature can be created using:

    (a) a *Cut* command   (b) a *Revolve* command

    (c) an *Extrude* command   (d) either (b) or (c)

100. When using a 2D *Sketch* command in the *Part* module, the first step is to:

    (a) use a *Line* command

    (b) define a sketch plane

    (c) rotate the part to view the sketch location

    (d) create a new part

101. 2D sketches cannot be used as input for the *Extrude* command. (T/F)

102. When drawing profiles, relationships are used to:

    (a) constrain the profile shape

    (b) point from one sketch element to another

    (c) identify the source of the sketch

    (d) reference the parent of the current element

103. In the sketch profile shown (Figure 171), the red cross pointed to by the green arrow:

    (a) shows location of the line's center

    (b) provides the handle by which the line can be moved

    (c) indicates that this line is constrained to remain horizontal

    (d) both (a) and (b)

FIGURE 171. A sketch profile.

104. In the sketch profile shown (Figure 172), the red square pointed to by the green arrow:

    (a) highlights the corner of the rectangle

    (b) provides the handle by which the corner can be moved

    (c) indicates that the end of one line must remain connected to the end of the other line

    (d) highlights the end of the line segment

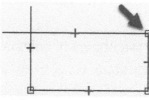

FIGURE 172. A sketch profile.

105. In the sketch profile shown (Figure 173), the line segment indicated by the green ellipse can be removed by:

    (a) selecting the line and pressing the *Delete* key

    (b) using the *Cut* command

    (c) using the *Trim* command

    (d) using the *Split* command

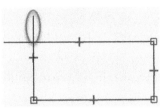

FIGURE 173. A sketch profile.

106. In the sketch profile shown (Figure 174), the red circle pointed to by the green arrow:

    (a) indicates the end of an arc

    (b) indicates the arc and the line must remain tangent

    (c) provides a handle for resizing of the arc

    (d) both (a) and (c)

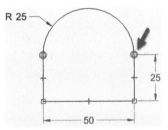

FIGURE 174. A sketch profile.

107. In the sketch profile shown in Question 106, the red color of dimensions 25 and 50:

    (a) highlights these as the critical dimensions

    (b) means these are both driven dimensions

    (c) makes the dimensions easier to see

    (d) means these are both driving dimensions

108. In the sketch profile shown in Question 106, the blue color of dimension R25:

    (a) means this is a radius dimension

    (b) means this is a driven dimension

    (c) indicates a call-out dimension style

    (d) indicates its location on a different layer

109. In the sketch profile shown in Question 106, if the red color dimension 50 is changed to 60, the R25 dimension will:

    (a) change to R30   (b) not be affected

    (c) increase by 10   (d) change color to red

110. The most efficient method to create the set of holes in the part shown (Figure 175) is by:

    (a) creating a Macro that will repeat the set of commands used to create each hole

    (b) copying and pasting the first hole created

    (c) using the *Pattern* command and choosing the *Circular* option

    (d) using the *Circumference Pattern* command

FIGURE 175. A part with the set of holes.

111. Using the *Pattern* command to create a set of 12 equally spaced holes arranged along the circumference of a full circle, you should choose the following option:

(a) fill   (b) fixed   (c) full   (d) fit

112. Using the *Pattern* command to create a set of equally spaced holes arranged along the circumference of a full circle and separated by a 45-degree angle, you should choose the following option:

(a) fill   (b) fixed   (c) full   (d) fit

113. Using the *Pattern* command to create a set of 5 equally spaced holes arranged along the circumference of a circle and separated by a 30-degree angle, you should choose the following two options:

(a) fit and full circle   (b) fixed and partial circle
(c) fill and partial circle   (d) fixed and full circle

114. When a *Swept protrusion* is defined, one or more:

(a) cross sections is swept along one or more path curves
(b) circles are swept along one or more lines
(c) lines are swept along one or more curves
(d) rectangles are moved along a circle

115. A line or an arc designated as a *Construction* element will not be used directly to create a profile for an extrusion or a cut. (T/F)

116. If two assembled parts need to be positioned so that a planar face on one part faces a planar face on another part, the correct assembly relationship to use is:

(a) insert   (b) mate   (c) planar align   (d) axial align

117. If two assembled parts need to be positioned so that a cylindrical feature on one part has its axis aligned with a hole axis on another part, the correct assembly relationship to use is:

(a) insert   (b) mate   (c) planar align   (d) axial align

118. If two assembled parts need to be positioned so that a planar face on one part is parallel to a planar face on another part, with both facing the same direction, the correct assembly relationship to use is:

(a) insert   (b) mate   (c) planar align   (d) axial align

119. If a bolt needs to be positioned in an assembly so that it sits in a hole with the bolt's head in contact with the part's surface, the best assembly relationship to use is:

(a) insert   (b) mate   (c) planar align   (d) axial align

120. When creating part views in the *Draft* module, to create drawings using third-angle projection, the setting shown in Figure 176 is defined by going to:

(a) *Tools / Settings / Drawings*

(b) *File / Options / Drawing Settings*

(c) *Application Button / Settings / Options / Drawing Standards*

(d) *View / Settings / Drawing Views*

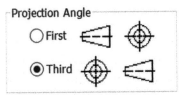

FIGURE 176. Projection angles.

121. When creating a section view using the *Section* command, the *cutting plane line* is defined by selecting a view where it will be drawn and then drawing a line in it. (T/F)

122. Any number of section views can be associated with the same cutting plane line. (T/F)

123. An auxiliary view can be created by first selecting an edge which is:

(a) parallel to the fold line   (b) either (a) or (c), as required

(c) perpendicular to the fold line   (d) horizontal

Answer Key									
1. (d)	**2.** (d)	**3.** (c)	**4.** (d)	**5.** (b)	**6.** (a)	**7.** (d)	**8.** (c)	**9.** (d)	**10.** (a)
11. T	**12.** (a)	**13.** (d)	**14.** (c)	**15.** (d)	**16.** (a)	**17.** (c)	**18.** (a)	**19.** (b)	**20.** (a)
21. (a)	**22.** (b)	**23.** (a)	**24.** (d)	**25.** (a)	**26.** (c)	**27.** (d)	**28.** (b)	**29.** (a)	**30.** (c)
31. (d)	**32.** (c)	**33.** (d)	**34.** (a)	**35.** (d)	**36.** (c)	**37.** (d)	**38.** (d)	**39.** (d)	**40.** (b)
41. (c)	**42.** (b)	**43.** (c)	**44.** (b)	**45.** (d)	**46.** (b)	**47.** T	**48.** (d)	**49.** (a)	**50.** (a)
51. (c)	**52.** (a)	**53.** (b)	**54.** F	**55.** (a)	**56.** (d)	**57.** (c)	**58.** (c)	**59.** F	**60.** F
61. T	**62.** F	**63.** T	**64.** F	**65.** F	**66.** F	**67.** T	**68.** F	**69.** (b)	**70.** (c)
71. (b)	**72.** (b)	**73.** (b)	**74.** (d)	**75.** (a)	**76.** F	**77.** (a)	**78.** (b)	**79.** (c)	**80.** (d)
81. (c)	**82.** (b)	**83.** (c)	**84.** (d)	**85.** (b)	**86.** (d)	**87.** (a)	**88.** (b)	**89.** (c)	**90.** (d)
91. (b)	**92.** (c)	**93.** F	**94.** (b)	**95.** (c)	**96.** (c)	**97.** (a)	**98.** (b)	**99.** (d)	**100.** (b)
101. F	**102.** (a)	**103.** (c)	**104.** (d)	**105.** (c)	**106.** (b)	**107.** (d)	**108.** (b)	**109.** (a)	**110.** (c)
111. (d)	**112.** (a)	**113.** (b)	**114.** (a)	**115.** T	**116.** (b)	**117.** (d)	**118.** (c)	**119.** (a)	**120.** (c)
121. F	**122.** F	**123.** (b)							

GOOD PRACTICES AND LEAN SIX SIGMA IMPLEMENTATION

Good Practices and Lean Six Sigma Implementation

The term *best practices* is a well-known expression in a variety of engineering disciplines in which the product CDIO (Conceive, Design, Implement, and Operate) life cycle concept is used. When working on a model, analysis, or process of any kind, a variety of techniques may be employed, revised, and expanded upon. Hence, the author does not necessarily agree with the expression *best practices*; instead, there are *good practices*, those which are more likely to lead to a useful outcome. Nevertheless, the challenge endures to make them better. This is to ensure improving performance and eliminating waste (i.e., *Muda*; a Japanese word for Futility, Uselessness, and Wastefulness), focusing on critical-to-quality characteristics [8]. You are to make sure that the difference between the potential and actual capability, which is to reflect the entitlement capability, is minimized to the extent possible. In other words, the difference between the entitlement and actual capabilities is the determining factor in defining the quality. Despite the common perception that the world is getting more complex as time passes due to the technological enhancements, people remain the driving force—they are the ones who use their faculties to create a better world. As is the case in many other life experiences, direct and clear communication is a must.

There are often preferred approaches to engineering sciences, which may fit into these categories: (a) technique and (b) approach. The techniques and engineering sciences improve or get revised; as science develops, new technologies become available, and new measuring and sensing tools are introduced or old ones are improved. For the purpose of standardization, these techniques are usually based on the recommendations made by a designated body and are to be followed by the members who are part of a bigger community (e.g., the American Society for Testing and Materials—ASTM). These are the responsibilities that come with being part of something bigger. These regulations are tested, proven, and accurate to the dots of the "i's" and crosses of the "t's." They are not to be treated casually.

The approaches, on the other hand, may vary from one specialist to another. The concept of the *judgment priority* cannot be underestimated, the purpose of which is to avoid chaos in design and implementation. Lack of engagement by the technical staff due to the *extreme* respect for the authority or in environments in which saying "yes, sir" or "yes, ma'am" blindly is a tradition does not usually produce safe products. In his popular book, *Outliers*, Malcom Gladwell tells a story of the tragic accident with a Korean airliner [9]. Upon reviewing the black box recordings, the investigators concluded that one of the main causes was the lack of engagement of the flight crew due to the *extreme* respect for the first-in-command (FIC), the authority, linked to the culture.

Although it may appear that independent thinking is discouraged in a strictly controlled environment such as that experienced in flight, you need to remember that in the end, the FIC, and the second-in-command (SIC), are both responsible for the aircraft control. For example, the air traffic controller (ATC) may be advising you to divert to the Northwest during your cross-country flight while you are flying inbound from the East, but at the same time you see the dark gathering clouds in that area. Based on the conditions, you decide to continue the route inbound and descend, since it is clearer and safer, and communicate your decision with the ATC. Would you expect to be blamed later on for not following the instructions? You should not be, since the ATC would not have been aware of the local conditions.

Your learning experience is not very different from other life experiences; either you do it because you like studying or view it as a "hobby" [10, 11, 12]. The latter applies to some, who require to show their documents so that they can take a job or assume the responsibility of a firm after their close relatives. Some choose to outsource their tasks by hiring test-takers such as what the Pakistani pilots did. After a fatal plane crash in Karachi in late May 2020, as part of a preliminary report to investigate the root causes, the aviation minister in Pakistan announced that almost 1 in 3 pilots in Pakistan had fake flight licenses; they had not taken the the flight tests themselves and paid others to sit for the test on their behalf. Reportedly, this trend is seen across the entire Pakistani aviation industry [13].

Implement the 5S methodology (Sort, Set in order, Shine, Sustain, and Standardize) in your learning and writing test activities. Standardization is particularly important; the challenge of any activity is how to standardize the steps and sustain the developed good standards. It is all about implementing the good practices in knowing, planning, executing, and reporting the learning-related tasks. This is a systematic approach to all the knowledge tests.

Recall the laws of learning—Intensity, Primacy, Readiness, Effect, Freedom, Exercise, Requirement, and Recency (IPREFERR) [14]. Note that the author chooses to place the *Effect* before the *Exercise*, because the principle of the *Effect* is directly related to the emotional reaction of the student and the motivation. It states that learning is reinforced when accompanied by a pleasant or satisfying feeling. Whatever the learning situation, it must include the basics that influence the students positively and give them a feeling of satisfaction. Additionally, learning is weakened when associated with an unpleasant feeling (e.g., repetitive demeaning and teasing). The student will endeavor to continue doing whatever provides an enjoyable outcome to continue learning. Positive reinforcement is more apt to result in success and motivate the learner, so the instructor should recognize and praise the improvement if they care for students' success. This also affects the *Readiness*, in other words, the degree of focus and eagerness—physically, mentally, and emotionally [15, 16]. The *Readiness* is best achieved through practice; *practice makes perfect*. The level of proficiency and skill is mostly achieved by persisting in learning and exercising it.

The author believes that Chris Austin Hadfield, a Canadian engineer, pilot, and astronaut, used this technique that combines knowledge with repetition to accomplish the most challenging tasks and to even overcome his fear of heights. In his memoirs, he tells a story that happened to him while doing one-on-one combat training in a CF-18 jet fighter. He found that his G-Suit's hose was accidentally disconnected during maneuvers through the movement of his elbow [17]. A G-Suit is worn when pilots experience high G-loads (acceleration) that may cause a sudden rush of blood to or away from the brain—also known as redout or black-out [18]. The G-Suit helps to stabilize the blood pressure by pressing against the body and therefore delaying these adverse effects. As a result of the G-Suit's malfunction, Hadfield became unconscious for sixteen seconds while his friend was trying to communicate with him; he called upon his operational awareness and got back on the ground first before trying to find what happened in the air. His experience resulted in modifications to the G-Suit connection in the CF-18 to improve safety. He recommends not to focus on the bee in the helmet but instead on the task at hand. He is referring to the unavoidable circumstances arising under unknown conditions. Self-reportedly, he is also scared of heights, a sensation that he experienced from a young age when he used to fly in his father's biplane. Knowledge and experience have helped him realize that he is not going to fall and that he is not helpless, and in fact he does have some control. You will feel the nerves and stress in a "high-stakes situation"—as he puts it—but should never be terrified. Were you aware of an astronaut who was scared of heights until now? Apparently, they are not the superheroes the author thought them to be, but only human beings, with a range of emotions, capabilities, and occasional insecurities.

Advances have been made in the aerospace industry, but still people are to *learn to fly* and occasionally walk in space as part of their regular extravehicular activity (EVA). The more complex a task is, the more dedication it may require to master it; anything from baking gluten-free breads with an automatic bread-maker to walking in space. You would know you made it when the smell of the bread is spread in the environment and the bread rises to the occasion, or when the hatch is closed behind you as you re-enter the space station at the end of an eight-hour EVA.

Making ethical decisions is part of living in this world, with the responsibility for executing it being deeply felt even more when peer pressure encourages the opposite. This is where Lean Six Sigma comes into play—the methodology to improve efficiency and effectiveness of activities while reducing wastage. As part of being human, behaving more accountably, responsibly, and respectfully toward our surroundings—including natural resources and life in general—is the true creation purpose. Note the difference between accountable and responsible. The former is individual-based while the latter is collective-based. Being accountable makes the individual answerable to actions. Responsibility also relates to any consequences due to the actions before or after the action has taken place. Given the world's complexities, adhering to Lean Six Sigma rules is a lifestyle to be thought about and promoted.

Engineering ethics principles require professional engineers who are working as designers to protect the public and to follow the guidelines that arose from human moral codes. Being part of this collective agreement, one promises to be faithful and true to the plans made, designs created, and words spoken. That promise affirms the commitment to the highest and truest expression of oneself via the *Golden Rule—do unto others as you would have them do unto you.* This is the basis for the ethical principles that were once intertwined with the scientific ones, when the scientists and mathematicians were also philosophers. This vision is the reason for many creations that we see around us. For reasons higher than only adhering to the *Golden Rule*—for the sake of us and nature—we are *accountable as an individual* and *responsible together* as *a clan* with respect to the things we create, from thoughts to words to actions. Think about this every time you use your magic of creation.

Assessment Tools

Assessment Tools

Writing a test is often a challenging task, be it for competitive reasons, such as to enter the job market, an educational institution, a trade school, or for personal evaluation—assessing one's learning or retention level. In most cases, the results depend to a large extent on the capability of the exam taker. Nevertheless, there are cases when other factors may result in unpredictable outcomes, either pleasant or otherwise, generating black swan scenarios. These factors may involve personal experiences surrounding the test conditions, such as the location and environment of the test, the policies of the organization that is administering the test, or the proctor's attitude and manners.

There are other assessment tools such as Learning Analytics, an online tool provided by Pearson Publishing, which the educator may employ in order for the student to be assessed in real time during the lecture [19]. Usually, several questions are asked to assess the students' learning and retention of the newly taught material. Most frequently these are multiple-choice questions. The students use either their mobile devices or laptops—anything that can connect to the Internet—to choose their response. The results are then analyzed and shared by automatically generated charts. Based on the responses, the educator concludes if the concepts have been grasped correctly. If not, further explanation is then given.

There are cases in which the trust in the ability and the kindness shown by a responsible teacher, a conscientious educator, has saved the day. Hadfield reports in his memoir that the first time he attempted to take his flight test, he did not pass it, even though he was well-prepared [17]. The examiner reviewed his records, however, and decided to give him a chance to retake the test and said "I'm going to chalk it up to a bad day. No re-ride." Academic failure was new to Hadfield. He studied further and became more familiar with the plane, almost sleeping in it to master the controls. Next time, he was better prepared and passed the test. If the examiner were to follow the "common" procedure, he would have reported a re-ride for him and Hadfield would not have made it, even though he could technically still try and make the next attempt to retake the test. This was due to the environment Hadfield was part of, the air force academy. He would start receiving humiliating looks and cold shoulders from his own peers. The peer pressure was often so severe that similar experiences would have reportedly resulted in abandoning the program. The decision of the examiner was a life-changing moment in the life of Hadfield. If he had not given Hadfield the "benefit of doubt," Hadfield may not have become an astronaut.

Let us examine an example in which miscalculations result in unforgiving circumstances. A pilot needs to include wind factor when flying in a windy condition. If she uses the same airspeed and bearing as if there were no wind, the ones she used when weather was calm, she would end up being deviated from the destination by a factor that depends on the atmospheric conditions. To further clarify this, in a calm day, if one deviates by one degree from her original bearing, in a 60-mile horizontal distance, she would be one mile away from her original destination (1.85 km). Viktor Emil Frankl, an Austrian neurologist and psychiatrist, presents an interesting analogy during one of his widely attended lectures. In his mid-late life, he passionately shares his attempt to *learn to fly*. His analogy mainly focuses on the drift correction principle used in navigation. His analogy is on the philosophical interpretation of Johann Wolfgang von Goethe, a German writer and statesman in the nineteenth century, which is human curiosity and search for meaning. If the person is not recognized for his thirst for betterment, he becomes bored and dull, and the oppressor contributes to his frustration. There must always be a spark of search for meaning. This presumption elicits all the goodness in the person so he can become *what he in principle is capable of becoming*. As an educator, if you take a person as he is, you make him become worse, but if you take him as he should be, you make him capable of becoming what he can be. In many ways, to be a good educator is to be an idealist.

One wonders if we need to have tests at all. How well do the written or oral tests evaluate the capability of the test takers, assess their knowledge and their level of commitment to learning and education? Furthermore, what do the test results really represent? Would everybody need to care to get an *A* in the exams they take or would passing them be sufficient? There was a music professor who would give an *A* to all his students to ensure that they were at ease about their music learning experience and not too worried about the outcome. In return, the students worked hard enough to actually earn the *A* they had already assumed, to repay the trust they were given! Would this approach work for other fields as well?

Consider the Ph.D. comprehensive tests. Some schools give these tests to doctoral students near the start of studies to assess their ability to undertake the studies. Failure (after two tries) means expulsion from the program. These tests can be written, oral, or a combination. Some of the test takers freeze when standing in front of the examiners. Many of these students are international students, and not familiar with the culture of the hosting institutions, not to mention having to use a foreign language. Their examiners' attitudes may be very different from what they are used to. If the test taker is coming from a culture where the people in authority are highly respected, they may feel uncomfortable to talk back and present a counter-argument. And one's foreign language skills are often diminished under extreme pressure, leading nearly to a complete lack of comprehension by a terrified student. So, the questions to be asked are: Is there anything useful determined by these tests and are they fair to all test takers?

A good practice is to use the resources that are available to you. Sometimes you will make use of the resources and experiences that on the surface are only very remotely connected to your preparing for and taking tests. The last suggestion is that you should always try to think ahead of the test. Ask yourself: What would be a good question to ask? What other related topics may help you understand the concept better? Eat lightly, wear comfortable outfits, arrive early enough to be able to relax and rest before the test, read the questions carefully and thoroughly, and ask for help if needed.

When taking comprehensive tests, written and particularly the oral ones, be prepared to defend your point of view, approaches you take in order to solve the problems, simplifying assumptions, input-output data and their accuracy, and validation methods. Present them as precisely as possible. Listen carefully to what the examiners ask. Ask questions to clarify anything you do not understand about the question. Do not rush. Think about the solution. Although you are not expected to be perfectly organized, you should attempt to have an organized method of thinking and presentation. Identify the knowns and unknowns on the board, express your thoughts as if you were having a conversation with yourself, explaining the concepts to the examiners, and hope for the best!

The *black swan* was a concept introduced by Nassim Nicholas Taleb, the Lebanese-American scholar and mathematical statistician, in his 2007 book of the same name [20]. It is an allegory for the events that are highly unlikely and unpredictable. It relates to a story that, at some point in the past, Europeans thought that swans could only be white and nobody thought otherwise, until one day black swans were discovered in Australia. The book's thesis is that these events simply cannot be predicted. One thus cannot plan for them. The only thing one can do is to improve the resilience of anything or anyone that may at some point be exposed to such an event. Thus, better preparation will allow you to meet that "black swan" question, or any other challenge, with confidence and succeed at overcoming it.

Taking tests and giving tests of any kind is often an experience filled with both happiness and sadness, and generally, it should be challenging mainly in the matter of the test subjects, not the unrelated distractions. At the end of the day, you are to take responsibility for your decisions and therefore their consequences. So, take charge of your own learning, no matter how hard it may seem!

Thermodynamic Tables And Diagrams

Thermodynamic Tables and Diagrams

TABLE 73. Saturated liquid - water - temperature.

Saturated Liquid - Water - Temperature Table

Item	Temperature T (°C)	Pressure P (Mpa)	Specific volume (m³/kg)			Specific energy (kJ/kg)			Specific enthalpy (kJ/kg)			Specific entropy (kJ/kgK)		
			vf	vfg	vg	uf	ufg	ug	hf	hfg	hg	sf	sfg	sg
1	0	0.00	0.00100	205.99	205.99	0.00	2,374.90	2,374.90	0.00	2,500.90	2,500.90	0.0000	9.1555	9.1555
2	10	0.00	0.00100	106.30	106.30	42.02	2,346.58	2,388.60	42.00	2,477.20	2,519.20	0.1511	8.7487	8.8998
3	50	0.01	0.00101	12.03	12.03	209.33	2,233.37	2,442.70	209.30	2,382.00	2,591.30	0.7038	7.3710	8.0748
4	100	0.10	0.00104	1.67	1.67	419.06	2,086.94	2,506.00	419.20	2,256.40	2,675.60	1.3072	6.0469	7.3541
5	150	0.48	0.00109	0.39	0.39	631.66	1,927.44	2,559.10	632.20	2,113.70	2,745.90	1.8418	4.9953	6.8371
6	200	1.55	0.00116	0.13	0.13	850.47	1,743.73	2,594.20	852.30	1,939.70	2,792.00	2.3305	4.0997	6.4302
7	250	3.98	0.00125	0.05	0.05	1,080.80	1,521.00	2,601.80	1,085.80	1,715.10	2,800.90	2.7935	3.2786	6.0721
8	300	8.59	0.00140	0.02	0.02	1,332.90	1,230.70	2,563.60	1,345.00	1,404.60	2,749.60	3.2552	2.4507	5.7059
9	350	16.53	0.00174	0.01	0.01	1,642.10	776.00	2,418.10	1,670.90	892.70	2,563.60	3.7784	1.4326	5.2110
10	374	22.06	0.00311	0.00	0.00	2,015.70	0.00	2,015.70	2,084.30	0.00	2,084.30	4.4070	0.0000	4.4070

TABLE 74. Saturated liquid - water - pressure.

Saturated Liquid - Water - Pressure Table

Item	Pressure P (Mpa)	Temperature T (°C)	Specific volume (m³/kg)			Specific energy (kJ/kg)			Specific enthalpy (kJ/kg)			Specific entropy (kJ/kgK)		
			vf	vfg	vg	uf	ufg	ug	hf	hfg	hg	sf	sfg	sg
1	0.0	6.97	0.00100	129.18	129.18	29.30	2,355.20	2,384.50	29.30	2,484.40	2,513.70	0.1059	8.8690	8.9749
2	0.0	45.81	0.00101	14.67	14.67	191.80	2,245.40	2,437.20	191.80	2,392.10	2,583.90	0.6492	7.4996	8.1488
3	0.1	99.61	0.00104	1.69	1.69	417.40	2,088.20	2,505.60	417.50	2,257.40	2,674.90	1.3028	6.0560	7.3588
4	1.0	179.88	0.00113	0.19	0.19	761.40	1,821.30	2,582.70	762.50	2,014.60	2,777.10	2.1381	4.4469	6.5850
5	4.0	250.35	0.00125	0.05	0.05	1,082.50	1,519.20	2,601.70	1,087.50	1,713.30	2,800.80	2.7968	3.2728	6.0696
6	8.0	295.01	0.00139	0.02	0.02	1,306.20	1,264.30	2,570.50	1,317.30	1,441.40	2,758.70	3.2081	2.5369	5.7450
7	12.0	324.68	0.00153	0.01	0.01	1,473.10	1,041.20	2,514.30	1,491.50	1,193.90	2,685.40	3.4967	1.9972	5.4939
8	16.0	347.35	0.00171	0.01	0.01	1,622.30	809.50	2,431.80	1,649.70	931.10	2,580.80	3.7457	1.5006	5.2463
9	20.0	365.75	0.00204	0.00	0.01	1,786.40	508.60	2,295.00	1,827.20	585.10	2,412.30	4.0156	0.9158	4.9314
10	22.1	373.95	0.00311	0.00	0.00	2,015.70	0.00	2,015.70	2,084.30	0.00	2,084.30	4.4070	0.0000	4.4070

TABLE 75. Superheated vapor - water.

	Superheated Vapor - Water						
Item	**Pressure**	**Temperature Saturated**	**Temperature**	**Specific volume**	**Specific energy**	**Specific enthalpy**	**Specific entropy**
	P (Mpa)	T (°C)	T (°C)	v (m³/kg)	u (kJ/kg)	h (kJ/kg)	s (kJ/kgK)
1			250.00	24.14	2,736.10	2,977.40	9.1020
2	0.01	45.8	450.00	33.37	3,050.30	3,384.00	9.7580
3			700.00	44.91	3,480.80	3,929.90	10.4060
4			250.00	1.20	2,731.40	2,971.20	7.7100
5	0.2	120.2	450.00	1.67	3,048.50	3,381.60	8.3730
6			700.00	2.24	3,479.90	3,928.80	9.0220
7			250.00	0.23	2,710.40	2,943.10	6.9270
8	1.0	179.9	450.00	0.33	3,040.90	3,371.30	7.6200
9			700.00	0.45	3,476.20	3,924.10	8.2760
10			250.00	0.09	2,663.30	2,880.90	6.4110
11	2.5	224.0	450.00	0.13	3,026.20	3,351.60	7.1770
12			700.00	0.18	3,469.30	3,915.20	7.8460
13			300.00	0.05	2,699.00	2,925.70	6.2110
14	5.0	264.0	450.00	0.06	3,000.60	3,317.20	6.8210
15			700.00	0.09	3,457.70	3,900.30	7.5140
16			375.00	0.01	2,650.40	2,858.90	5.7050
17	15.0	342.2	450.00	0.02	2,880.70	3,157.90	6.1430
18			700.00	0.03	3,409.80	3,839.10	6.9570

TABLE 76. Compressed liquid - water.

	Compressed Liquid - Water						
Item	**Pressure**	**Temperature Saturated**	**Temperature**	**Specific volume**	**Specific energy**	**Specific enthalpy**	**Specific entropy**
	P (Mpa)	T (°C)	T (°C)	v (m³/kg)	u (kJ/kg)	h (kJ/kg)	s (kJ/kgK)
1			60.00	985.30	250.30	255.40	0.8287
2	5	264.0	120.00	945.50	501.90	507.20	1.5236
3			240.00	815.10	1,031.60	1,037.70	2.6983
4			60.00	989.60	248.60	263.70	0.8234
5	15	342.2	120.00	950.40	498.50	514.30	1.5148
6			240.00	825.00	1,021.00	1,039.20	2.6774
7			60.00	995.80	246.10	276.30	0.8156
8	30	-	120.00	957.40	493.70	525.00	1.5020
9			240.00	838.40	1,006.90	1,042.70	2.6491

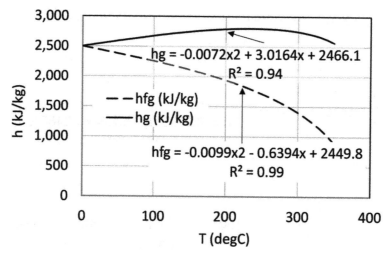

FIGURE 177. Specific enthalpy versus temperature for saturated liquid - water.

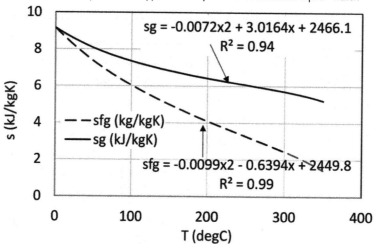

FIGURE 178. Specific entropy versus temperature for saturated liquid - water.

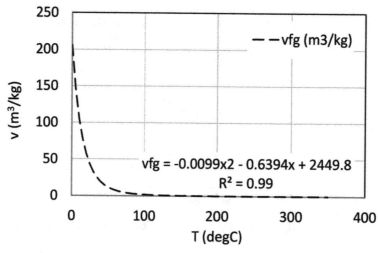

FIGURE 179. Specific volume versus temperature for saturated liquid - water.

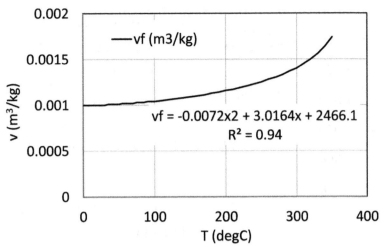

FIGURE 180. Specific volume versus temperature for saturated liquid - water.

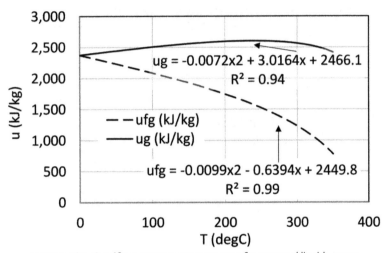

FIGURE 181. Specific energy versus temperature for saturated liquid - water.

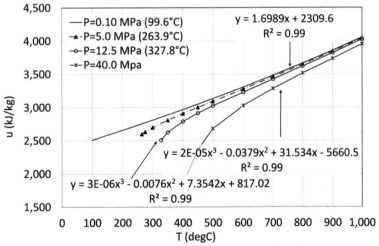

FIGURE 182. Specific energy versus temperature for superheated vapor - water.

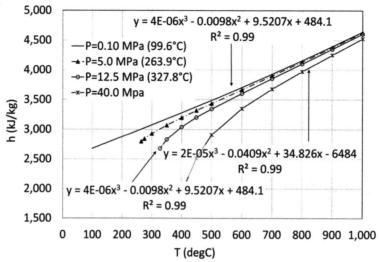

FIGURE 183. Specific enthalpy versus temperature for superheated vapor - water.

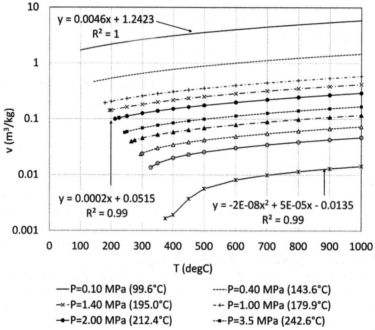

FIGURE 184. Specific volume versus temperature for superheated vapor - water.

BIBLIOGRAPHY

1 Layla S. Mayboudi, *Heat Transfer Modelling Using COMSOL®: Slab to Radial Fin (Multiphysics Modeling Series)*, pp. 272, Mercury Learning and Information, 2018

2 Layla S. Mayboudi, *Geometry Creation and Import with COMSOL Multiphysics®*, pp. 260, Mercury Learning and Information, 2019

3 Layla S. Mayboudi, *Flight Science Mathematics Techniques Sensibility*, pp. 500, Mercury Learning and Information, 2019

4 Layla S. Mayboudi, *COMSOL® Heat Transfer Models (Multiphysics Modeling Series)*, pp. 400, Mercury Learning and Information, 2020

5 https://www.canada.ca/en/public-health/services/diseases/2019-novel-coronavirus-infection.html

6 https://www.firestonecompleteautocare.com/tires/tire-pressure/inflation/

7 https://tiresize.com/tires/Ferrari/488/2020/Pista/

8 Lean Six Sigma approach focuses on improving the bottom line through better performance and by eliminating waste (i.e., *Muda*; a Japanese word for Futility, Uselessness, and Wastefulness), focusing on critical-to-quality characteristics. The training for Lean Six Sigma is provided through the belt-based training system. The belt personnel are designated as white, yellow, green, black, and master black belts, similar to judo. https://en.wikipedia.org/wiki/Lean_Six_Sigma

9 Malcolm Gladwell, *Outliers: The Story of Success*, pp. 336, Back Bay Books; Illustrated edition, 2011

10 https://www.cbc.ca/news/world/indian-students-aided-by-wall-climbing-parents-expelled-for-cheating-1.3003096

11 https://www.bbc.com/news/world-asia-31998343

12 https://www.theglobeandmail.com/opinion/in-india-exam-cheating-is-just-a-symptom/article23669759/

13 https://www.cnn.com/2020/06/25/business/pakistan-fake-pilot-intl-hnk/index.html

14 Author's suggestion (IPREFERR) as a memory aid—"I Prefer Recency."

15 Alfred H. Fuchs, Katharine S. Milar, Psychology as a Science, In *Handbook of Psychology*, I. B. Weiner (Ed.), First Edition, John Wiley and Sons, 2003

16 https://en.wikipedia.org/wiki/Principles_of_learning

17 Chris Hadfield, *An Astronaut's Guide to Life on the Earth*, pp. 320,, First Edition, Random House Canada, 2013.

18 Unlike a G-Suit, a Space-Suit is to protect you against extreme temperature and zero-pressure conditions as well as protect against UV radiation and collision with high velocity small objects while flying in space.

19 https://www.pearson.com/us/higher-education/products-services-teaching/learning-engagement-tools/learning-catalytics.html

20 Nassim Nicholas Taleb, *The Black Swan: The Impact of the Highly Improbable*, Incerto, pp. 366, Annotated Illustrated Edition, Random House, 2007